T0137829

Advances in Intelligent Systems and Computing

Volume 349

Series editor

Janusz Kacprzyk, Polish Academy of Sciences, Warsaw, Poland
e-mail: kacprzyk@ibspan.waw.pl

About this Series

The series "Advances in Intelligent Systems and Computing" contains publications on theory, applications, and design methods of Intelligent Systems and Intelligent Computing. Virtually all disciplines such as engineering, natural sciences, computer and information science, ICT, economics, business, e-commerce, environment, healthcare, life science are covered. The list of topics spans all the areas of modern intelligent systems and computing.

The publications within "Advances in Intelligent Systems and Computing" are primarily textbooks and proceedings of important conferences, symposia and congresses. They cover significant recent developments in the field, both of a foundational and applicable character. An important characteristic feature of the series is the short publication time and world-wide distribution. This permits a rapid and broad dissemination of research results.

Advisory Board

More information about this series at http://www.springer.com/series/11156

Radek Silhavy · Roman Senkerik
Zuzana Kominkova Oplatkova · Zdenka Prokopova
Petr Silhavy
Editors

Software Engineering in Intelligent Systems

Proceedings of the 4th Computer
Science On-line Conference 2015
(CSOC2015), Vol 3: Software
Engineering in Intelligent Systems

 Springer

Editors
Radek Silhavy
Faculty of Applied Informatics
Tomas Bata University in Zlín
Zlín
Czech Republic

Zdenka Prokopova
Faculty of Applied Informatics
Tomas Bata University in Zlín
Zlín
Czech Republic

Roman Senkerik
Faculty of Applied Informatics
Tomas Bata University in Zlín
Zlín
Czech Republic

Petr Silhavy
Faculty of Applied Informatics
Tomas Bata University in Zlín
Zlín
Czech Republic

Zuzana Kominkova Oplatkova
Faculty of Applied Informatics
Tomas Bata University in Zlín
Zlín
Czech Republic

ISSN 2194-5357   ISSN 2194-5365   (electronic)
Advances in Intelligent Systems and Computing
ISBN 978-3-319-18472-2   ISBN 978-3-319-18473-9   (eBook)
DOI 10.1007/978-3-319-18473-9

Library of Congress Control Number: 2015938581

Springer Cham Heidelberg New York Dordrecht London

Springer International Publishing AG Switzerland is part of Springer Science+Business Media
(www.springer.com)

Preface

This book constitutes the refereed proceedings of the Software Engineering in Intelligent Systems Section of the 4th Computer Science On-line Conference 2015 (CSOC 2015), held in April 2015.

The volume Software Engineering in Intelligent Systems brings 36 of the accepted papers. Each of them presents new approaches and methods to real-world problems and exploratory research that describes novel approaches in the field of cybernetics and automation control theory.

Particular emphasis is laid on modern trends in selected fields of interest. New algorithms or methods in a variety of fields are also presented.

CSOC 2015 has received (all sections) 230 submissions, 102 of them were accepted for publication. More than 53% of all accepted submissions were received from Europe, 27% from Asia, 10% from America and 10% from Africa. Researches from 26 countries participated in CSOC 2015 conference.

CSOC 2015 conference intends to provide an international forum for the discussion of the latest high-quality research results in all areas related to Computer Science. The addressed topics are the theoretical aspects and applications of Computer Science, Artificial Intelligences, Cybernetics, Automation Control Theory and Software Engineering.

Computer Science On-line Conference is held on-line and broad usage of modern communication technology improves the traditional concept of scientific conferences. It brings equal opportunity to participate to all researchers around the world.

The editors believe that readers will find the proceedings interesting and useful for their own research work.

March 2015

Radek Silhavy
Roman Senkerik
Zuzana Kominkova Oplatkova
Zdenka Prokopova
Petr Silhavy
(Editors)

Organization

Program Committee

Program Committee Chairs

Zdenka Prokopova, Ph.D., Associate Professor, Tomas Bata University in Zlin, Faculty of Applied Informatics, email: prokopova@fai.utb.cz

Zuzana Kominkova Oplatkova, Ph.D., Associate Professor, Tomas Bata University in Zlin, Faculty of Applied Informatics, email: kominkovaoplatkova@fai.utb.cz

Roman Senkerik, Ph.D., Associate Professor, Tomas Bata University in Zlin, Faculty of Applied Informatics, email: senkerik@fai.utb.cz

Petr Silhavy, Ph.D., Senior Lecturer, Tomas Bata University in Zlin, Faculty of Applied Informatics, email: psilhavy@fai.utb.cz

Radek Silhavy, Ph.D., Senior Lecturer, Tomas Bata University in Zlin, Faculty of Applied Informatics, email: rsilhavy@fai.utb.cz

Roman Prokop, Ph.D., Professor, Tomas Bata University in Zlin, Faculty of Applied Informatics, email: prokop@fai.utb.cz

Program Committee Members

Boguslaw Cyganek, Ph.D., DSc, Department of Computer Science, University of Science and Technology, Krakow, Poland.

Krzysztof Okarma, Ph.D., DSc, Faculty of Electrical Engineering, West Pomeranian University of Technology, Szczecin, Poland.

Monika Bakosova, Ph.D., Associate Professor, Institute of Information Engineering, Automation and Mathematics, Slovak University of Technology, Bratislava, Slovak Republic.

Pavel Vaclavek, Ph.D., Associate Professor, Faculty of Electrical Engineering and Communication, Brno University of Technology, Brno, Czech Republic.

Miroslaw Ochodek, Ph.D., Faculty of Computing, Poznan University of Technology, Poznan, Poland.

Olga Brovkina, Ph.D., Global Change Research Centre Academy of Science of the Czech Republic, Brno, Czech Republic & Mendel University of Brno, Czech Republic.

Elarbi Badidi, Ph.D., College of Information Technology, United Arab Emirates University, Al Ain, United Arab Emirates.

Luis Alberto Morales Rosales, Head of the Master Program in Computer Science, Superior Technological Institute of Misantla, Mexico.

Mariana Lobato Baes,M.Sc., Research-Professor, Superior Technological of Libres, Mexico.

Abdessattar Chaâri, Professor, Laboratory of Sciences and Techniques of Automatic control & Computer engineering, University of Sfax, Tunisian Republic.

Gopal Sakarkar, Shri. Ramdeobaba College of Engineering and Management, Republic of India.

V. V. Krishna Maddinala, Assistant Professor, GD Rungta College of Engineering & Technology, Republic of India.

Anand N Khobragade, Scientist, Maharashtra Remote Sensing Applications Centre, Republic of India.

Abdallah Handoura, Assistant Prof, Computer and Communication Laboratory, Telecom Bretagne - France

Technical Program Committee Members

Ivo Bukovsky
Miroslaw Ochodek
Bronislav Chramcov
Eric Afful Dazie
Michal Bliznak
Donald Davendra
Radim Farana
Zuzana Kominkova
    Oplatkova
Martin Kotyrba
Erik Kral

David Malanik
Michal Pluhacek
Zdenka Prokopova
Martin Sysel
Roman Senkerik
Petr Silhavy
Radek Silhavy
Jiri Vojtesek
Eva Volna
Janez Brest
Ales Zamuda

Roman Prokop
Boguslaw Cyganek
Krzysztof Okarma
Monika Bakosova

Pavel Vaclavek
Olga Brovkina
Elarbi Badidi

Organizing Committee Chair

Radek Silhavy, Ph.D., Tomas Bata University in Zlin, Faculty of Applied Informatics, email: rsilhavy@fai.utb.cz

Conference Organizer (Production)

OpenPublish.eu s.r.o.
Web: http://www.openpublish.eu
Email: csoc@openpublish.eu

Conference Website, Call for Papers

http://www.openpublish.eu

Contents

Physicians' Perspectives in Healthcare Portal Design

Petr Silhavy, Radek Silhavy, and Zdenka Prokopova

Tomas Bata University in Zlin, Faculty of Applied Informatics, nam. T.G.M. 5555,
Zlin 76001, Czech Republic
{psilhavy,rsilhavy,prokopova}@fai.utb.cz

Abstract. In this contribution the web-based portal for healthcare electronic communication is discussed. Authors offer a system design, which reflect a trustworthiness and reliability. The paper deals with system design from the perspective of physicians and offers functional analysis, which is proposed by system design. The construction itself is derived from requirements, which were gathered by literature review.

Keywords: Healthcare electronic communication, System design, Use case modelling.

1 Introduction

Web-based portals take an ever-more important role in the medical informatics. Electronic information sharing is increasing rapidly as the web is becoming an appropriate eco-system for medical information portals and information sharing platform.

The web is only one of many possible ways of delivering information; in this paper we discuss a position for delivery service portal, which can be used for patients-provider communication [1]. Discussion between medical professionals and academic researchers about the advantages and disadvantages of this approach is ongoing, mainly about if such a system can be trusted. Rauer's [2] investigations indicated that trust in electronic records varies, with patients having a greater tendency to trust them than professionals.

The most important advantage [1] of a web-based portal is the possibility of connecting with personal computers at home or in the office. The web platform allows to communication of abroad set of patients and providers, thus for using only spread technology is required.

Bashshur [3] describes telemedicine in healthcare. These results should be used for Internet service in general. Kohane et al. [4] describe methods for building an electronic health records system.

These findings are useful for an understanding of web technologies in healthcare, but they only describe one part of a web-based portal. Therefore, their research is expanded in this paper and further discussion is engaged.

Provider's needs may vary in relatively huge interval terms. There are systems for clinical care, information exchange, and service request functions, whose integration

© Springer International Publishing Switzerland 2015
R. Silhavy et al. (eds.), *Software Engineering in Intelligent Systems,*
Advances in Intelligent Systems and Computing 349, DOI: 10.1007/978-3-319-18473-9_1

into existing patient-care and communication workflows requires careful planning. Later in the paper, the authors address patient requirements. Modern Internet users are looking for user-friendly and information intensive systems, which they are used to in other sectors. Providers are looking for modern and secure systems, which can help cost-management and can help to build a cost-effective healthcare system at the national level.

According to study [5], reducing face-to-face communication is one possible approach, which increases the efficacy of the healthcare's systems, for example, by allowing those patients who need complex monitoring or regularly examination to ask for a remote communication solution. In [6–8], the authors present the fact that patients can omit some important information or they can fail to remember what the physician told them.

The aim of this paper is to present a system design showing by use case models. The authors believe that, by using the results of this investigation, it is possible to improve web-based frameworks for portals in the healthcare sectors. The organization of this contribution is as follows: Section 2 describes the problem formulation. Section 3 describes the Effective Electronic Methodology, which were adopted. Section 4 brings a results, which are represented by system models and provides a discussion of those results. Finally, Section 5 gives the conclusions.

2   Problem Formulation

The system architect has design the system with emphasis on process or procedural part of the system. The architecture that is proposed in this article is built according to user needs, which were ascertained from a user survey. Healthcare providers are trying to both improve communication ability and reduce loading capacity. They have therefore begun by implementing provider-patients web-based portals [7–11].

The communication methods, which are typically involved, include: e-mail, appointment scheduling, and medication refill requests. In addition, the provider's information system supports transfers of clinical data. In the healthcare sector, the FIFO-based system is the most popular approach, i.e., using the chronology of the attendance of waiting patients. This causes long queues in the waiting rooms. There is no chance to control either the time of the patients, nor that of the providers. Although there is the possibility of telephone orders, but very often, patients miss their examination appointments.

The electronic prescription is a specific form of provider-patient communication. This form is based on an electronic request for a medicament prescription. Nowadays, typically only a physical visit to the health-care provider's office leads to the prescription of medicaments. This is highly time-consuming. The second aspect is, of course, the control of interference with or misuse of the medicaments prescribed.

The possibility of electronic consultation is the third load-bearing pillar in using web portals and technology. Web portals—as a usual web-based application, allow the transfer of multimedia information—e.g., photographs, or text information only.

In the written form, both patient and provider should have the opportunity to better rethink their questions and answers.

Some studies [1, 5–10] present the results of the patient's requests:

1. Electronic consultations
2. Shortened waiting time for an examination
3. Simplification of prescriptions
4. Reminder services and invitations for preventive examinations

It is critically important to solve the first requirement precisely. Both Physician and Patients should understand this requirement in similar way. Mainly, therefore this is taken as a key component of the system. Electronic communication use can be restricted by process regulations, which will declare that the system cannot be used in an emergency.

The second requirement should be achieved by a built-in scheduler, which allows one to choose the time of the examination/appointment, which is the best for both patient and physician. If adequate time for examination is reserved, the number of patients in the queue will rapidly decrease.

The third, simplification of the prescription process, is based on electronic prescription requests. The prescription should be delivered to the selected chemist or it should be collected in the health-care provider's office.

The last patient's request is connected to the automation of sending invitations to attend preventive examination appointments, for vaccinations or as a reminder service. This request, probably, is based on patients' inability or the impossibility of tracking these important tasks, but patients do feel the obligation to take part in these processes. Physician profits too, therefore each examination or consultation, which is missed by patient, has influence to the overall costs.

As can be seen in [1, 7,11–17], describing the electronic system, such electronic visits can potentially improve communication, reduce frustration, and enhance relationships between patients and physicians (or the provider itself).

Furthermore, overall satisfaction is increased significantly in situations when non-personal communication can help with information sharing. Physician should significantly improved time management. According [1] many tasks can be done remotely in a form of web messaging, combined with a secure form of communication over a web portal.

3   Effective Electronic Methodology

The proposed approach allows patients and providers, separated by both space, and time, to share information asynchronously in the timeslots that are usually unused for daily scheduled tasks. This is an important fact, resulting from prior study [5,7,11]. Traditional electronic communication (electronic mail) cannot be used.

The basics blocks are described in the form of user cases. Based upon the results of their investigations, the authors have clustered the functionality of the basic blocks by system actors.

In Fig. 1, the use case model [1], representing the overview of the functionality can be seen. The basic functionalities, described by the design, can be found here. The system is designed as open, and therefore, it can be expanded as needed in various differing healthcare systems. Four actors, which are triggers for the basic functionality of the system, are modelled (see Fig. 1).

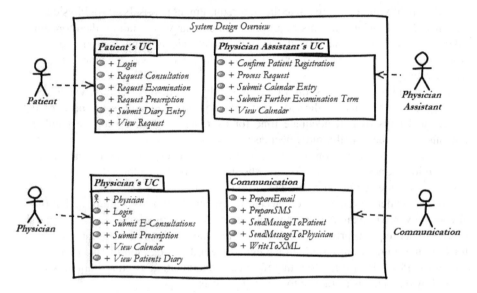

Fig. 1. Use Case Model [1]

The system design is based on shared communication using a web service (Central Layer, see Fig. 2) [1]. Healthcare providers use their own information system, to which the data exchange layer is added. A generic Internet connection is described as a data exchange layer.

This leads to an extremely cost-effective solution. The system is based on methodology proposed in [1] and is based on creation of a Central Layer, which will act in the role of Service Provider.

The authors propose leaving data to be stored in individual information systems, on the provider side (i.e., Provider Layer). The security and accessibility of the data is better than that in a central data store. Moreover, the system is resistant to network failure.

The Central Layer contains all of the use cases shown in Fig. 1. The authors propose these blocks, which are the result of an analysis of users' requirements.

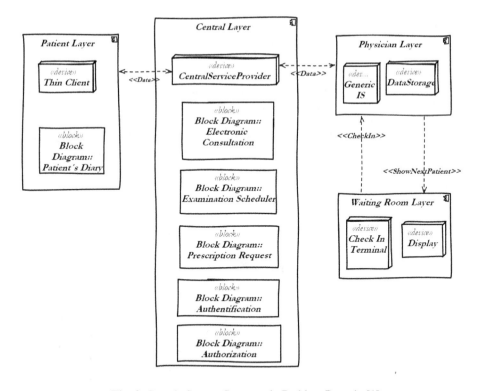

Fig. 2. Generic System Structure in Problem Domain [1]

This architecture implements the most common parts, which are described as basic building blocks. The authentication and authorization blocks implement Rights Management and are able to grant access to specific functions that the individual healthcare centre wants to offer their patients. Furthermore, the design is based on parity—a 1:1 relation between the physical patient and users of the system. Therefore, it is possible to share data among healthcare centres or among independent providers. This results that one person (e.g. patient) is represent as entity by using unique identification (e. g. birth number in Czech Republic).

The blocks represent system components. Therefore, the communication block can be understood as a broker, who offers message-composing services. Individuals—i.e., patients, providers or healthcare centres, can configure different methods of data transfer. The web-based services are used as an inbound channel. Therefore, various information systems can be connected to a central server.

We can see the similarity to a common-object-request broker architecture, which represents an independent layer.

4   Results

Physicians are offered to use a several use-cases in our system design. These use cases can be seen in Fig. 3.

View Patient's Diary supports Submit E-Consultations. This is the task of the Patient Diary block (see Fig. 2). This is not a regular medical record, which are stored in local Datastore in Physician Layer block (Fig. 2), but takes the form of relatively free-form notes of personal feeling and issues, which are described by the patients; these notes can be transfer as enclosed document together with electronic consultation requests. View Calender is a use case, which demonstrates how Physician works with internal time organizer.

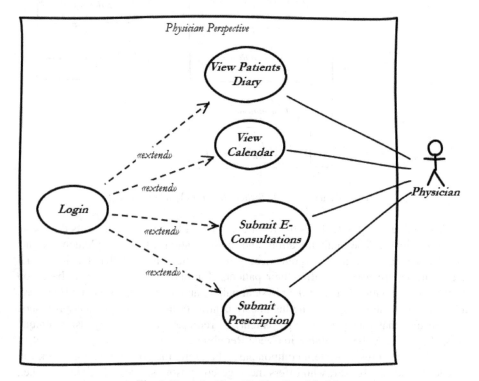

Fig. 3. Physician Specific Use Case Model

The Submit E-Consultation use case brings a scenario for solving less-serious patient's needs. The Physician tasks, mainly those related to chronically-ill patients can be performed by using remote monitoring or by electronic communication.

Physician should be authorized for selected task. After the authorization is granted, the list of electronic consultation request is displayed (see Fig. 4). Physician can choose a request, which need to be handled. System wills look-up for patient's notes according PatientID attribute. As notes system recognizes textual notes submitted by patients, data files from sensors and laboratory data files.

Physician will evaluate all submitted information and add response. When response is added, selected diary's entry are locked. Optionally using Examination Scheduler and its PlanExamination() operation physician-patient meeting is planned.

Finally email message is generated and send by Communication block to patients preferred email address. Optionally information about examination plan is included to the email message.

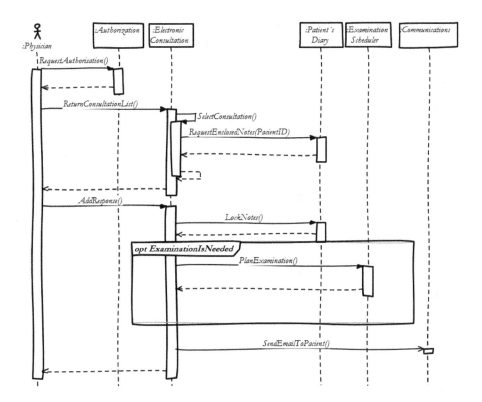

Fig. 4. Physician handles Electronic Consultation Process

Patients should be able to ask for a prescription—as described in the Prescription Request block, and choose if they will pick-up the prescription personally or if it should be sent to a local pharmacy.

The methodology [1] also contains an attribute, which is used for storing the date of the last examination. Therefore, the system knows if the patient should ask for a prescription by wire or not.

Physician handles an electronic prescription requests as follows. The Prescription Request Block (see Fig. 5) is performed. The authorization process is used, which the same operation RequestAuthorization() from Authorization block as in previous example, and the following process starts.

If Authorization is granted, the system will open a list of the Prescription Requests. The system will be looking-up for the specific patient in local Datastore block and check a cross-prescription by using a communication block.

The physician chooses the diagnosis and system will be checked issues in prescription or multiply prescription among more healthcare providers.

If the prescription limit has not been reached, then the patient's request is processed and physician send prescription to selected chemistry.

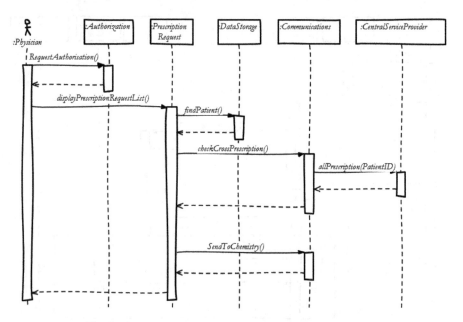

Fig. 5. Physician handles Electronic Prescriptions Requests

5   Conclusion

The main contribution of the system design can be found in its creating a description and system design of implementation of principles of Effective Electronic Methodology [1]. The physician s key tasks were described in the section 3; sequence diagram were used for process demonstration. The system design resolves the complex set of tasks for physician, which is presented as use case model.

The described use case model brings an overview of commonly asked task in electronic healthcare, as resulted from literature view. Moreover, existing systems for sharing medical records can be integrated by using the proposed set of web services, which adopted form described methodology.

The physician part of system, which was described herein, represents a complex set of interconnected use cases, which offer a synergistic effect.

References

1. Silhavy, P., Silhavy, R., Prokopova, Z.: Patients' Perspective of the Design of Provider-Patients Electronic Communication Services. International Journal of Environmental Research and Public Health 11(6), 6231–6245 (2014)
2. Rauer, U.: Patient Trust in Internet-based Health Records: An Analysis Across Operator Types and Levels of Patient Involvement in Germany. Social Sciences Division, Oxford Internet Institute, St. Cross College, p. 66. University of Oxford, Oxford (2011)
3. Sun, S., Zhou, X., Denny, J.C., Rosenbloom, T., Xu, H.: Understanding patient-provider communication entered via a patient portal system. Proc. Amer. Soc. Inform. Sci. 49, 1–4 (2012)
4. Bashshur, R.L.: Telemedicine and the Health Care System. In: Bashshur, R.L., Sanders, J.H., Shannon, G.W. (eds.) Telemedicine Theory and Practice, pp. 5–35. Thomas Publisher Ltd., Springfield (1997)
5. Kohane, I.S., Greenspun, P., Fackler, J., Cimino, C., Szolovits, P.: Building national electronic medical record systems via the World Wide Web. J. Amer. Med. Inform. Assoc. 3, 191–207 (1996)
6. Silhavy, P., Silhavy, R., Prokopova, Z.: Web-based Service Portal in Healthcare. In: Elleithy, K. (ed.) Advanced Techniques in Computing Sciences and Software Engineering, pp. 269–272. Springer, The Netherlands (2010)
7. Vaart, M.R.V.D.: Development and Evaluation of a Hospital-based Web Portal with Patient Access to Electronic Medical Records. In: Proceedings of European League against Rheumatism Congress, Madrid, Spain, June 12-15 (2013)
8. Hassol, A., Walker, J.M., Kidder, D., Rokita, K., Young, D., Pierdon, S., Deitz, D., Kuck, S., Ortiz, E.: Patient experiences and attitudes about access to a patient electronic health care record and linked web messaging. J. Amer. Med. Inform. Assoc. 11, 505–513 (2004)
9. Chen, C., Garrido, T., Chock, D., Okawa, G., Laing, L.: The Kaiser permanent electronic health record: Transforming and streamlining modalities of care. Health Affair 28, 323–333 (2009)
10. Ralston, J.D., Martin, D.P., Anderson, M.L., Fishman, P.A., Conrad, D.A., Larson, E.B., Grembowski, D.: Group health cooperative's transformation toward patient-centered access. Med. Care Res. Rev. 66, 703–724 (2009)
11. Harris, L.T., Martin, D.P., Haneuse, S.J., Ralston, J.D.: Diabetes quality of care and outpatient utilization associated with electronic patient–provider messaging: A cross-sectional analysis. Diabetes Care 32, 1182–1187 (2009)
12. Katz, S.J., Moyer, C.A.: The emerging role of online communication between patients and their providers. J. Gen. Intern. Med. 19, 978–983 (2004)
13. Houston, T.K., Sands, D.Z., Jenckes, M.W., Ford, D.E.: Experiences of patients who were early adopters of electronic communication with their physician: Satisfaction, benefits and concerns. Amer. J. Manage. Care 10, 601–608 (2004)
14. Lin, C.T., Wittevrongel, L., Moore, L., Beaty, B.L., Ross, S.E.: An internet-based patient-provider communication system: Randomized controlled trial. J. Med. Internet Res. 7 (2005), doi:10.2196/jmir.7.4.e47
15. Sittig, D.F.: Results of a content analysis of electronic messages (email) sent between patients and their physicians. Med. Inform. Decis. Making 3 (2003), doi:10.1186/1472-6947-3-11

16. Leong, S.L., Gingrich, D., Lewis, P.R., Mauger, D.T., George, J.H.: Enhancing doctor-patient communication using email: A pilot study. J. Am. Board Fam. Med. 18, 180–188 (2005)
17. Ralston, J.D., Carrell, D., Reid, R., Anderson, M., Moran, M., Hereford, J.: Patient web services integrated with a shared medical record: Patient use and satisfaction. J. Amer. Med. Inform. Assoc. 14, 798–806 (2007)
18. Liederman, E.M., Morefield, C.S.: Web messaging: A new tool for patient-physician communication. J. Amer. Med. Inform. Assoc. 10, 260–270 (2003)

Applied Least Square Regression in Use Case Estimation Precision Tuning

Radek Silhavy, Petr Silhavy, and Zdenka Prokopova

Tomas Bata University in Zlin, Faculty of Applied Informatics,
nam. T.G.M. 5555, Zlin 76001, Czech Republic
{rsilhavy,psilhavy,prokopova}@fai.utb.cz

Abstract. In the presented paper the new software effort estimation method is proposed. The Least Square Regression is used to predict a value of correction parameters, which have a significant impact. The accuracy estimationis of 85% better than the convectional use case points methods in tested dataset.

Keywords: Use case points, Least square regression, Software effort estimation, Achieving accuracy, Software project planning.

1 Introduction

In Software Engineering field a project planning and time consumption is crucial task. Its importance dates from the sixties and seventies of the last century, when a software crisis phenomena, begun. Thus, the investigation of software effort estimation takes an important role in software planning and project development.

One of the most common and easier technique is the use case points (UCP) method, which was developed by Gustav Karner [1]. His perspective addresses object-oriented design and is thus widely used. The method is based on use case models, which are widely used as functional descriptions of proposed systems or software. A number of use case steps were initially involved in the estimation process. There have been several modifications of the original principles, such as use case size points [2], extended use case points [3], use case points modified [4], adapted use case points [5], and transaction or path analysis [6].

Karner's basic method is based on assigning weights to clustered actors and use cases. Karner's UCP method identifies three clusters – simple, average and complex. The sum of weighted actors creates a value called unadjusted actor weights (UAW), and in the same sense, the unadjusted use case weights (UUCW) value is calculated. Two coefficients, technical factors and environmental factors, are used to describe the project, related information and development team experience. Summing UAW and UUCW and then multiplying this value by the technical and environmental factor coefficients obtain the number of UCP. Finally, Karner [1] works with the productivity factor, i.e., the number of man-hours per point. This factor is one of the greatest issues with the methodology, as this final operation can change the estimate considerably.

© Springer International Publishing Switzerland 2015

R. Silhavy et al. (eds.), *Software Engineering in Intelligent Systems*,
Advances in Intelligent Systems and Computing 349, DOI: 10.1007/978-3-319-18473-9_2

Use case size points is investigated in [2]. The authors emphasize the internal structure of the use case scenario in their method. The primary actors take roles and classifications based on the adjustment factor. Fuzzy sets are used for the estimation.

The second modification described here is prepared by Wang [4] and is called extended UCP. This approach employs fuzzy sets and a Bayesian belief network, which is used to set unadjusted UCP. The result of this approach is a probabilistic effort estimation model.

Diev [5] notes that if the actors and use cases are precisely defined, then unadjusted UCP can be multiplied by technical factors. Technical factors are used to establish a coefficient of base system complexity. According to [5], the effort of further activities must be added. These further activities are supportive tasks, such as configuration management or testing.

The next interpretation is called adapted UCP [5]. UCP are adapted to incremental development for large-scale projects. In the first increment, all actors are classified as average and all use cases as complex. The authors also propose the decomposition of use cases to smaller ones, which are classified into the typical three categories.

Use case-based methods of estimation have certain well-known issues [6]. Use cases are written in natural language, and there is no rigorous approach to allow for a comparison of the use case quality or fragmentation.

The aim of this paper is to present a new version of estimation equation, which will applicable for Use Case Points precise tuning. The authors believe that, by using the using least square regression a minimalizing brings a minimal magnitude of minimal error.

The organization of this contribution is as follows: Section 2 describes the use case points method. Section 3 describes the Least Square Regression and how it is used in our approach. Section 4 brings a results and a discussion of those results. Finally, Section 5 gives the conclusions.

2   Methods

2.1   Use Case Points Definition

In the UCP method [1, 3, 8], actors and use cases are classified as simple, average or complex [2]. A simple actor typically represents an application programming interface (API) or, more generally, a non-physical system user. A system interconnected by network protocol is classified as average, and finally, physical users interacting with the system through a graphical user interface are complex actors. These factors are summarized in Table 1.

Table 1. Actor Classification

Actor Classification (AC)	Weighting Factor (WFa)
Simple	1
Average	2
Complex	3

The total unadjusted actor weights (TUAW) are calculated according to the following formula:

$$TUAW = \sum AC \times WFa \qquad (1)$$

Use cases are classified in a similar manner (see Table 2). The complexity of use cases is based on the number of steps. The step counting process is not clear from the inline view. The original method counts steps in a primary scenario and in alternative scenarios.

Table 2. Use Case Classification

Use Case Classification (UCC)	Number of Steps	Weighting Factor (WFb)
Simple	(0,4)	5
Average	<4,7>	10
Complex	(7, ∞)	15

The unadjusted use case weights (UUCW) are calculated according to the following formula:

$$UUCW = \sum UCC \times WFb \qquad (2)$$

Technical factors are then applied to the unadjusted use case points (UUCP). The second adaptation is based on environmental factors. Table 3 presents the technical factors, and Table 4 presents the environmental factors, as they are known in UCP.

Table 3. Original Technical Factors

Factor ID	Description	Weight (WFc)	Significance (SIa)
T1	Distributed System	2	<0,5>
T2	Response adjectives	2	<0,5>
T3	End-User Efficiency	1	<0,5>
T4	Complex Processing	1	<0,5>
T5	Reusable Code	1	<0,5>
T6	Easy to Install	0.5	<0,5>
T7	Ease to Use	0.5	<0,5>
T8	Portable	2	<0,5>
T9	Easy to Change	1	<0,5>
T10	Concurrent	1	<0,5>
T11	Security Feature	1	<0,5>
T12	Access for Third Parties	1	<0,5>
T13	Special Training Required	1	<0,5>

Table 4. Original Environmental Factor

Factor ID	Description	Weight (WFd)	Significance (SIb)
E1	Familiar with RUP	1.5	<0,5>
E2	Application Experience	0.5	<0,5>
E3	Object-oriented Experience	1	<0,5>
E4	Lead Analyst Capability	0.5	<0,5>
E5	Motivation	1	<0,5>
E6	Stable Requirements	2	<0,5>
E7	Part-Time Workers	-1	<0,5>
E8	Difficult Programming Language	2	<0,5>

Factors T1-T13 and E1-E8 have fixed weights. Moreover, for each factor, the significance can be set in the interval from 0 to 5, where 0 indicates no impact, 3 indicates an average impact, and 5 indicates a strong impact. The technical complexity factor (TCF), that is, the first correction coefficient, can be calculated according to the following formula [10]:

$$TCF = 0.6 + (0.01 \times \sum_{T13}^{T1} WFc \times SIa) \tag{3}$$

The second correction coefficient is called the environmental complexity factor (ECF). The formula for calculation is similar to the above:

$$ECF = 1.4 + (-0.03 \times \sum_{E8}^{E1} WFd \times SIb) \tag{4}$$

The final result of the estimation is called adjusted UCP (AUCP) and represents the project (system or software) size in points. The unadjusted value of points from actors and points from use cases are summarized, and then, the TCF and ECF are applied. The following formula is used [1, 2, 4, 7, 10]:

$$AUCP = (TUAW + UUCW) \times TCF \times ECF \tag{5}$$

The AUCP value represents a system size, but a man-hour value is used to measure the work effort. Therefore, AUCP is typically multiplied by 20 [1]. This value is a productivity factor [10] of the project developers.

3   Applying Least Square Regression

Appling Least Square Regression [9] is motivated by investigating the possibility of solving issues with the classical mathematics approach. As the basic for optimization the original version of the Use Case Points method is used, as was previously described in the text.

First, the AUCP is calculated for each new project. Each analytical individual uses a slightly different estimation style. Least squares regression is applied in the second phase to perform a precise tuning. The modified AUCP equation (5) is as follows:

$$MAUCP = a_1(TUAW \times TCF \times ECF) + a_2(UUCW \times TCF \times ECF) \quad (6)$$

The precise tuning is provided by setting a_1 and a_2 as two correction coefficients. These values are calculated by least squares regression according to the following formulas:

$$\binom{y_1}{y_n} = \begin{pmatrix} x_{11} & x_{12} \\ x_{n1} & x_{n2} \end{pmatrix} \times \binom{a_1}{a_2} \Rightarrow \binom{a_1}{a_2} = (X^T \times X)^{-1} \times (X^T \times Y) \quad (7)$$

$$x_{i1} = (TUAW_i \times TCF_i \times ECF_i) \quad (8)$$

$$x_{i2} = (UUCW_i \times TCF_i \times ECF_i) \quad (9)$$

4   Results

Evaluation was perform by using original Use Case Poitns [1] as was previously declared and new approach designed as modified MAUCP.

We evaluated the proposed algorithm using datasets prepared by Ochodek [7], which contain 14 projects data. These projects have a real effort from 13,85 to 179,65 points. We use a constant 20 man-hours per AUCP for our comparison and evaluation, which allows for the recalculation of the real effort to UCP and for a comparison of the effectiveness of methods with the productivity factor omitted.

In Fig. 1 there is a summarization of the raw project data, which are used to test the new algorithm.

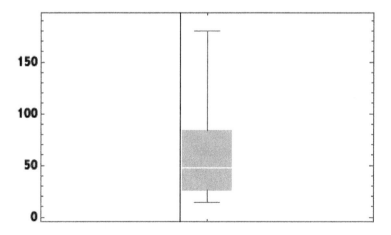

Fig. 1. Real Effort Visualization [7]

In Fig. 2 there can be a visualization of AUCP values calculated according original UCP method.

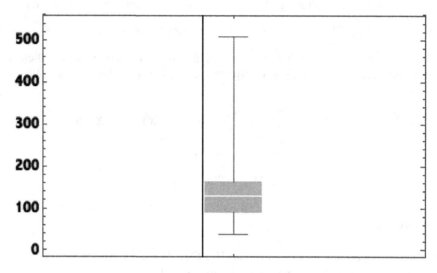

Fig. 2. AUCP values visualized from [7]

Modified AUCP equation produces significantly better estimates than AUCP the mean MRE is only 0.23 in the case of MAUCP, which is more than 85% better than that obtained with the UCP method. In Fig. 3. comparison of MRE is shown. The left plot represents modified version of AUCP equation

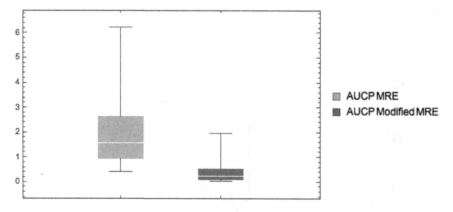

Fig. 3. MRE Comparison

Overall, Modified AUCP equation produces consistent and predicable effort estimations and is more suitable than the Use Case Points algorithm considered here.

5   Conclusion

In this paper, we presented a modified version of ACUP equation, which is more applicable for Use Case Points estimation. Setting a_1 and a_2 as obtained via least squares regression provides precise tuning.

The modified AUCP equation is significantly better than previously used estimation method, with a mean MRE that is 85% better than that obtained with UCP.

A comparison drawn with regard to a number of descriptive characteristics has shown that datasets can influence the quality of the estimation. Therefore, in our future research, we will focus on developing a complex estimation algorithm, which will be based on historical data and will be dived in two-phase estimation algorithm.

References

1. Karner, G.: Metrics for objectory, Diploma, University of Linkoping, Sweden, No. LiTH-IDA-Ex-9344. 21 (December 1993)
2. Braz, M.R., Vergilio, S.R.: Software effort estimation based on use cases. In: 30th Annual International Computer Software and Applications Conference, COMPSAC 2006. IEEE (2006)
3. Wang, F., et al.: Extended Use Case Points Method for Software Cost Estimation, pp. 1–5 (2009)
4. Diev, S.: Use cases modeling and software estimation. ACM SIGSOFT Software Engineering Notes 31(6), 1 (2006)
5. Mohagheghi, P., Anda, B., Conradi, R.: Effort estimation of use cases for incremental large-scale software development, pp. 303–311 (2005)
6. Azevedo, S., et al.: On the refinement of use case models with variability support. Innovations in Systems and Software Engineering 8(1), 51–64 (2011)
7. Ochodek, M., Nawrocki, J., Kwarciak, K.: Simplifying effort estimation based on Use Case Points. Information and Software Technology 53(3), 200–213 (2011)
8. Ochodek, M., et al.: Improving the reliability of transaction identification in use cases. Information and Software Technology 53(8), 885–897 (2011)
9. Pindyck, R.S., Rubinfeld, D.L.: Econometric models and economic forecasts. Irwin/McGraw-Hill, Boston (1998)
10. Silhavy, R., Silhavy, P., Prokopova, Z.: Requirements Based Estimation Approach for System Engineering Projects. In: Innovations and Advances in Computing, Informatics, Systems Sciences, Networking and Engineering, pp. 467–472. Springer International Publishing (2015)

Functional and Non-functional Size Measurement with IFPUG FPA and SNAP — Case Study

Mirosław Ochodek and Batuhan Ozgok

Poznan University of Technology,
Faculty of Computing, Institute of Computing Science,
ul. Piotrowo 2, 60-965 Poznań, Poland
Miroslaw.Ochodek@cs.put.poznan.pl

Abstract. Software size measures are probably the most frequently used metrics in software development projects. One of the most popular size measurement methods is the IFPUG Function Point Analysis (FPA), which was introduced by Allan Albrecht in the late-1970's. Although the method proved useful in the context of cost estimation, it focuses only on measuring functional aspects of software systems. To address this deficiency, a complementary method was recently proposed by IFPUG, which is called Software Non-functional Assessment Process (SNAP). Unfortunately, the method is still new and we lack in-depth understanding of when and how it should be applied.

The goal of the case study being described in the paper was to investigate how FPA and SNAP measurement methods relate to each other, and provide some early insights into the application of SNAP to measure the non-functional size of applications.

The results of the study show that SNAP could help mitigating some well-known deficiencies of the FPA method. However, we have also identified some potential problems related to applying SNAP in a price-per-size-unit pricing model.

Keywords: IFPUG, Function Points, FPA, SNAP, size measurement.

1 Introduction

Software size measures are probably the most frequently used measures in software development projects. They are, for instance, used as independent variables in many models of effort estimation and pricing. They also constitute a natural normalizing factor for many indirect measures.

In the late-1970's, Allan Albrecht of IBM introduced one of the most well recognized sizing methods, which is called Function Point Analysis (FPA) [2,4]. FPA measures the functionality which was requested and received by the user, independent of the technical details involved. Since 1986, the International Function Point Users Group (IFPUG) has continued to maintain Albrecht's method, which has become a de facto industrial standard.

In 2007 IFPUG started to work on developing a method complementary to FPA that could handle measuring non-functional aspects of software systems.

© Springer International Publishing Switzerland 2015
R. Silhavy et al. (eds.), *Software Engineering in Intelligent Systems,*
Advances in Intelligent Systems and Computing 349, DOI: 10.1007/978-3-319-18473-9_3

The first version of the new method, called Software Non-functional Assessment Process (SNAP), was released in 2010. The second release of the method was published in 2013[5], preceded by beta tests that gave promising results [10]. Since then, IFPUG started widely promoting the method.

Having both functional and non-functional size would definitely provide a more exhaustive picture of software development. Unfortunately, SNAP is a new method, and users have yet to learn its potential advantages. The next step which must be taken is to understand how FPA and SNAP be can be used together. This question is important from the practical perspective, because many projects use the price per Function Point model (a certain amount of money is agreed upon per Function Point). [1] Extending this model, by inclusion of an additional price per SNAP Point model, would require lowering the price of a Function Point in order to keep the price of the product at the same level. Although SNAP is beginning to be used in industry, we still lack an in-depth understanding on how these two measures could be applied conjointly.

The goal of the study presented in the paper was to collect and disseminate early insights on applying FPA and SNAP to measure the functional and non-functional size of software applications. We focused particularly on observing what benefits could the addition of SNAP to FPA bring in respect to the completeness of the size measurement and pricing model. We also wanted to investigate the pitfalls related to the SNAP method's application.

2 Methods

2.1 Objectives and Research Questions

We defined the following research questions in order to explore practical aspects of applying IFPUG FPA and SNAP methods to measure the size of software applications:

- RQ1: How can SNAP be applied to measuring user-level non-functional requirements?
- RQ2: Does the inclusion of SNAP into size measurement with FPA bring any added value?

In order to answer these questions, we decided to conduct a *case study*. A case study is an empirical method that "*aims at investigating contemporary phenomena in their context*" [11]. Its flexibility allows combining different sources of information and performing different types of analyses (both qualitative and quantitative). Focusing on a particular case helps in obtaining in-depth insights; unfortunately, it comes with the trade-off of reducing experimental control when compared to more traditional types of empirical studies, such as controlled experiments.

2.2 Case Description and Selection Strategy

We decided to follow the *embedded case study* design. The *case* was the application of IFPUG and SNAP methods to measure the size of a software application.

The *units of analysis* were three in-house software development projects (A, B, and C) developed at the Poznan University of Technology within the Software Development Studio framework [8] that aimed at delivering IT services for the University administration. All projects followed the fixed–price budgeting model.

We decided to choose these projects because of the similarities between them, and the fact that we had in-depth knowledge about them (including full access to all produced artifacts).

The projects were very similar in respect to the actual effort spent on development. In addition, they were developed in nearly the same environment. Such homogeneity made it possible for us to better isolate and focus on analyzing differences between the products they delivered, i.e. variation in non-functional requirements and types of functionality.

In addition to the main characteristics of the projects under study, presented in Table 1, we would like to briefly describe their goals, and the products they delivered.

Table 1. Main characteristics of the projects under study

	Project A	Project B	Project C
Actual effort [h]	1216	1260	1230
#User-level use cases	10	13	8
#User-level NFRs	28	38	58
Process	The process was organized according to the XPrince methodology [9], which combines PRINCE2 and eXtreme Programming.		
Team	Development team consisting of 6 people (4 developers, project manager and analyst). It does not include people involved in PRINCE2 roles such as Executive, Senior user, Senior supplier.		
Technologies	Java, Ninja fram. PostgreSQL DBMS	Java, SmartGWT, PostgreSQL DBMS	PHP, Moodle, PostgreSQL DBMS

Project A. The goal of the project was to develop an application that would collect, aggregate and report bibliometric information for University employees. It is a standalone web application that heavily depends on a legacy system which already stores and provides basic information about researchers, their publications and citations. Application A fetches data (using screen scrapping) from the legacy system, aggregates it (e.g., it calculates the measures that are used by Polish Ministry of Science and Higher Education to evaluate research units) and provides reports to users. It is worth mentioning that the legacy system is closed for modification and often contains incomplete or inconsistent data.

Project B. The goal of the project was to deliver an application that enables monitoring the assignment of organizational duties at the University (i.e. duties not related to the educational process and research). The created web-application enables assigning organizational duties to employees and monitoring the effort spent by individuals on these types of activities.

Project C. The goal of the project was to develop a plugin to the Moodle e-learning platform that enables surveying students and alumni. The requested functionality differed from the available surveying plugins for Moodle. For instance, the application should make it possible to aggregate and compare responses from different surveys. The plugin had to integrate with the existing web services at the University to fetch information about students.

2.3   Theory — IFPUG FPA and SNAP in a Nutshell

Function Point Analysis (FPA) is a method dedicated to measuring the functional size of an application, expressed in Function Points (FP). The SNAP method measures non-functional aspects of an application and expresses size in SNAP Points (SP). Therefore, when using both methods, a single application receives two size measures.

Figure 1 presents the main components of both methods. It is worth mentioning that both methods share some of the concepts and measurement procedure steps. For instance, they both require determining the purpose of the count, its scope and type (new development, application, or enhancement).

The next common step of measurement procedures is to determine so-called ① boundary, which is the border between the application being sized and its users (either people or other software systems). It is important to keep in mind that the boundary is determined from the user perspective, and not according to technicalities (e.g., architecture of a system). In order to distinguish technical components of application SNAP introduces the concept of ② partition.

Functional size is measured for data and transactional functions. Data functions relate to logical data that are identifiable by the users of a system. They are either stored within the system or retrieved from external sources. There are two types of *data functions*: ③ Internal Logical Files (ILF) (data maintained by application) and ④ External Interface Files (EIF) (data read from external sources). The complexity of each data function is determined based on the number of Data Element Types (DETs), which are unique attributes, and the number of Record Element Types (RET), which are logically related sub-groups of DETs. There are three complexity levels in the FPA method—low, average, and high. The size of the data function measured in FP is determined based on its type and complexity.

The interactions between the application and its users are measured using the concept of *transactional function*. Each transactional function has to be a separate *elementary process*—it is the smallest, self-contained and consistent unit of activity that is meaningful to the user. There are three types of functional transactions defined in FPA: ⑤ External Input (EI) (data coming from the outside of

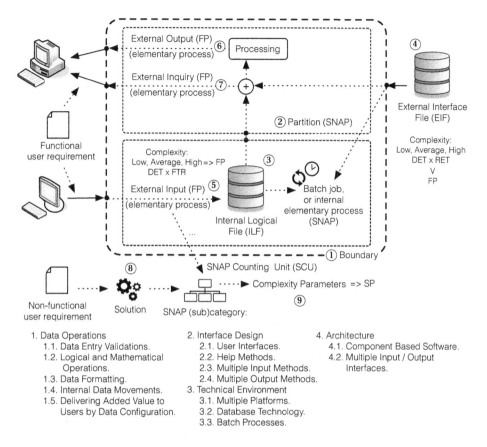

Fig. 1. The main components of IFPUG FPA and SNAP

an application to modify data or change state) , ⑥ External Output (data are processed within and sent outside boundary), and ⑦ External Inquiry (data are sent outside boundary without processing). The complexity of a transactional function is determined based on the number of DETs crossing the boundary and the number of logical files that are read or written into (FTRs). Again, the number of FP is determined based on the type of the transactional function and its complexity.

The base functional size can be adjusted by the so-called value adjustment factor (VAF). It is calculated based on the assessment of 14 general system characteristics (GSCs), which correspond to non-functional aspects of the system. The application of VAF can modify the initial size by +/-35%. SNAP was, in fact, created to replace GSCs and VAF.

SNAP can be used independently of FPA, however it easier to apply the method if data and transactional functions were previously identified and measured. One of the key steps in the measurement process is associating non-functional user requirements with SNAP (sub)categories ⑧. In most cases it cannot be done directly. First, one must determine how the requirement is going to be implemented. Knowing the solution, we can try to find an appropriate SNAP (sub)category. SNAP defines 4 categories, which are further divided into 14 sub-categories (see Figure 1). Each sub-category defines a SNAP Counting Unit (SCU). The SCU is a component or activity in which complexity and non-functional size is assessed. Most categories define SCU as an elementary process.

The number of SP is calculated for each SCU taking into account complexity parameters defined by the selected sub-category ⑨. For instance, category 1.1 Data entry validation defines SCU as an elementary process. Therefore, if there is a non-functional requirement related to entry validation, we have to identify all transaction functions it affects (they are identified as SCUs). Sub-category 1.1 defines two complexity parameters — nesting level and the number of DETs used for validation. They have to be calculated for each SCU. Finally, one has to apply appropriate rules to obtain their non-functional size in SP.

2.4   Study Procedure

During the case study we decided to perform the following steps:

1. Size measurement — this step involved measuring the functional (FPA) and non-functional (SNAP) size of applications A, B, C. We performed measurements based on the Software Requirements Specification (SRS) document, user manuals, accessing working software, and technical documentation. During this step, we also collected practical insights related to calculating Function and SNAP Points.
2. Analysis of mapping between non-functional user requirements and SNAP (sub)categories — we explored issues related to mapping between the non-functional requirements defined in SRS and SNAP (sub)categories.
3. Analysis of measurement results — in this step we focused on the analysis of measurement results, including the analysis of relationships between SNAP and FPA.

3   Results

3.1   Results of Functional Size Measurement

We identified and measured 5 data functions in each of the applications, while the number of transactional functions differed from 14 to 23. The final functional size of the applications ranged from 92 to 131 Function Points (see Table 2 for details). The functional size of applications A and C turned out to be almost equal (92 and 98 FP), while application B was visibly larger (131 FP). Taking into account only the applications's functional size, the Product Delivery Rate (PDR) would range from 9.6 to 13.2 hours per FP depending on the project (A=13.2; B=9.6; C=12.6).

Finally, we assessed GSCs and calculated VAFs for the projects: A=1.04; B=1.06; C=1.10.

Table 2. Summary of FPA calculations for applications A, B, and C

Project	Function type	#Low	#Average	#High	FP
			Complexity		
Project A	ILF	5	0	0	35
	EIF	0	0	0	0
	EI	4	1	1	22
	EQ	4	1	0	16
	EO	0	1	2	19
			Functional size		92
Project B	ILF	5	0	0	35
	EIF	0	0	0	0
	EI	4	3	5	54
	EQ	6	3	0	30
	EO	0	1	1	12
			Functional size		131
Project C	ILF	4	0	0	28
	EIF	1	0	0	5
	EI	9	0	0	27
	EQ	5	2	1	29
	EO	1	1	0	9
			Functional size		98

3.2 Results of Non-functional Size Measurement

The first major step of calculating SNAP Points was the analysis of non-functional user requirements and mapping them into SNAP (sub)categories. It is worth mentioning that, in all projects, requirements were elicited using so-called SENoR workshops [7]. Such a workshop has the form of a brainstorming session, where participants analyze each of the ISO 25010 [6] quality (sub)characteristics in order to identify potential non-functional requirements.

During the analysis we filtered out requirements related to development process or environment (ca. 15% of requirements), because they could not be measured in SNAP. We were able to establish mappings between 29–48% of the remaining non-functional user requirements and SNAP (sub)categories (A=48%; B=34%; C=29%). Table 3 presents the number of non-functional requirements that were successfully associated with SNAP (sub)categories, grouped by ISO/IEC 25010 characteristics. In addition, we observed that 6–22% non-functional user requirements were fully measured in FPA rather than SNAP (e.g., functions related to authentication, authorization, and managing data access permissions).

Table 3. Number of non-functional user requirements mapped to SNAP categories in comparison to the total number of non-functional requirements; grouped by ISO/IEC 25010 characteristics

Category	Project A	Project B	Project C
Compatibility	0 / 1	3 / 5	3 / 6
Functional suitability	1 / 2	1 / 1	2 / 2
Maintainability	0 / 3	0 / 7	0 / 12
Performance efficiency	1 / 2	0 / 1	0 / 4
Portability	1 / 1	1 / 6	1 / 3
Reliability	2 / 4	2 / 3	1 / 3
Security	1 / 4	1 / 5	1 / 6
Usability	4 / 6	2 / 4	4 / 11
Documentation*	2 / 2	1 / 1	2 / 2

This category is not an ISO 25010 (sub)characteristic.

In the next steps we identified SNAP Counting Units for each associated non-functional requirement, and finally calculated SNAP Points. The summary of SNAP counts for the applications is presented in Table 4.

3.3   Relationship between FP and SP

At the application level, the ratio between SP and FP was, on average, 3.6 and varied visibly between the applications (A=4.1; B=3.2; C=7.4).

The relationship between FP and SP can be also considered at the elementary process level, because it is the most frequently defined type of SCU in the method. The ratios between SP and FP calculated for transactional functions were on average equal to 5.49, however they also differed visibly between projects (A=4.46; B=3.34; C=8.84). Figure 2 presents distribution of the SP/FP ratio for all applications.

As the next step, we investigated whether the ratio between SP and FP depends on the type of transactional function (the observed mean SP/FP ratio: EI=7.11; EQ=3.44; EO=5.65). We decided to use the Kruskal-Wallis test, which is a non-parametric version of ANOVA, because we suspected, having analyzed QQ-plots and performing Shapiro-Wilk tests, that the normality assumption may be violated. We were able to reject the null hypothesis about the equality of medians with the presupposed significance level, α equal to 0.05 (p-value = 0.005).

Proceeding to the next step of analysis, we performed the post-hoc pairwise comparison of subgroups according to the procedure proposed by Conover [3]. According to the results of the analysis, the following differences between the median number of SNAP Points per transactional function (μ_{SP}) seemed significant: $\mu_{SP}(EI) > \mu_{SP}(EO) > \mu_{SP}(EQ)$.

Finally, we investigated whether there is a correlation between the functional and non-functional size of transactional functions measured with FPA and

Table 4. Summary of SNAP calculations for applications A, B, and C

	Project A			Project B			Project C		
SNAP Category	#SCU	SP	%	#SCU	SP	%	#SCU	SP	%
1. Data Operations	–	48	12.7	–	192	45.3	–	168	23.2
1.1 Data Entry Validation	7	48	12.7	9	44	10.4	9	36	5.0
1.2 Logical and Mathematical Operations	0	0	0	0	0	0	0	0	0
1.3 Data Formatting	0	0	0.0	2	6	1.4	1	4	0.6
1.4 Internal Data Movements	0	0	0.0	7	64	15.1	2	32	4.4
1.5 Delivering Added Value to Users by Data Configuration	0	0	0.0	3	78	18.4	3	96	13.2
2. Interface Design	–	182	48.1	–	174	41.0	–	123	17.0
2.1 User Interface	11	139	36.8	15	156	36.8	4	45	6.2
2.2 Help Methods	1	31	8.2	1	12	2.8	1	78	10.8
2.3 Multiple Input Methods	0	0	0.0	0	0	0.0	0	0	0.0
2.4 Multiple Output Methods	2	12	3.2	1	6	1.4	0	0	0.0
3. Technical Environment	–	148	39.2	–	58	13.7	–	434	59.9
3.1 Multiple Platforms	0	0	0.0	0	0	0.0	18	360	49.7
3.2 Database Technology	6	48	12.7	10	54	12.7	11	66	9.1
3.3 Batch Processes	2	100	26.5	1	4	0.9	2	8	1.1
4. Architecture	–	0	0.0	–	0	0.0	–	0	0.0
4.1 Component Based Software	0	0	0.0	0	0	0.0	0	0	0.0
4.2 Multiple Input / Output Interfaces	0	0	0.0	0	0	0.0	0	0	0.0
Non-functional size	–	378	–	–	424	–	–	725	–

SNAP. The calculated Kendall's rank correlation τ was equal to 0.035, therefore it is rather unlikely that such correlation exists (p-value = 0.74).

4 Discussions

4.1 Adequacy of Functional Size Measurement

In our opinion, the obtained functional size measurement results correctly characterized the applications.

Looking from the end-user perspective, the functionality of application A seemed to be the simplest among the systems being analyzed. In essence, it provided access to on-line reports concerning individual researchers and research teams, presenting various bibliometric measures associated with them. It also had a limited number of input and administrative functions. The functional size

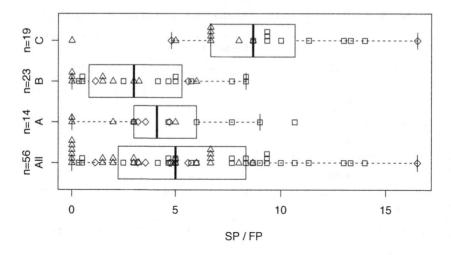

Fig. 2. Ratio between SNAP and Function Points calculated for transactional functions (□ — External Input; ◊ — External Output; △ — External Inquiry)

of application C seemed similar, however it offered many more input functions (e.g., editing and responding to surveys). It also seemed to have more less complex functions. The functionality of application B could be characterized as a combination of CRUDL (Create, Retrieve, Update, Delete, Listing) and workflow operations. Different actors are allowed to modify different portions of data in the course of the business process. These modifications often trigger changes in object states; this results in many simple input functions. In addition, the application has an administrative panel which allows modifying most objects by entering SQL queries.

4.2 Adequacy of Non-functional Size Measurement

Prior calculating SNAP, we perceived application A as the most complex. Surprisingly to us, the SNAP measurement results yield contradictory results. They indicated application A as being the smallest one.

In our opinion, the reason for that was that we based our subjective assessment of non-functional size on the perceived difficulty of implementing non-functional requirements, while SNAP measures *non-functional aspects of functional requirements* (with one exception—sub-category 2.2 Help methods). We believe that this is one of the key issues needing to be understood in order to correctly interpret measurement results. For instance, a single non-functional requirement stating that application C should support three web browsers (IE 7.0+, Firefox 15, Opera 12) generated nearly 50% of SP (sub-category 3.1 Multiple Platforms), because it affected nearly all elementary processes. It is worth emphasizing that the addition of a new elementary process which interacts with web browsers will automatically increase the non-functional size of the application by 20 SP. If we resigned from this requirement, the size of application C would drop to 365 SP, making it the smallest in the analyzed data set.

We were able to capture some specific non-functional aspects of the applications under study. For instance, because application A would not be able to process data on-line, a special internal batch job was implemented to periodically re-calculate aggregated values (sub-category 3.3 Batch processing). Some of the reporting functions of applications B and C were supposed to be delivered using the existing Business Intelligence platform. However, in order to create a new type of report, the platform had to be configured to establish connection and fetch data (1.5 Delivering added value to users by data configuration). In addition, we were able to reflect the need to integrate applications B and C with a special e-mail delivery service, which we placed in a separate partition (sub-category 1.4 Internal data movements).

4.3 Requirements and Solutions: Customer vs. Developer Perspective

SNAP APM states that *SNAP sub-categories do not define or describe non-functional requirements; they classify how these requirements are met within the software product.* [5] We felt that there is a gap between the non-functional user requirements which are defined in SRS and the technical solutions that can be measured with SNAP. We believe that, from a practical point of view, bridging the gap by associating requirements with SNAP categories is a focal point of the method. It seems especially important if we are to use the price-per-SP pricing model.

In some cases, we were able to easily associate non-functional user requirements with SNAP sub-categories. For instance, we found it trivial to associate non-functional requirements related to user documentation and built-in help (sub-category 2.2 Help methods). The same relates to the majority of requirements concerning user error protection (sub-category 1.1 Data entry validation). Unfortunately, we also observed that associating some of the other non-functional requirements may be a cumbersome task. Especially when we need to identify the way that the non-functional requirement is going to be met. For instance, we observed that it was the case for requirements concerning performance. Looking at performance from the user perspective, it is important how quickly the system is going to respond to stimuli, what the transaction throughput will be, etc. Unfortunately, in order to measure requirements in SNAP, one has to decide how the requested performance-level is going to be achieved. The most obvious SNAP sub-categories that seem adequate in this respect are 3.2 Database technology, 4.2 Multiple I/O interfaces and, to some extent, 3.3 Batch processing (e.g., internal, periodic processes used to calculate aggregated values). However, the list of possibilities is not fully covered by SNAP sub-categories. For instance, if performance is going to be achieved by writing optimized, well-designed code, it will not be possible to measure this type of requirement using SNAP. We believe that dependency on solutions can, in many cases, negatively affect the transparency of the measurement process for both the customer and developer.

For instance, the developer could claim that it is necessary to add large number indices to a database in order to achieve the requested performance-level. It would be very difficult, or even impossible, for the customer to verify if the step is really necessary. On the other hand, the developer could select very sophisticated technology which allows meeting the performance requirements out-of-the-box; it might again be impossible to measure the requirement in SNAP. The aforementioned problem is probably less visible if SNAP is used internally by a software development organization.

In order to solve that problem we recommend creating a dedicated manual on top of SNAP APM, which should contain detailed rules concerning how to associate non-functional requirements with SNAP (sub)categories. For instance, a good example would be category 2.1 User interfaces. It is worth discussing in the project the types of UI elements which are present in the system, what properties they poses and how they are going to be configured to meet the requirements (e.g., UI elements have white background by default, which suits customer). In the projects under study we identified 21 requirements related to usability. However, we found it difficult to determine which properties were actually configured to meet usability requirements, which were left with default configuration, and which were configured by developers because of reasons other than meeting non-functional user requirements.

4.4   Measurement Deferred in Time

During the requirement analysis, we observed that some of the non-functional user requirements could not be measured when they are defined, because there were no SCUs available at the time. It can be seen when the customer requests to prepare infrastructure for a functionality which is going to be implemented sometime in the future. Therefore a change of this sort could be measured only when SCUs are available in the forthcoming releases. Similarly, if only a subset of potential SCUs is defined at the moment of the non-functional user requirement being construed, it can then only be partially measured and the rest of the measurement will be deferred until the remaining functions are available. For instance, if the customer requests creating a database infrastructure for logging events (code data) in the system, but the functionality that is going to be included in the logging process is yet to come, the result of the measurement in the sub-category 3.2 Database operations will be equal to zero.

4.5   Maintainability and Design Constraints

The largest group among the not-mapped non-functional user requirements was related to the maintainability characteristic (22 requirements). The requirements were related to the requested properties of code, e.g., coding standards or the language used for writing comments within the code. There were also requirements stating that it needs to be possible to reconfigure the application to run in a special safe mode (to ease development and testing new functions), which, instead of communicating with other systems, uses pre-prepared mock interfaces delivering exemplary data. In application C there were even requirements to prepare infrastructure for implementing a certain set of potential enhancements.

4.6   Combining FPA and SNAP

SNAP Added Value to FPA. We observed that combining SNAP and FPA for pricing may be beneficial. It enables pricing requests for change which would not be possible to measure using standard FPA. For instance, in applications A and B there were requirements stating that some of the reports should be available in two formats—web pages and spreadsheet files. If only FPA rules were applied, we would count only one transactional function per report no matter how many different output formats were requested. This is because only unique elementary processes[1] are taken into account when calculating FP. Therefore, requesting to handle new output methods for an existing elementary process would not increase the functional size of application. This is often difficult to accept when FPA is used for pricing. SNAP introduces sub-categories 2.3 and 2.4 that could be used to measure requirements concerning additional input and output methods.

In our opinion, SNAP might be especially useful in enhancement projects, because certain frequently requested types of modifications of applications could not be counted in FPA. For instance, requesting changes in the graphical user interface without changing DETs, FTRs or the processing logic of transactional functions cannot be measured in FPA, however they can be measured in SNAP sub-category 2.1 User interfaces. The same relates to the modification of database schema without changing DETs or RETs of logical files (sub-category 3.2 Database technology).

Relationship between SP and FP. Before the tandem of SNAP and FPA is going to become broadly accepted by industry, we need to better understand how both measures relate to each other. Due to the small sample size analyzed in this case study, we were not able to make strong conclusions concerning the relationship between SP and FP at the application level. However, we can state that one should expect the number of SP to be at least several times greater than the number of FP (we suspect that for complex systems it could be even an order of magnitude).

Our observation from this case study is that, in order to predict the potential number of SP, one should focus on identifying non-functional requirements that affect many elementary processes. In our case it was a requirement related to supporting multiple web browsers, however we believe that it could concern other sub-categories as well. For instance, if we introduce additional partitions, we should identify elementary processes that cross the boundaries of partitions, because they may generate a large number of SP in sub-category 1.4 Internal data movements.

---

[1] An elementary process is not unique if there is another elementary process that require the same set of DETs, FTRs and types of processing logic.

Another practical insight from the case study was that we did not observe any visible correlation between SP and FP at the transactional function level. Therefore, if we wish to estimate the number of SP based on the number of transactional functions, we need not necessarily consider the complexity of these functions. Still, it may be valuable to identify types of transactional functions. We should expect to have more SP calculated for External Inputs.

SNAP vs. GSC. Interestingly, both SNAP and GSC indicated the same order of applications in respect to non-functional size ($C > B > A$). However, the differences between VAFs were minor — ranging from 0.02 to 0.06.

4.7   Threats to Validity

The main threats to validity relate to the correctness of size measurement, which is by its nature subjective. In order to mitigate this threat, the measurement was perform by one of the authors who is an IFPUG Certified Function Point Specialist (CFPS), and had participated in one of the IFPUG's workshops on SNAP. In addition, the authors had deep knowledge of the projects, which limits the possibility of introducing bias in the requirement analysis stage.

Another important group of threats relate to external validity—ability to generalize results onto the bigger population. In case studies we usually focus on analyzing a limited number of cases, but we achieve deeper understanding of the phenomenon in a given context. Unfortunately, it limits the possibility of generalizing the results beyond these circumstances. One important threat to validity regards the size of the analyzed projects. We analyzed projects with actual an effort value of around 1K man-hours. It is difficult to state how the application size influenced the observations being made, however we suspect that increasing size should not have significant impact on our observations, as it would lead to a potential increase in the number of SCUs for non-functional requirements. However, if function is more complex in respect to its number of DETs, it may have visible impact on those categories which use DETs as attributes (e.g., 1.1 Data entry validation).

5   Conclusions

In the study presented in this paper, we measured the functional and non-functional size of three applications in order to gain insight into combining the well-known IFPUG FPA and the relatively new SNAP method. We focused particularly on investigating the problems and benefits of applying the latter one.

We observed that FPA and SNAP synergize and are able to better characterize applications when used together. In addition, we collected a set of observations that may help new SNAP users to better understand the measurement method. First of all, we noted that certain non-functional user requirements could not be measured in SNAP (e.g., requirements related to maintainability). We further discovered that in some cases non-functional requirements could not be measured at the time of their definition, because SCUs were not yet available. We also investigated the relationship between SP and FP; we can conclude here that it is rather unlikely that the non-functional size of transactional functions correlates with their functional size. On the other hand, we believe it is highly probable that a transactional function's type influences its size measured in SP. Finally, we found that a single non-functional requirement can visibly influence the size of an application if it affects a large number of elementary processes (e.g., a requirement concerning the necessity to support many web browsers).

References

1. Aguiar, M.: When metrics mean business. In: Proceedings of the 9th Software Measurement European Forum, pp. 1–12 (2012)
2. Albrecht, A.J.: Measuring application development productivity. In: Proceedings of the Joint SHARE/GUIDE/IBM Application Development Symposium, pp. 83–92 (October 1979)
3. Conover, W.J.: Practical Nonparametric Statistics, 3rd edn. John Wiley & Sons (1999)
4. IFPUG. Function Point Counting Practices Manual, Release 4.3.1 (2010) ISBN 978-0-9753783-4-2
5. IFPUG. Software Non-functional Assessment Process (SNAP) — Assessment Practices Manual, Release 2.2 (2014) ISBN 978-0-9830330-9-7
6. ISO/IEC. 25010: 2011: Systems and software engineering–Systems and software Quality Requirements and Evaluation (SQuaRE)–System and software quality models (2011)
7. Kopczynska, S., Nawrocki, J.: Using non-functional requirements templates for elicitation: A case study. In: 2014 IEEE 4th International Workshop on Requirements Patterns (RePa), pp. 47–54. IEEE (2014)
8. Kopczynska, S., Nawrocki, J., Ochodek, M.: Software development studio— Bringing industrial environment to a classroom. In: 2012 First International Workshop on Software Engineering Education based on Real-World Experiences (EduRex), pp. 13–16. IEEE (2012)
9. Nawrocki, J., Olek, Ł., Jasinski, M., Paliświat, B., Walter, B., Pietrzak, B., Godek, P.: Balancing agility and discipline with XPrince. In: Guelfi, N., Savidis, A. (eds.) RISE 2005. LNCS, vol. 3943, pp. 266–277. Springer, Heidelberg (2006)
10. Tichenor, C.: A new metric to complement function points. CrossTalk—The Journal of Defense Software Engineering 26(4), 21–26 (2013)
11. Yin, R.K.: Case Study Research: Design and Methods. SAGE Publications (2003)

A Models Comparison to Estimate Commuting Trips Based on Mobile Phone Data

Carlos A.R. Pinheiro[1,2], Véronique Van Vlasselaer[1], Bart Baesens[1],
Alexandre G. Evsukoff[3], Moacyr A.H.B. Silva[2], and Nelson F.F. Ebecken[3]

[1] KU Leuven, Research Center for Management Informatics,
Naamsestraat 69, Leuven, Belgium
`{carlos.pinheiro,bart.baeses,`
`veronique.vanvlasselaer}@kuleuven.be`
[2] Getúlio Vargas Foundation, School of Applied Mathematics,
Praia de Botafogo 190, Botafogo, Rio de Janeiro, Brazil
`moacyr.silva@fgv.br`
[3] Federal University of Rio de Janeiro, Department of Civil Engineering,
Centro de Tecnologia, Cidade Universitária, Rio de Janeiro, Brazil
`alexandre.evsukoff@coc.ufrj.br, nelson@ntt.ufrj.br`

Abstract. Upon an overall human mobility behavior within the city of Rio de Janeiro, this paper describes a methodology to predict commuting trips based on the mobile phone data. This study is based on the mobile phone data provided by one of the largest mobile carriers in Brazil. Mobile phone data comprises a reasonable variety of information about subscribers' usage, including time and location of call activities throughout urban areas. This information was used to build subscribers' trajectories, describing then the most relevant characteristics of commuting over time. An Origin-Destination (O-D) matrix was built to support the estimation for the number of commuting trips. Traditional approaches inherited from transportation systems, such as gravity and radiation models – commonly employed to predict the number of trips between locations(regularly upon large geographic scales) – are compared to statistical and data mining techniques such as linear regression, decision tree and artificial neural network. A comparison of these models shows that data mining models may perform slightly better than the traditional approaches from transportation systems when historical information are available. In addition to that, data mining models may be more stable for great variances in terms of the number of trips between locations and upon different geographic scales. Gravity and radiation models work very well based on large geographic scales and they hold a great advantage, they are much easier to be implemented. On the other hand, data mining models offer more flexibility in incorporating additional attributes about locations – such as number of job positions, available entertainments, schools and universities posts, among others –and historical information about the trips over time.

Keywords: human mobility behavior, trips prediction, transportation models, pattern recognition.

© Springer International Publishing Switzerland 2015
R. Silhavy et al. (eds.), *Software Engineering in Intelligent Systems*,
Advances in Intelligent Systems and Computing 349, DOI: 10.1007/978-3-319-18473-9_4

1   Introduction

Human mobility analysis has received reasonable attention in recent years. However, it is not a straightforward and trivial task to collect data which allows performing human mobility studies on large scale. Many studies show relevant results by tracking mobile phone data to learn about human mobility behavior [1] [2] [3] [4] [9] [12] [14] [18] [21]. Other approaches are based on the collection of transactions by using cell phone applications and GPS devices [1] [2] [6] [22]. Several studies are based on the mobile phone records and the trip prediction modeling is developed under this approach. These studies raise general knowledge about human motion considering distinct geographic scales [1] [2] [8] [18] [21] [22].

A frequent focus in human mobility analysis is associated to understand migration trends in large geographic scale and to predict the number of trips between locations [7] [14]. Many studies have been carried out on human mobility in developed countries [4] [12] [13] [14]. However, less is known about human mobility in developing countries like Brazil.

Trip prediction models perform well when they are applied on large geographic scales. However, their accuracy decreases when geographic scale is lower [15] [17]. Traditional approaches applied to solve problems in transportation systems are the gravity and radiation models [10]. Statistical and data mining models are approaches commonly deployed in machine learning solutions. Linear regression, decision tree and artificial neural network are frequently applied in a variety of real world problems, performing very well in most of cases. Data mining can also be used to understand frequent trajectories and human mobility behavior [11] [20] [22]. In this paper we compare the performance of statistical and data mining models in addition to the transportation methods.

2   Methodology and Data Preparation

Mobile phone data contains transactional records about caller and callee information, associating each call or text message to the corresponding cell tower that is spread out through the metropolitan areas. These cell towers process the incoming and outgoing calls, as well as text and multimedia messages sent and received by subscribers, providing relevant information about their geographic locations at particular points in time. This geographical information basically consists of the latitude and longitude, the radius covered by the cell towers and information about the physical addresses such as street, neighborhood, city and state. Even though such locational information only approximates real human mobility, recent studies [5] [19] show that by using the appropriate techniques, mobile phone data may offer the possibility to statistically characterize human trajectories and journeys on an urban area scale.

We used mobile data provided by one of the largest telecommunications company in Brazil. The models were developed considering six months of call detail records, comprising 3.1 billion records, 2.1 million subscribers, 1,500 cell towers, 416 neighborhoods and 26 counties.

2.1   Presumed Domiciles and Workplaces

Mobile carriers have an unbalanced distribution of prepaid and postpaid phones. For the mobile carrier in this study, it is 85% for prepaid and 15% for postpaid. Due to this distribution, there is a lack of information about subscribers' home address. As information about home address and workplace has an indispensable value in building commuting trips, we computed *presumed domiciles* and *workplaces* for all subscribers, irrespective whether they are prepaid or postpaid.

The presumed domicile is the most frequently visited location for each subscriber between 07:00 pm and 07:00 am the day after. Analogously, the presumed workplace is the most visited location for each subscriber during the period between 09:00 am and 05:00 pm. The periods of 07:00 am and 09:00 am and from 05:00 pm to 07:00 pm were considered as *commuting periods*, which represent the time when people are moving from home to work or from work back to home.

Two additional constraints were imposed to identify the presumed domiciles and workplaces. First, a location can be considered a presumed domicile or workplace only if it is visited at least three times by the subscriber over the entire period of analysis. This threshold was chosen based on the subscribers' distribution considering all locations over time. The value of three is assigned to the percentile p10 (90% of the subscribers have more than 3 visits). Moreover, for each analyzed time slot, a location considered as a presumed domicile or workplace needs to be at least 20% more frequently visited than the second most frequent visited location. This threshold was chosen based on the distribution of the difference between first and second most visited locations. The value of 20% corresponds to the percentile p25 (75% of the subscribers visit the first most frequently visited location more than 20% of the times than they visit the second one).

In order to validate the method to compute presumed domiciles, we compare the population obtained to the official neighborhoods' population provided by official reports of Rio de Janeiro's administration offices. The presumed domiciles explain more than 88% of the observed population in the city according to official reports. There is no similar report available to validate the presumed workplaces.

2.2   Clustering the Cell Towers

In order to compare the performance of the transportation and data mining models upon different geographic scales, we are going to use the cells towers – the lowest granularity in terms of geographic areas –, the neighborhoods – an aggregation of cell towers according to the administrative boundaries in space – and a fixed number of cell towers clusters.

The clustering method used in this study is based on distance. In order to find the best number of clusters, we computed the Cubic Clustering Criterion (CCC) under the uniform null hypothesis [27]. A preliminary clustering was performed to find a possible optimal number of clusters upon the coordinates of the cell towers. This clustering was based on Ward's method [28], defining the distance between two clusters as sum of squares between the two clusters added up over all the variables.

Then a second clustering process was performed upon the fixed number of clusters (61) found in the first step. This second method was based on a disjoint cluster analysis on the distances computed from two quantitative variables, the latitude and the longitude of the tower cells.

3   Gravity and Radiation Models to Predict Commuting Trips

The gravity model tracks its origin from the gravitational law. Two bodies are attracted to one another with a force that is proportional to the product of their masses and inversely proportional to the square of their distance. Gravity models are mapped to human mobility by replacing bodies by locations and masses by importance. Importance can be measured in terms of population, but can also incorporate other attributes like the number of jobs, gross domestic product, public facilities, transportation resources, traffic routes, among others.

The *gravity model* usually incorporates parameters to define constraints to paths and displacements followed by people, such as the cost to travel [10]. The cost to travel consists of several attributes like the distance, the resources to cover the path, the number of people travelling, etc. In particular, it assumes that the commuting activity between two distinct locations is proportional to the product of the population of these two locations and inverse proportional to the distance between these two locations.

Individuals are attracted to other locations as a function of the distance between two different places and the cost of travel between them. The gravity model considers that individuals are more attracted to close locations than to long-distance locations. This last hypothesis is based on the natural limited resources to travel between locations and the higher cost involved in long distances. In problems related to transportation systems, the distance to travel is a crucial factor in users' decision making process when they have to commute between locations. Trips between two locations with the same distance may have different costs, for instance based on possible routes, traffic jam, public transportation resources etc. The population of the locations involved in the trip is also important to predict trips within geographic areas. Large number of people associated to origin and destination locations may imply more trips.

The gravity model is expressed as a simple function including the population in the origin location, the population in the destination location, and the distance between both locations. This is expressed in the following formula.

$$T_{ij} = k \, \frac{P_i P_j}{d_{ij}} \tag{1}$$

where T_{ij} is the number of trips between locations i and j, P_i is the population of location i, P_j is the population of location j, d_{ij} is the distance between locations i and j, and k is a scaling factor. The scaling factor is a problem-specific parameter, and differs between geographical regions. This factor is estimated by linear regression models.

The *radiation model*, on the other hand, tracks its origin based on theories about diffusion dynamics, where particles are emitted at a given location and have a certain probability of being absorbed by surrounding locations.

It uses the spatial distribution for the population as input, not needing any other additional parameters. It basically depends on the populations of the locations involved in the trips and their distances [10].

As the radiation model is parameter-free, the model can be much easier implemented in mobility behavior analyses and trip prediction models, especially when using mobile phone data. Although the radiation model may not seem sufficient to predict human mobility in low geographic scales, this model is successfully applied in reproducing mobile patterns at large spatial scale [16] [17] [18]. As a result, this type of model can reasonably forecast mobility trends and the number of trips in great metropolitan areas, big cities and even countries, particularly when long-distances travels are involved.

Accordingly, the radiation model is expressed as a function of the population of the origin location, the population of the destination location, and by the *radius population* which is the population comprised within a circle where the origin location is the center and the radius is the distance between the origin and destination locations (discarding the populations of origin and destination locations). The radiation model also considers the number of travels initiated in the origin location, considering all possible destinations. This attribute may increases the number of trips between any two locations, for instance when the origin location works as a hub. This is expressed by the following formula.

$$T_{ij} = T_i \frac{P_i P_j}{(P_i + P_{ij})(P_i + P_j + P_{ij})} \tag{2}$$

where T_{ij} is the number of trips between locations i and j, T_i is the number of trips started at location i (not matter the destination), P_i is the population in location i, P_j is the population in location j, and P_{ij} is the total radius population in the circle of radius r_{ij} (the distance between locations i and j) centered at location i (excluding the populations of locations i and j).

4   Statistical and Data Mining Models to Predict Commuting Trips

4.1   Linear Regression Models

Considering the same geographic scales, we developed linear regression models and applied them on our mobility dataset in order to predict the commuting trips. We choose to estimate two types of linear regression models: a Quantile Regression and a Robust Regression. We found that those two models performed slightly better than the models inherited from physics – i.e. the gravity and radiation model. For comparison purposes, the mobility's attributes used to feed these statistical models were the same as used in the gravity and radiation models. These attributes include the distance between the two locations involved in the trip, the number of trips originated from the origin location (regardless the destination), the population of the origin location, the population of the destination location and the radius population (the population in the circle between origin and destination locations).

Both statistical models estimate the dependent variable Y by linear predictor functions of regression variables X, including estimated intercepts β_n, as presented in the following formula:

$$Y = \beta_0 + \beta_1 * Dist + \beta_2 * PopDst + \beta_3 * PopOrg + \beta_4 * PopRad + \beta_5 TripsIni + \varepsilon \quad (3)$$

with Y the estimated number of trips, β_i the parameters assigned to the mobility attributes used to predict the number of trips between locations. In general, we use five independent variables to estimate   linear regression formula: the distance between origin and destination locations *(Distance)*, the population of destination location *(PopDst)*, the population of origin location *(PopOrg)*, the population within the circle between origin and destination locations *(PopRad)* and the number of trips initiated in origin location *(TripsIni)*.

The first model is based on Quantile Regression [23], which generalizes the concept of a univariate quantile to a conditional quantile given one or more covariates. This method models the effect of covariates on the conditional quantiles of a response variable by means of quantile regression. Ordinary least squares regression models the relationships between the covariates X and the conditional mean of the response variable Y, given $X = x$. The Quantile Regression extends the regression model to conditional quantiles of the response variable such as the median or the 90th percentile.

The second model is a linear regression which detects and considers outliers, and provides resistant and stable results, even in the presence of unexpected observations. The Robust Regression [24] limits the influence of outliers' presence by using a high breakdown value method based on the least trimmed squares estimation. The breakdown value is a measure of the contamination's proportion, and then an estimation method can withstand and still maintains its robustness.

The variation of the number of trips between two distinct locations in the city of Rio de Janeiro is significantly high – varying from 2 to 148K, with a coefficient of variation of 738. As some two locations have few trips between them, others have a huge number. Particular pairs of locations seem to behave as outliers, as there are many trips between those locations. The Quantile Regression and the Robust Regression can handle such nonlinear distribution of the number of trips between two locations better than the regular linear regression models.

4.2   Decision Tree Models

A decision tree is a supervised learning model for classification problems [25]. Each input variable may corresponds to a node in the tree – if it increases the classification rate. Otherwise an input variable may be discarded. Each possible value for the input variable corresponds to edges to split nodes. In the case of continuous values the algorithm estimates cut-off values to properly create the edges. Each leaf node represents a value of the target variable given the values of the input variables, represented by the path from the root of the tree to the leaf. Afterwards, the algorithm prunes away some of the created paths in order to avoid *overfitting*. A tree learns by splitting the input data set into subsets by testing the attributes' value. The model evaluates the data on the target variable by selecting the most promising independent variable to distinguish between the values of the target variable. This process is recursively repeated on each derived subset. This process is called *recursive partitioning*, and it ends when the subset at a node has all the same value of the target variable, or when splitting no longer adds value to the predictions. In data mining, a decision tree can be described as the combination of mathematical and computational techniques to describe, categorize and generalize a particular data set. In this case the data comes in the following form.

$$(x, Y) = (x_1, x_2, x_3, \dots, x_{k,}, Y) \tag{4}$$

The dependent variable Y is the target variable and the vector x is composed of the input variables, $x_1, x_2, x_3, \dots, x_k$. Once again, the dependent variable, or the target variable, is the number of trips estimated within the timeframe of analysis. The independent variables are the ones assigned to the distances and populations involving the locations travelled by the subscribers throughout the city of Rio de Janeiro.

4.3 Artificial Neural Network Models

The other data mining model deployed in this study is the artificial neural network (ANN). This technique is inspired by the central human nervous systems. In an ANN, neurons are connected together to form a network which simulates a biological neural network. This network consists of sets of adaptive weights associated to each neuron, represented by numerical parameters adjustable by a learning algorithm. These neurons should be able to approximate non-linear functions of their inputs. The weights are activated during training and prediction phases [26].

An ANN is a model formed by the interconnection of basic processing units, called artificial neurons. In the mathematical model of the biological neuron, the output Y is calculated from inputs u_i as showed in the following formula:

$$Y = f(\textstyle\sum_i u_i \theta_i) \tag{5}$$

where the parameters θ_i represent the connection weights and f is the activation function. The output value is also called neuron activation value.

Different models are obtained according to the way the neurons are connected, that is, the topology of the network. A neural network can consist of one or multiple hidden layer. An ANN with multiple hidden layers is called a Multilayer Perceptron (MLP) and is the most popular topology in ANN applications. The MLP topology is formed by neurons organized in layers, in a way that the neurons of one layer are connected to the ones of the following layer. In most applications, a three layer structure is used. In general, more intermediary layers can be placed between the input and the output layers. The input layer distributes the entries – mobility attributes – to the neurons of the first layer. In the intermediary layer, the entries of each neuron are associated to a group of parameters, which indicate the weights of the respective connections with the units of the input layer. The outputs of the intermediary layer feed the output layer of the network, which produces the answers of the network for the problem studied.

5   Models' Performance in Estimating Commuting Trips between Locations

All models, including gravity and radiation, linear regression (quantile and robust) and the data mining (decision tree and artificial neural networks) were developed by using three distinct datasets, one for training, one for validation and one for test, 40% for training, 30% for validation and 30% for testing.

The modes were developed considering three different geographic scales, cell towers, neighborhoods and clusters of cell towers. The dataset assigned to the cell towers contains 827,759 records, which means distinct commuting trips between two cell towers. The dataset assigned to neighborhoods contains 15,850 records, and the dataset assigned to clusters contains 3,470 records.

We can observe that as greater the geographic scale better the performance or the accuracy of the model. The Decision Tree has presented the best performance in predicting the number of commuting trips for the geographic scale based on cell towers. The Artificial Neural Network has presented the best performance for neighborhoods clusters of cell towers. A models comparison is presented in Table 1 according to the R^2, the coefficient of determination.

Table 1. The performance of all models in predicting the number of trips between two distinct locations considering different geographic scales

Model	Cell Tower	Neighborhood	Cluster
Gravity Model	0.3823	0.6247	0.6716
Radiation Model	0.3215	0.5937	0.6548
Quantile Linear Regression	0.2871	0.6753	0.7091
Robust Linear Regression	0.2256	0.6439	0.6544
Decision Tree	0.4246	0.6910	0.8392
Artificial Neural Network	0.2124	0.7539	0.8687

6   Conclusions

Nowadays, human mobility behavior is a hot topic in network science. This subject comprises many possibilities to better understand patterns in mobility behavior, like people's frequent trajectories and migration trends. Such analyses may be applied on different geographic scales, such as cell tower level, neighborhoods, cities, states or even countries. One of the major sources of information about human mobility is mobile phone records from telecommunication companies. As we gather more and more data from mobile carriers, and consider the mobile phones dissemination within big cities, we would be able to closely approximate populations' mobility behavior based on subscribers' mobile behavior. As a result, we should raise several insights, for different industries and business scenarios.

Mobile phone records are used by data scientists to aggregate raw information about space and time of subscribers' mobile activities, and create the basis for a set of human mobility studies. There are several approaches in analyzing human mobility. These approaches vary from exploratory analysis – explaining the overall human mobility behavior within geographic locations – to trip prediction models – estimating the number of trips between any two locations.

To fine-tune the above methods for trip prediction, we can add further information about locations, aiming to improve general model performance. The gravity and radiation models have a negative correlation between the number of trips and the mobility attributes. Additional information like the number of companies, universities, restaurants, etc. associated with the locations involved in the travels may increase the overall performance when predicting the number of trips. Such information can better describe why certain trips are more frequently used than others. Also, including historical trip information to the statistical and data mining models, will create a feasible opportunity to increase the overall performance of these models. Finally, a next study should apply and compare all those models on distinct geographic scales, considering cell towers, neighborhoods, counties and cities.

References

[1] González, M., Hidalgo, C., Barabási, A.: Understanding individual human mobility patterns. Nature 453, 779–782 (2008)
[2] Simini, F., González, M., Maritan, A., Barabási, A.-L.: A universal model for mobility and migration patterns. Nature 484, 96–100 (2012)
[3] Rubio, A., Sanchez, A., Martinez, E.: Adaptive non-parametric identification of dense areas using cell phone records for urban analysis. Engineering Applications of Artificial Intelligence 26, 551–563 (2013)
[4] Liu, F., Janssens, D., Wets, G., Cools, M.: Annotating mobile phone location data with activity purposes using machine learning algorithms. Expert Systems with Applications 40(8), 3299–3311 (2013)
[5] Candia, J., González, M., Wang, P., Schoenharl, T., Madey, G., Barabasi, A.-L.: Uncovering individual and collective human dynamics from mobile phone records. Journal of Physics A: Mathematical and Theoretical 41(224015) (2008)
[6] Schneider, C., Belik, V., Couronné, T., Smoreda, Z., González, M.: Unraveling daily human mobility motifs. Journal of The Royal Society Interface 10(84), 20130246 (2013)
[7] Yan, X.-Y., Zhao, C., Fan, Y., Di, Z., Wang, W.-X.: Universal predictability of mobility patterns in cities. Physics and Society, arXiv:1307.7502 (2013)
[8] Park, J., Lee, D., González, M.: The eigenmode analysis of human motion. Journal of Statistical Mechanics: Theory and Experiment 2010 (2010)
[9] Jiang, S., Fiore, G., Yang, Y., Ferreira, J., Frazzoli, E., González, M.: A review of urban computing for mobile phone traces: current methods, challenges and opportunities. In: Proceedings of the ACM SIGKDD International Workshop on Urban Computing (2013)

[10] Masucci, A., Serras, J., Johanson, A., Batty, M.: Gravity vs radiation model: on the importance of scale and heterogeneity in commuting flows. arXiv:1206.5735 (2012)

[11] Lee, A., Chen, Y.-A., Ip, W.-C.: Mining frequent trajectories patterns in spatial-temporal databases. Information Sciences 179, 2218–2231 (2009)

[12] Järv, O., Ahas, R., Witlox, F.: Understanding monthly variability in human activity spaces: A twelve-month study using mobile phone call detail records. Transportation Research Part C 38, 122–135 (2014)

[13] Sun, J.B., Yuan, J., Wang, Y., Si, H.B., Shan, X.M.: Exploring space-time structure human mobility in urban space. Physica A 390, 929–942 (2011)

[14] Zong, E., Tan, B., Mo, K., Yang, Q.: User demographics prediction based on mobile data. Pervasive Mobile Computing 9(6), 823–837 (2013)

[15] Makse, H.A., Havlin, S., Stanley, H.E.: Modelling urban growth patterns. Nature 377, 608–612 (1995)

[16] Bettencourt, L.M.A., Lobo, J., Helbing, D., Kühnert, C., West, G.B.: Growth, innovation, scaling, and the pace of life in cities. Proceedings of the National Academy of Sciences of the United States of America 104, 7301–7306 (2007)

[17] Batty, M.: The size, scale, and shape of cities. Science 319, 769–771 (2008)

[18] Becker, R., Cáceres, R., Hanson, K., Isaacman, S., Loth, J.M., Martonosi, M., Rowland, J., Urbanek, S., Varshavsky, A., Volinsky, C.: Human mobility characterization from cellular network data. Communications of the ACM 56(1), 74–82 (2013)

[19] Balcan, D., Colliza, V., Bruno, G., Hu, H., Ramasco, J.J., Vespignani, A.: Multiscale mobility networks and the spatial spreading of infectious diseases. Proceedings of the National Academy of Sciences of the United States of America 106(51), 21484–21489 (2009)

[20] Wang, L., Hu, K., Ku, T., Yan, X.: Mining frequent trajectory pattern based on vague space partition. Knowledge-Based Systems 50, 100–111 (2013)

[21] Bayir, M.-A., Demirbas, M., Eagle, N.: Mobility profiler: A framework for discovering mobility profiles of cell phone users. Pervasive and Mobile Computing 6(4), 435–454 (2010)

[22] Lin, M., Hsu, W.-J.: Mining GPS data for mobility patterns: A survey. Pervasive and Mobile Computing (Available online July 8, 2013)

[23] Koenker, R.: Quantile Regression. Cambridge University Press (2005)

[24] Andersen, R.: Modern Methods for Robust Regression. Sage University Paper Series on Quantitative Applications in the Social Sciences, 07-152 (2008)

[25] Howard, R.-A.: The Foundations of Decision Analysis. IEEE Transactions on System Science and Cybernetics SSC–4(3), 211–219 (1968)

[26] Bishop, C.-M.: Neural Networks for Pattern Recognition. Oxford University Press (1995)

[27] Sarle, W.S.: Cubic Clustering Criterion. SAS Technical Report, vol. 108 (1983)

[28] Ward, J.H.: Hierarchical grouping to optimize an objective function. Journal of the American Statistical Association 58, 236–244 (1963)

Mobile Learning Systems and Ontology

Sergey Rodzin and Lada Rodzina

Southern Federal University, Nekrasovsky Lane 44, 347928, Taganrog, Russia
lada.rodzina@gmail.com

Abstract. The paper proposes a scenario model of learning and the open architecture of context-based mobile learning system. Developed structure of a content management system is based on semantic web. The structure of the content management system contains the following main elements: the ontology metadata, ontologies particular domain, which describes the structure of indexing resources, and, finally, models of training scenarios and adaptive selection of learning resources. The model is proposed for building a content management system and based on probabilistic automata. Context-sensitive learning system should be able to personalize the best learning style. For this purpose we propose to use the apparatus of Bayesian networks and evolutionary computation.

Keywords: Mobile learning, context-aware system, content management, ontology, probabilistic automaton, script learning, Bayesian network.

1 Introduction

Modern learning should be more integrated into our everyday life, and, preferably, available upon request and without the need for planning or scheduling in the months ahead. We could hear about advantages and prospects of mobile learning already a couple of years ago. According to the development of mobile networks and increasing opportunities for data transfer (especially video), mobile learning is becoming one of the most convenient and promising ways of acquiring knowledge. All the signs show that mobile devices will support many educational activities. The devices that are capable of receiving, storing and transmitting information, such as: cameras, smart phones, handheld and tablet computers. Mobile learning (or, m-learning) is a completely new approach with its own basic principle: the learning process is "free" format that adapts to the listener and not vice versa. The term "mobile" is primarily characterized an access to training tools and forms of implementation of educational interaction. More integrated education involves the development of convenient alternative scenarios of distance learning that use smartphones and laptops. And this is not limited by classroom space or capabilities of usual computer. The constant availability of information (educational segments) is extremely important. It gives an opportunity and motivation for studying anywhere, as you like, and whenever you want. Thus, the article is devoted to the development of appropriate scenarios and architectures of mobile learning system, according to the modernization of the educational programs in the format of the mobile structure in the near future.

© Springer International Publishing Switzerland 2015
R. Silhavy et al. (eds.), *Software Engineering in Intelligent Systems*,
Advances in Intelligent Systems and Computing 349, DOI: 10.1007/978-3-319-18473-9_5

2   Methods

2.1   Mobile Learning Scenario

In recent years, a new principle of learning systems was formulated: the learning process is considered as a process of knowledge management of a trainee [1]. The different research and development are performed under this approach. They are focused on creation of adaptive learning systems that support personalized learning approach. As well as it provides a possibility of contextual usage of educational resources, mobility, and deep personalization of educational services [2]. Context is one of the key issues for the individualization of learning. Context-Aware System should be able to analyze the state of the user, his physical environment, and adapt its behavior accordingly. Creation of context-aware adaptive learning systems is an interdisciplinary problem. Fundamental importance here is the internal logic of the learning process. This logic is reflected in the pedagogical and technological scenarios that cut across a set of techniques, processes, procedures and training.

Scenario is methodically targeted sequence-built methods and technologies to achieve their learning goals. And for each student we need a different scenario. In computer science is called "curse of dimensionality". It means that we can have a great quantity of the scenarios, and, consequently, a serious problem of their classification, rationalization, and organization into a coherent structure. The purpose of the scenario is to describe the learning process and activities focused on the acquisition of knowledge. Scenario is defined by characteristics such as: structure, coordination, typology of activities, and distribution of roles between students, teachers and computer systems.

Used scenarios in the majority are not context sensitive and adaptive to different audiences. In [3] argues that there are hundreds of different pedagogical models and training scenarios. In [4] a general abstract scenario was proposed to represent different pedagogical models. It is determined by the theme of training, students, integrable knowledge, teacher, used recourses (communication and information technologies and technical tools), pedagogical and didactic models of learning, and some other elements. This scenario provides only a very limited ability for adaptation with rules "if-then-else" [5]. Educational resources are defined in a priori and cannot be changed. The script also does not provide management tools for changing domain knowledge or usage of contextual learning technologies [6].

For knowledge integration envisaged by a scenario it requires a single conceptual description of knowledge through ontology reflecting the subject area [7] formalizes the structure of the ontology learning process from the perspective of competencies formed [8, 9], a repository of learning objects, objects of research and project activities, public information and educational resources. This will allow increasing the relevance of selection of studied objects according to the individual characteristics of students.

2.2   Architecture of Mobile Learning System

The purpose of developing the architecture of information learning systems is to set at a high level abstraction framework for understanding certain types of systems, subsystems and their interactions with other systems. Over the last decade learning infomation systems have evolved from centralized systems on dedicated computers for distance learning system with a distributed architecture "client-server".

Disadvantages of a centralized architecture are obvious: it is difficult for deploing, expensive to maintain and difficult to adapt to the constant changes of the educational process. Such systems depend on private user tools and imposed by developers of educational resources. As a result we have an environment that does not take into account any differences in the task and the user level, or changes of the educational market's needs, or conditions of education. The situation may be improved by the Internet-, Java- and other web-based technology that has already demonstrate it in handling themselves as effective tools in developing applications for any purpose. Informational and educational systems of distance learning require specific architectural solutions. This type of architecture should be based on web and telecommunication technologies, as well as able to provide the optimum combination of performance, functionality, and powerful management tools. Such decisions, unfortunately, currently do not exist. However, the real benefits of web-based technologies are allowing the creation of fundamentally new application architecture that is based directly on the Internet and mobile telecommunications technology. Clearly, the characteristics of a learning system and its functionality are dependent on the capabilities and limitations of the architectural pattern. The proposed model is the open architecture of adaptive context-sensitive mobile learning system and it is shown in Figure 1.

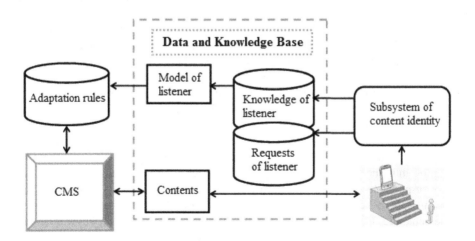

Fig. 1. Mobile learning system architecture

Database and knowledgebase include the following information: contextual data and knowledge, student's profile, and module of knowledge control. Context data contains information about the place, time, mobile learning session, as well as information about training materials. Student's profile contains personal information about him, his queries, his knowledge level, as well as disposable time for a learning session. The control module knowledge includes test items and the results of previous monitoring inspections. Contextual information includes student's requests and information about the level of his knowledge. Contextual information obtained from the students' request indicates its location (campus, home, house), disposable student's training interval and level of concentration. Each location has a specific contextual factor that affects the educational activity (the level of concentration, time for learning, etc.). The lower this factor, the higher its impact, and vice versa. Free time frames, which is available to spend on lesson by the listener, may be different, for example, 15, 30, 45, or 60 minutes. Concentration levels can also be described by discrete values, such as 1, 2, or 3 (low, medium or high). Information about the level of knowledge defined by the results of surveys and tests can be evaluated, for example, on a 5-point scale.

Mobile content can include full-text versions of textbooks, interactive simulations, tests, glossary search, a variety of educational games, etc. Content management system provides access and management of those objects, planning of training, knowledge control. Content described in a hierarchical tree structure with educational themes as vertices. Certainly, the textbook view on mobile devices is not comfortable yet: constantly you have to scroll the screen, both vertically and horizontally, and images are not displayed completely. The learner has to distract from time to time by the navigation instead of catch on the material. This problem can be solved by creating of programs with feature of "customizing" of training volume segment to fit the screen of your smartphone.

Some defined themes could be provided in the student's model and are necessary for studying. The choice is made according to the request. Thus, the routes for studying are selected on the tree structure. Student's model is the base for adaptive selection of the course content according to all the contextual factors described above. All links of contextual factors identified in the tree model. "if-then-else" rules for adaptive selection of educational resources are built on this basis. Their goal is to help the user, for example: to show information window for the current task, to restruct hyperspace to help navigate and move around in it, or to provide additional explanations about some of the training concept, etc. Adaptation rules determine which component of adaptation must be selected according to the model of the listener. These rules are mainly responsible for content adaptive presentation and adaptive navigation. The following approaches are used for implementation of adaptation mechanisms: based on the semantic domain concepts and semantic indexing of content, keywords, and automatic indexing content-based information retrieval, as well as social mechanisms, such as navigation on the basis of history and collective filtering.

3   Results

3.1   Ontological Structure Content Management Software Package

Distinctive features of mobile learning is usage of mobile devices and receiving training regardless of location using portable technologies. Mobile learning reduces the restrictions on obtaining education at the location with the help of portable devices. In the field of mobile learning we can recognize the following problems:

- technical, such as connectivity, screen size, the battery life, numerous standards, operating systems;
- social and educational, for example, the availability and price for the end-user support, training in different situations, the development of relevant theories of learning for the mobile age, tracking results and the proper use of this information.

Modification of learning process to mobile format structure requires the development of suitable systems content management. The proposed learning model is adaptive, it uses an approach based on Semantic Web) [10]: the learning environment includes a set of resources, ontologies and tools that enable the flexibility to choose the appropriate resources for a specific audience and actual learning situation. Figure 2 shows the structure of complex content management software based on the Semantic Web.

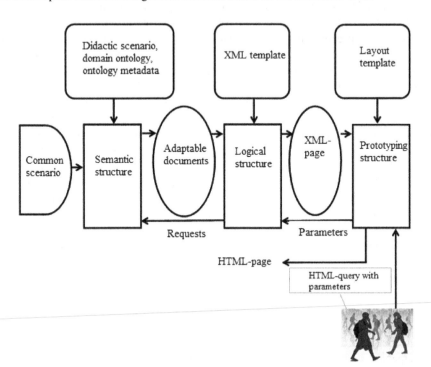

Fig. 2. Structure of the complex content management

Common scenario of learning situation is the input specification for software complex content management. Structure of a content management system includes four main elements: ontology metadata, model training scenarios, adaptive selection of learning resources, and, finally, ontologies particular domain that describes the structure of resource indexing. Metadata is the information contained on the web page of data. Metadata is an important component of distributed learning systems, enabling reuse of educational materials in various educational institutions, fast and efficient retrieval of educational materials in the Internet both teachers and students, copyright protection, etc. Metadata that is structured as a hierarchy represents itself as the ontology, for example, XML-schema. Domain ontology is a formal description of a domain, which presents and defines the concepts and terminology database domain. Learning scenario model is a directed graph that represents the basic concepts of the hierarchical model of tasks and communication of various types, depending on the application. Existing models of scenarios are not context-dependent. That is why the challenge is to formalize the context model so that the general scenario of the training system could "calculate on the fly" specific scenario tailored to the students' individuals and the current learning situation. The process of content management systems can be divided into several stages: semantic selection, prototyping and assembly of logic. The process ends by giving an HTML-document from XML (display or rendering). Analysis of the educational needs of users in terms of the complexity of their treatment allows them to allocate the following types: simple requests (to define a new concept to explain using examples, etc.), the study of a single issue, the study theme, the study section of the course, the study of the course; inquiry level educational program consisting of a plurality of interconnected courses. With regard to the educational process service request, it involves an iterative refinement of the educational needs and requests detail and customize the training program. As a result it should be built as an individual training program consisted from ontology concepts. The coating is carried out after. This program is made up of concepts that are available in the educational space of learning resources. However, in an open educational environment available a large number of educational resources, and for each concept there are many coverage options. The following additional restrictions of users should be used for reducing the running: on the form of presentation, learning strategies, time and financial resources, etc. The result of this step is a training program consisting in real learning objects. Presented in [11] didactic approaches to adaptation for a perfect training system is proposed to use for building a content management system. The reason is it lets you personalize and optimize the process of mobile learning in context (user preferences and learning objectives). A content management system provides for the functions related to the alert the user to the academic calendar (lectures, tests, homework, etc.) depending on the external conditions, the current situation, including in the context of the free time and location.

An important role is also played by context-sensitive communication: asynchronous (using email, discussion boards) and synchronous (online chat) to exchange messages between teachers and students, or between students. For building a content management system is proposed to use a model based on probabilistic automata [12, 13]. In probabilistic automata transition from one state to another happens according to the random input signals or depending on the sequence of previous states. Usually probabilistic automata are used to demonstrate the behavior of systems, the reactions, which are difficult to predict. In this case, it is assumed that the listener behaves as probabilistic automaton. The work algorithm of probabilistic automaton appears as stochastic graph with set of vertices connected by edges that correspond to the probabilities of transitions from one state to another. The input function of a probabilistic automaton has the form:

$$In(t) = [SS(t), UD(t), SI(t), SM(t)],$$

where $SS(t)$ – state of listener, $UD(t)$ – state of learning activity, $SI(t)$ – state of infrastructure, $SM(t)$ – state of environment.

Output function has the following form:

$$Out(t+1) = [UD(t+1), SI(t+1)],$$

where $UD(t+1)$ presents adapted state of educational activity at the time $(t+1)$, $SI(t+1)$ - adapted state of environment at the time $(t+1)$.

Suppose at time t automata with probability $p_m(t)$ is in state $UD(t) = UD_m$, and state $IS(t) = IS_n$ with probability $p_n(t)$. Set of state probabilities $UD(t) = \{UD_1(t), UD_2(t),..., UD_m(t)\}$, set of state probabilities $IS(t) = \{IS_1(t), IS_2(t),..., IS_n(t)\}$.

Machine learning occurs by the method of rewards and sanctions [11] according to the rules:

- suppose that in time t, $UD(t) = UD_m$ with probability $p_m(t)$. If a result of learning is "good" (for example, listener is satisfied), then probability $p_m(t)$ increases and decreases the probability of selecting all the other states $UD_s(t)$. Otherwise, if the listener is not satisfied, conversely, the probability $p_m(t)$ decreases and increases the probability of selecting all the other states $UD_s(t)$.
- Suppose that in time t, $IS(t) = IS_n$ with probability $p_n(t)$. If a result of learning is "good" (for example, listener is satisfied), then probability $p_n(t)$ increases and decreases the probability of selecting all the other states $IS_s(t)$. Otherwise, if the listener is not satisfied, conversely, the probability $p_n(t)$ decreases and increases the probability of selecting all the other states $IS_s(t)$.

The example uses a linear law of promotion / punishment, but training can be performed using other laws and depending on the situation.

3.2   Usage of Bayesian Network for Adaptation to the Learning Style

Each pedagogical scenario describes a typical situation in the training system with specially formulated goal. Each scenario has a name, parameters and purpose. Achieving the goal involves one or more students (agents, individuals) in one or more processes. Scenario describes a series of actions and communications of agents to achieve a specific goal.

Ontology of pedagogical scenarios allows to build rapport between the participants of the educational process, re-use previously created knowledge facilitates the understanding of the subject area in terms of the tasks and functions to ensure the interaction of different applications to model the semantic content of web pages, provide an unambiguous behaviour training system. Creating a directory of training scenarios support the construction of new training scenarios.

Individualization of the learning process is mainly achieved through changes in his scenario, depending on the category of students, the available educational resources and the form of training. These features are crucial for building adaptive learning environment. The question is to develop a common scenario that would allow coping with a wide range of individual situations in the learning process.

It is suggested that the production of a common scenario is going from several stages. At the first step, you create the initial version based on the recommendations of expert teachers. In the next step scenario is refined and modified using the theory of didactic anthropology of knowledge [14]. Then formalization of the hierarchical model of the problem is carried and typology of learning tasks and the possibility of adaptation is built. Consider these steps in more detail. The initial version of the general scenario of learning SC0 is defined by two sets:

SC0 = <Ph1, Ph2>,

Where Ph1 – set of recommendations for the didactic training, Ph2 – study plan. The set Ph1 includes learning tasks, learning resources, explanation of approaches to the problem, etc. The set Ph2 includes didactic description of the method of solving the problem for actions, etc. In the next step the script is specified in terms of praxeology (the various steps or set of actions in terms of establishing their effectiveness) [15]. Essence of praxeological approach is finding, selecting and implementing educational practices in a variety of resources needed for its implementation from the standpoint of such categories as "rationality", "efficiency".

In our case, praxeology learning system helps to clarify and structure-centred type of the problem, methods for its solution, options interaction scenarios, determined by the triple <T, M, D>, where T – the type of problem to be solved, M – methods for solving it, D – discourse (screenplay interaction).

Signature <T, M> has a hierarchical structure. In other words, the problem can be decomposed into a number of subtasks, the solution of which is achieved by the method M and operators such as sequence, alternative choice and parallel execution.

Context-sensitive learning system should be able to personalize the best learning style. For this purpose, you can use the device Bayesian networks and evolutionary computation [16-18]. Bayesian network, according to J. Pearl, is a probabilistic model, which is a set of variables and their probabilistic dependencies. Formally, the Bayesian network - a directed acyclic graph, each vertex of which corresponds to that which some variable and the arcs of the graph encode the conditional relationship of independence between these variables. Vertices may be variables of any type, be weighted parameters, hidden variables or hypotheses. Effective methods that have been successfully used for computing and learning Bayesian networks. If you specify a probability distribution on the set of variables corresponding to the vertices of the graph, the resulting network is a Bayesian network. On the network can use Bayesian inference to calculate the probability of consequences of events.

For testing a developed mobile application in Java was created a questionnaire. Surveyed a group of 40 students who used the application to their mobile phone. The average score on a 5-point scale is 3.9. Note that at this stage of development model listener is not able to consider the whole context. Noticeable problem is the mapping of fragments of content on mobile phones, as well as the need to organize the search adapted to the learning style of teaching materials listener. Promising direction of solving this problem is the use of multi-agent technology [19 - 20].

4   Discussion

The key issue is not just about studying constantly. Education should be mobile and demand. In other words, the gap between receiving knowledge in the learning process and the time when new knowledge is needed must be reduced. From a user perspective, the knowledge gained at the first request and for the hours until their usage in practice is far more relevant than the information obtained in the weeks, months or years before it will be in demand. It is clear, that the traditional learning formats such as "learn and forget" cannot give such interactivity. Moreover, the main thing in studying is the development of new knowledge. It does not matter how they are received as long as they were relevant.

Additionally, we note the high innovation potential of mobile devices and technologies. In the educational process should be used as a new form of educational activity (interactive slide lectures, webinars, training, and computer simulations, telecom discussions with experts from domestic and foreign universities), and new types of tasks and exercises (teaching and training assignments; slide presentations, web projects). In addition, mobile learning will help to overcome the destructive impact of information and communication technology on cognitive and social activities, such as: downloading finished articles and abstracts from the internet to perform tasks, ignoring the rules of copyright, usage of any mobile device as cribs.

An important aspect is the development and implementation of mobile technology in education as an opportunity to study people with disabilities. In addition, mobile learning is economically justified, because of training materials easily propagate between users through the modern wireless technologies (WAP, GPRS, EDGE, Bluetooth, Wi-Fi); information in multimedia format promotes better assimilation and memorization of the material, increasing interest in the educational process, and, finally, youth technically and psychologically ready for the usage of mobile technologies in education.

Acknowledgements. The study was performed by the grant from the Russian Science Foundation (project № 14-11-00242) in the Southern Federal University.

References

1. Koulopoulos, T.M., Frappaolo, K.: Smart Things to Know about Knowledge management. Capstone, Oxford (1999)
2. Grachev, V.V., Sitarov, V.A.: Personalization training: requirements for the content of education. J. Alma Mater. Journal of Higher School (8), 11–15 (2006)
3. Koper, R., Olivier, B.: Representing the Learning Design of Units of Learning. J. Educational Technology & Society 7(3), 97–111 (2004)
4. Nodenot, T.: Contribution à l'Ingénierie dirigée par les modèles en EIAH: le case des situations-problems cooperatives. Habilitation à Diriger des Recherchéen Informatique, Pau, Uni de Pau et des Pays de l'Adour (2005)
5. IMS Global Learning Consortium, http://www.imsglobal.org
6. Verbitsky, A.A.: Active learning in higher education: the contextual approach. High School, Moscow (1991)
7. Gruber, T.R.: Toward Principles for the Design of Ontologies Used for Knowledge Sharing. Int. J. of Human & Computer Studies 43(5/6), 907–928 (1993)
8. Kureichik, V.V., Bova, V.V., Rodzin, S.I.: Integrated development environment of innovation support educational processes. J. Open Education 4(81), 101–111 (2010)
9. Rodzin, S.I.: On the innovation component of engineering education programs. J. Innovations 5(92), 66–71 (2006)
10. Garlatti, S., Iksal, S.: A Flexible Composition Engine for Adaptive Web Sites. In: De Bra, P.M.E., Nejdl, W. (eds.) AH 2004. LNCS, vol. 3137, pp. 115–125. Springer, Heidelberg (2004)
11. Zarraonandia, T., Fernandez, C., Diaz, P., Torres, J.: On the way of an ideal learning system adaptive to the learner and her context. In: Fifth IEEE Int. Conf. on Advanced Learning Technologies, pp. 128–134 (2005)
12. Pospelov, D.A.: Probabilistic automata. Energy, Moscow (1970)
13. Economides, A.A.: Adaptive Mobile Learning. In: 4th Int. Workshop on Wireless, Mobile & Ubiquitous Technologies in Education, pp. 263–269 (2006)
14. Bim-Bad, B.M.: Pedagogical anthropology. URAO, Moscow (1998)
15. Grigoriev, B.V., Chumakov, V.I.: Praxeology or how to organize a successful business. School Press Publ., Moscow (2002)
16. Russell, S., Norvig, P.: Artificial Intelligence - A Modern Approach, 3rd edn. Prentice-Hall (2010)
17. Tulupyev, A.L., Nikolenko, S.I., Sirotkin, A.V.: Bayesian networks: logical-probabilistic approach. Science, St. Petersburg (2006)
18. Rodzin, S.I., Kureichik, V.V., Kureichik, V.M.: Theory of evolutionary computation. Fizmatlit, Moscow (2012)
19. Rodzina, L.S.: Applied multiagent systems. Programming Platform JADE. LAP LAMBERT Academic Publ. GmbH & Co., Saarbrucken, Germany (2011)
20. Rodzina, L., Kristofferson, S.: Context-dependent car Navigation as kind of human-machine collaborative interaction. In: Proc. of the 2013 Intern. Conf. on Collaboration Technologies & Systems – CTS 2013, San Diego, California, USA, May 20-24, pp. 253–259. Publ. of the IEEE (2013)

Synchronization Algorithm for Remote Heterogeneous Database Environment

Abdullahi Abubakar Imam, Shuib B. Basri, and Rohiza Binti Ahmad

CIS Department, Universiti Technologi PETRONAS Malaysia
Aiabubakar3@gmail.com,
{shuib_basri,rohiza_ahmad}@petronas.com.my

Abstract. Inconsistencies and lack of uniformity of data and its structure is a deep-seated problem of data synchronization in heterogeneous database environment. In this paper, we proposed a model and a unique algorithm to synchronize data between heterogeneous databases, in order to provide alternative synchronization paths, concurrency in operation and bidirectional mode of data exchange. To achieve this feat, three separate algorithms are generated. In addition, jQuery technology is adopted to determine the active path, and the data is transmitted via web service that operates on XML technology over HTTP protocol. The model is independent of any database vendor or proprietary solutions and can be adapted to any database or platform. While fulfilling the set objectives, our proposed algorithm managed to provide an efficient data exchange mechanism for distributed databases using 3 steps for proper consistency and synchronization along with network utilization when tested with three different databases.

Keywords: data synchronization, heterogeneous, remote databases, XML, Web-service, jQuery.

1 Introduction

Data sharing and synchronization have been a topic of research for quite some time in the area of databases and distributed databases [1]. Database synchronization is a record exchange between two different databases, when changes are made to a database, will appear the same way in another database on a different servers and networks transversally. Database synchronization can be a one-way or a two-way program and can also be immediate or episodic [2], that is, 1) *Synchronous* which is more suitable for LAN and 2) *Asynchronous* most suitable for WAN due its delayed technique.

Looking at the expansion of automated enterprise business processes and enterprise information systems advancements, the use of heterogeneous database become higher in all aspects. The importance and significance of data synchronization between these distributed heterogeneous databases cannot be over emphasized. Considering the differences of heterogeneous databases and their dependent functions in the distribution nodes like different period of development, running environment, structures have made data synchronization very complicated. Many solutions have been provided to address the aforementioned problem, however, most of the solutions are proprietary

© Springer International Publishing Switzerland 2015
R. Silhavy et al. (eds.), *Software Engineering in Intelligent Systems*,
Advances in Intelligent Systems and Computing 349, DOI: 10.1007/978-3-319-18473-9_6

and use databases dependent information such as timestamp and trigger. Other solutions have one communication route while some are one directional (Master to Slave).

The objective of this paper is to provide solution to the problems mentioned above to enhance the quality of our day-to-day database drive software applications. In doing so, we proposed a unique approach that has the following characteristics.

✓ Routing Alternatives: for reliability provision to the system, we introduced a hybrid model to expedite possible routing substitutes for data exchange in case one path is down;

✓ No trigger or timestamp: in our proposed algorithm, the synchronization flows from the source to the destination and vice versa in which the use of database dependent information is eliminated and the solution is made generic, thus; heterogeneous.

✓ Bidirectional: the data exchange can be in all directions, i.e., all connected databases can send and also receive information.

✓ Parallelism: each process is handled individually which makes them operate at the same time without a single intervention. That is to say both the sending and the receiving of data can be done at the same time.

✓ XML: the system uses the standard XML for data transmission only.

The remainder of this document is organized into sections and subsections. Section 2: the related work is discussed, while section 3: the proposed approach is described. Section 4: the example (implementation) and section 5: conclusion and recommendations.

2   Related Work

A.   *Heterogeneous Database Synchronization*

In [3], a model was introduced which conceptually has a Data Exchange Center. In this system, the data exchange center must always be used to establish communication between two or more databases that need to synchronize information with each other. To utilize data exchange center, the databases have to use either of the two models to connect to each other [3]: The models are as shown below:

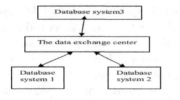

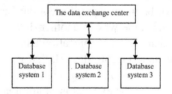

Fig. 1. Star linear   **Fig. 2.** Bus shape

Furthermore, the synchronization technique adopted in this paper was based on trigger mechanism [3]. even though many databases use trigger, some still do not support it [3] and [4]. The system uses the XML as the message form.

The solution is perfect for some areas of application such as LAN because it is based on synchronous mechanism, however, in order to improve communication frequent flaws and make the solution suitable for WAN there is need to build a hybrid model. To add, the model is linear and bus shape which has no alternative routing path incase the solution is down for any reason as it can be clearly seen in Fig. 1 and 2 above.

According to [5] Application data used by enterprise applications can be transmitted as XML files over HTTP. Moreover, the solution sends application data to target server identified by a given URL. The application synchronizes with the master data repository, and then master data repository notifies all subscribed data repositories (broadcast). In comparison, this solution is One-to-Many (master to slaves) and also uses broadcast notification mechanism, while in our approach the relationship is many-to-many, bidirectional and can be between the concerned databases only, thus; improve data privacy.

In [6], mobile agents was used as a connector for distributed databases. The communication between the systems is done via the mobile agents where aglet is appended with different queries and routs them to the intended destination. In this system although the queries are from different databases they are considered to be the same when retrieving the data. However, the solution focused on data selection from different sources while our emphasis is on the data exchange.

In [7], Only two components are used unlike others [6] [5] [3] and [4] that have three components involved. The two designated components connect and share information using the technology known as JSON. The process assumes that for an effective implementation of the algorithms, all devices have their respective times. These two timestamps are compared to initiate the process.

According to [8] and [9], believe that by introducing MSync system, possibilities to have data synchronized between portable devices, laptops and desktops with Oracle databases can be easily achieved[10]. While in [11], GAAMov was introduced to ease developers' work by generating mobile applications that require accessing a remote database. Java ME was used for client-side while PHP and MySQL on the server side.

In [4] and [12], the idea of global table as the central storage area was introduced. [4] Further explained that global table information can be stored in XML documents while [12] said that, the data copy in the cache should not be invalidated, instead it should just be updated in case of any change. However, the solution uses trigger which makes it dependent to some database vendors. Bus shape model is also adopted in [4] where only one possible route is expected.

In view of the above discussions, in order to essentially improve the existing data synchronization solutions between two, three or more heterogeneous databases it is indispensably imperative to understand the existing studies. The aforesaid synchronization solutions have one big disadvantage. Almost all of them are proprietary synchronization solution or are based on database dependent information such as timestamp and trigger which eliminate the heterogeneity concept in the solutions as other databases don't support trigger or timestamp, and also increase the efforts for server or PC application providers and different databases providers to make their respective products interoperable.

Furthermore, some solutions operate on master-to-slaves only, thereby making it difficult to communicate in in both directions. In addition, some solutions act as a mediator whenever two or more servers want to communicate. The question is what happens when the solution is down? It is therefore remained challenging since we are dealing with heterogeneous databases to have several alternative communication paths which improve reliability.

B.   Web Service

A *Web service* is a set of Web methods that are hosted on an application server. *Web methods* are methods that can be invoked remotely over a network or the internet. To invoke a Web method, a SOAP message is created and sent to the Web service via HTTP. It can be written in any suitable language. A SOAP message is an XML document that follows a specific standard [13].

C.   XML

According to [3], XML is a data description language and it is SGML abridge edition. XML is not a programming language. XML is a document that usually constitutes state, element, attribute and text. The text is the data which must be truly saved. But the attribute and the element are usually used to describe the nature of the text.

D.   jQuery

jQuery is not a language, but it is a well written JavaScript code [14]. As quoted on official jQuery website, "it is a fast and concise JavaScript Library that simplifies HTML document traversing, event handling, animating, and Ajax interactions for rapid web development". Prior to using jQuery, basics of JavaScript are expected [14]. Looking at the licensing, jQuery It is free and open source, it is also Dual-licensed under the MIT and General Public License GPL.

3   Proposed Approach

3.1   The System Model

In this system, different RDBMS, platform and structure is entirely not an issue. The process does not depend on any peculiar functionality. The synchronization can either be direct or indirect depends on the availability of the data exchange path.

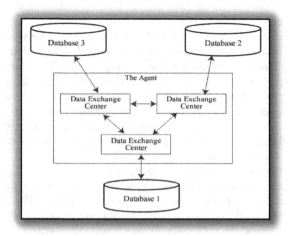

Fig. 3. The Hybrid Synchronization Model

The model is hybrid in nature, in the sense that, when data is to be exchanged can have several alternative routes, if one path (Data Exchange Center) is down; the other available route can be used to exchange data. This eliminates the issue of complete system failure in the cause of synchronization.

3.2 Data Synchronization Process

The process of data exchange and sharing between the heterogeneous databases begins by choosing the trail, and then proceed to send request to the database server where the original information is located. The continuation is based on the activeness of the path. If a given way is active, the source server prepares the required data and sends it back to the data manipulation module on the destination server. Also if the data is not available in the pointed database, another data source server is chosen. The scenario is shown in fig. 4 below:

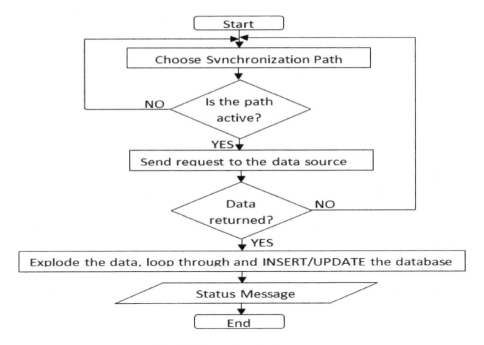

Fig. 4. Data synchronization process

3.3 System Architecture

The system architecture consists of three main parts: 1) the heterogeneous databases. 2) the agent where all the initial configurations are administered and managed. 3) the web service (with XML) that can be consumed by any databases server. A scenario is depicted in fig. 5 below.

The system operates on three main database actions and they are: INSERT UPDATE and DELETE.

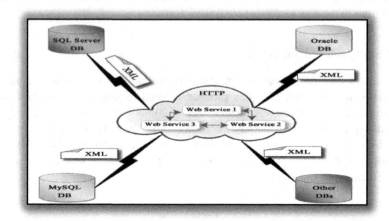

Fig. 5. System Architecture

The system adopts the use of XML as the message form. We use the standard SQL query to produce the data retrieved from the source local database and organize it in form of text (see A below) which is then hashed and packaged for transmission. The destination server then connects and grabs the file via the web service and transmits the data back to the destination server in an XML file through the same medium. When the destination server receives the file, it eliminates all the XML and HTML tags and use the row data as structured by the data-source-file (see C below). The parameters that are used to specify the number of records or group of records needed are transmitted via the web service in a jQuery data (see B below).

Additionally, the middle tier module receives such file (as described above) and passes it to the data processing module (DPM). The DPM uses simple algorithm (as shown in section C below) to explode the data, loop through it and perform any action defined by the user. This process makes our approach independent to any database vendor and also independent to any proprietary solution as it is not using any database dependent information such as trigger, stored procedure or timestamp etc.

In our approach, three major algorithms are necessary. One would be on the source server and the remaining two on the destination server. They are as follows:

A. Data-Source_Module_Algorithm (DSMA):

For an effective synchronization with the destination server, source server must have at least one file that prepares the data (at run time) retrieved from the source local database. The algorithm is as follows:

    1.0 Start.
    2.0 Establish connection to the databases. *//on the data-source server*
    3.0 Declare variable (*A1, A2---An*).

4.0 Variable ← get values from the URL *//usually from records destination server.*

5.0 QueryVariable ← select data from the database.

6.0 Results = Null.

7.0 Loop through the array of the queried data *While (row=ArrayOfData(*QueryVariable*))*

    7.1.1   Results ← Results + the concatenated columns in a row and a delimiter that marks the end of a row.

8.0 Display results.

B. *Data_Destination_Module_Algorithm (DDMA)—Middle Tier*

In order to send request to and to receive data from data source server, the destination server must have at least one file to be able to send request and also receive data prepared by *section A* above. Below is the algorithm:

1.0 Start.

2.0 Declare variable (*A1, A2---An*)

3.0 Variable ← get values from the user or controls *//may be used as parameters.*

4.0 DataSourceURL ← http://data_source_file.... *//prepared by sect A above.*

5.0 Send Ajax request to URL1 for data retrieval.

6.0 If Ajax request successful

    6.1 Datum_variable ← data from URL1.

    6.2 DataDestinationURL ← http://destination_file...   *//to prepare and place the data in the destination database appropriately.*

    6.3 Send Ajax request to the URL2 with the data in the Datum_variable *//for insert or updates.*

    6.4 If Ajax request successful

        6.4.1   display success message

    6.5 If Ajax request fails

        6.5.1   display failure message

    6.6 else

        6.6.1   display *no action performed*

7.0 Else *first Ajax request is not successful.*

    7.1 Go To 4.0 to choose the next available route.

8.0 Method call for the execution. *//using any event preferred by the user.*

C. *Data_Destination_Module_Algorithm (DDMA) ——Data Manipulations*

To be able to distribute data downloaded from the data-source server, the destination server should have a second file for query executions such as INSERT, UPDATE and DELETE. The *section B* above receives data and forwards it to this file for an appropriate action. The algorithm is below described:

1.0 Start.

2.0 Establish connection to the databases. *//on the destination server.*

3.0 Declare variable (*A1, A2---An*)

4.0 Variable ← get POST datum sent from *section B above* via URL2.

5.0 Trim the data in the *variable   //to remove white spaces and tags.*

6.0 Make the data in row-by-row format using the delimiter introduced in *section A* above.

7.0 Loop through the rows in the explode dataset. *for ($i=0; $i<count($records); $i++)*

8.0 Extract all columns from each row using the column delimiter introduced in *section A* above.

9.0 QueryVariable ← INSERT, UPDATE or DELETE the data in the destination database.

10.0If the query is successful.

  10.1Display success message.

11.0 Else

  11.1display error messages

D. XML:

This system message format is as follows:

< XML_ Data >
  <INFO>
    <Destination_database> </ Destination _database>
    < Destination _table_name> </ Destination _table_name>
  </INFO>
<DATA>
......
</DATA>
</ XML_ Data >

The explanations of the above sample XML code are:

  ✓ *< XML_ Data >   XML definition.*
  ✓ *< Destination _database>: caries the name of a database in with the DATA will be deposited.*
  ✓ *< Destination_table_name>: takes the name of a table in which the data will be structurally arranged.*
  ✓   *<DATA>: the actual data that really matters for synchronization.*

E. Message Digest

We conceptually describe message digest using the formula below:

  h = H (M) where:
  h = the hashed message
  H = the function used for hashing
  M = the actual row data.

To further explain the formula, the function will produce a fixed length h (message digest) when the row data is passed through the formula. This is to say, if the value of M (row data) changes, the value of h (message digest) will also be changed, thus track inconsistencies and data integrity.

4   Example

We selected three different databases to be used in our example to show the capabilities and the results of our approach. These databases are Oracle, SQL Server 2008 and MySQL 5.5. The tables constructed are *Students*, *applicants*, *application details and processing fee*. The table structures are below descried:

Table 1. Students (MySQL DB)

Description	field	Data type	Comment
Application number	App_number	Integer(10)	Public key
Student number	Reg_number	Varchar(15)	Primary key
Name	St_name	Varchar(50)	Not Null
Program admitted	Prog_name	Varchar(80)	
Payment status	P_status	Boolean	Paid OR Not Paid

Table 2. Applicants (Oracle DB)

Description	field	Data type	Comment
Application number	App_number	Integer(10)	Primary key
Name	St_name	Varchar(50)	Not Null
Applicant gender	sex	Varchar(1)	
Date of birth	dob	date	
Application status	App_status	Varchar (10)	Confirm or Not

Table 3. Application_details (Oracle DB)

Description	field	Data type	Comment
Application number	App_number	Integer(10)	Foreign key
Program applied	Prog_applied	Varchar(50)	Not Null
Level	Level_app	Integer (5)	
Previous qualification	Prev_qual	Varchar(50)	

Table 4. Processing_fee (SQL Server)

Description	field	Data type	Comment
Application number	App_number	Integer(10)	Foreign key
Amount	amount	Float	Not Null
Payment status	status	Varchar(5)	Not Null

We have taken a school as an instance where an applicant applies to become a member of the school. We consider student's portal (runs MySQL DB), application portal (Oracle DB) and payment portal (SQL Server DB). We inserted 10, 000 records in table 2 (oracle) and also in the corresponding tables (table 3 (oracle) and table 4 (SQL Server)). For the students' portal at this point it's still empty, therefore there is need to synchronize the records from other databases to make the school portal up-to-date. In doing so, we wrote some codes in Php (following the section 3.3 — A and C *algorithms*) and jQuery (based on section 3.3 — B algorithm).

In addition, MD5 was used to produce the message digest before the data transmission and is decrypted upon arrival at the destination. With this experiment, the following objectives are achieved:

✓ *Active Path:* using the model introduced, we were able to determine the route that is available using the response received from the jQuery. Another path is chosen when the path response is false.

✓ *Data availability:* if data is not available in one database, the algorithms choses another configured database for the data retrieval.

✓ *No trigger or timestamp*: the event that initiates the synchronization process is customizable. In this example, we adopt system-launch-based event to start the process.

✓ *Bidirectional*: table 1 can send data to table 4 on different server, and table 4 can also send data to table 1, not (master-to-slave) as other solutions.

✓ *Parallelism*: the sending and receiving data between table 4 and table 1 are done at the same time without waiting for sending process to finish before the receiving begins.

✓ *XML:* was used to package data for transfer.

5   Conclusion and Future Work

We have presented a model and an algorithm, for synchronizing data among heterogeneous database environment along with minimum involvement of database or server dependent information hence maximize independency of vendors thus maintaining the integrity of diverse database and server in an occasionally or persistent connected network. We have used jQuery-data-response mechanism to determine which server contains the data required by requester, thereby giving alternative paths to the agent in case the pointed server is down. Determining the active path at early stage of synchronization greatly reduces the occurrences of data lost during synchronization.

The algorithm suggested in this work allows bidirectional communication and data exchange. For instance, the server A can send information to server B and the vice versa. Moreover, parallelism is also a priority in our approach where the sending and the receiving of data occur at the same time on a given server without any collision.

Finally the event that triggers the process can be defined by the user during configuration, thereby giving preference to a user to determine which event will best activate the synchronization process. Our future concern will be the large data size where thousands or millions of records are expected to cross over the network without any distortion or distraction which brings the issue of data compression, transfer speed and data security.

References

[1] Wiesmann, M., Pedoney, F., Schiper, A., Kemmez, B., Alonsoz, G.: Database Replication Techniques: a Three Parameter Classification. In: Proc. 19th IEEE Symp. Reliab. Distrib. Syst. (2000)

[2] Lu, Z., Zhang, C., Liu, Z.: Optimization of Heterogeneous Databases Data Synchronization in WAN by Virtual Log Compression. In: Second Int. Conf. Futur. Networks, ICFN 2010, pp. 98–101 (January 2010)

[3] Geng, Y., Zhao, Z., Kong, X.: Synchronization System among Remote Heterogeneous Database Systems. In: 2008 IEEE Int. Symp. Knowl. Acquis. Model. Work. (2008)

[4] Zhang, Z., Lu, C.: Research and Implementation for Data Synchronization of Heterogeneous Databases. In: 2010 Int. Conf. on Comput. Commun. Technol. Agric. Eng. (CCTAE), vol. 3, June 12-13, pp. 464–466 (2010)

[5] Selman, I.D., Gb, S., Bergman, R., Us, C.O., Neil, E.K.O.: (12) United States Patent Selman et al. (10) Patent No.: (45) Date of Patent 2(12) (2008)

[6] Gajjam, N., Apte, S.S.: Mobile Agent based Communication Platform for Heterogeneous Distributed Database 2(9), 203–207 (2013)

[7] Sethia, D., Mehta, S., Chodhary, A., Bhatt, K., Bhatnagar, S.: MRDMS-Mobile Replicated Database Management Synchronization. In: 2014 Int. Conf. Signal Process. Integr. Networks, pp. 624–631 (2014)

[8] Stage, A.: Synchronization and replication in the context of mobile applications, pp. 1–16 (2005)

[9] Stage, A.: Synchronization and replication in the context of mobile applications. In: Second International Conference on ICFN 2010, pp. 98–101 (2012)

[10] Oracle®, Database Lite Developer's Guide 10g (10.0.0) Part No. B13788-0

[11] Vazquez-Briseno, M., Vincent, P., Nieto-Hipolito, J.I., Sanchez-Lopez, J.D.: Design and Implementation of an Automated Mobile Applications Generator for Remote Database Access 9(3), 351–357 (2011)

[12] Yang, G.: Data Synchronization for Integration Systems based on Trigger, pp. 310–312 (2010)

[13] Rayns, C., Burgess, G., Cooper, P., Fitzgerald, T., Goyal, A., Klein, P., Li, G.Q., Liu, S., Sun, Y.: Front cover Application Development for CICS. Contract (2015)

[14] Narayan, S.: What is jQuery and how to start using jQuery (2011),
http://www.codeproject.com/Articles/157446/
What-is-jQuery-and-how-to-Start-using-jQuery
(accessed: February 02, 2015)

Cognitive Framework for Intelligent Traffic Routing in a Multiagent Environment

Shailendra Tahilyani and Manuj Darbari

Babu Banarasi Das University,
Lucknow, U.P., India
stahil.ec@gmail.com, manuj_darbari@acm.org

Abstract. The paper highlights the use of two basic components: Cognitive Ra-dio and EMO in controlling Urban Traffic System. The coordination of the ac-tivities like: identification of route network, traffic jams and coordination of VMS for route diversion. The learning of central control system is achieved by software defined radios integrated with evolutionary Multiobjective optimiza-tion.

Keywords: Urban Traffic System, Cognitive Radio, EMO.

1   Introduction

Traffic management using Multi-Agent system[1,10] has grown significantly. Auton-omous agent provides excellent coordination capabilities while we are designing co-operative Multi Agent. Due to advancement in Urban Traffic there is an urgent need of developing a complex control system which can manage traffic efficiently. We focus on all the design aspect of Information Model which is divided into two parts namely Urban Decision Subsystem and Urban Physical Subsystem. The focus is on hierarchical structure for urban traffic control and management. In recent years, the evolving growth of UTS[11,12,13] application is generating an increased need for tools to aid in the design and assessment of the system. To obtain these objectives, there are various models to simulate the traffic and they have proved to be very cost effective. Various tools are required to improve the increasing complexity and swiftly weakening transportation systems of the present. The ability of traffic simulation models to quantify the urban transportation system is still missing even today. Cur-rently there are 80 simulation models available out of them CORSIM and ITEGRATION seem to have the highest success probability in the real world scenar-io. But, still both these models suffer from the disadvantage of high complexity of the system, which makes them difficult to be enhanced or customized according to speci-fied users requirements.

The main goal of this paper is to incorporate the unified architecture which allows to couple Transportation Systems and Cognitive extension[2,4,9] in addition to its potential application to the real world. The whole framework serves as an important tool to assess and validate innovative approaches to deploy sustainable transportation solutions. However, some challenges like coordination amongst agents and validation

© Springer International Publishing Switzerland 2015   67
R. Silhavy et al. (eds.), *Software Engineering in Intelligent Systems*,
Advances in Intelligent Systems and Computing 349, DOI: 10.1007/978-3-319-18473-9_7

of behavioral models is an actual issue which we have also incorporated in our work. Coupling multi-agent systems with intelligent transportation solutions have boosted our efforts to promote the quite recent area of Artificial Transportation Systems as one important instrument to better analyze sustainable transportation.

1.1   Motivation

The limited road infrastructure and exponentially increasing number of vehicles are major reasons for traffic congestion in urban cities. With these unavoidable limitations, the Artificial Intelligence approaches are being applied to deal with this problem. A new lane-by-pass approach has been investigated to deal with the problem of traffic congestion using cognitive radio in a multi-agent environment[14]. Genetic Algorithms are utilized for optimization of various parameters used in traffic routing mechanism. As mechanism involves generation of minor sublane and dynamic route in-formation focusing on three basic aspects:

(i) Velocity of information and
(ii) Quality of information
(iii) Dynamicity of the information

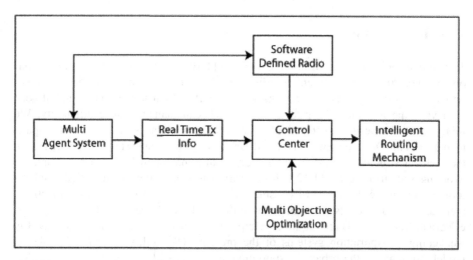

Fig. 1. MACNUTS(Multi Agent Cognitive Radio Network for Urban Traffic System)

Besides these it has got various factors like signal control from control centre (figure 1), these are controlled using the concept of Multi Objective Optimization[3,5,14]. Before applying this three major activities are to be performed:

(i)   Identifying the Route Networks in the city that represents the relationship between all the routes.
(ii) Identifying the Traffic Jams and alternative minor sub lane route information.
(iii) Synchronization between all the alternative traffic routes to ensure smooth flow of the traffic from the source to destination.

In an MOO problem space, a set of solutions optimizes the overall system, if there is no one solution that exhibits a best performance in all dimensions. This set, the set of non-dominated[17] solutions, lies on the Pareto-optimal front (hereafter called the Pareto front). All other solutions not on the Pareto front are considered dominated, suboptimal, or locally optimal. Solutions are non-dominated when improvement in any objective comes only at the expense of at least one and other objective.

2 MACNUTS (Multi Agent Cognitive Radio Network for Urban Traffic System) a Framework to Model UTS Intelligently

Our work proposes to generate various sets of subroutine program using Software Defined Radio(SDR) using XML. The Cognitive radios[7,8,15,16,18,19] should be able enough to inform other cognitive radios about their observation that may affect the performance of the radio communication channel. The receiver must be able to measure the signal properties and based upon that they should be able to make an estimation of what the transmitter intended to send. Apart from this, it also needs to be able to let know the transmitter how to change its waveform in a way that will repress interference. In short we can say that the cognitive radio receiver needs to convert this information into a transmitted message to send back to the transmitter. In Radio XML, Radio defines the domain of natural and artificial know-ledge and skill having to do with the creation, propagation, and reception of radio signals from sources natural and artificial. That's pretty much how RXML, defined in terms of the use of XML Tag as schema-schema, was envisioned, within an open framework for general world knowledge needed for AACR. RXML recognizes critical features of micro-world not openly addressed in any of the e-Business or semantic web languages yet:

(i)   Knowledge often is procedural.
(ii) Knowledge has a source that often establishes whether it is authoritative or not, or its degree of attributed voracity.
(iii) Knowledge takes computational resources to store, retrieve, and process.
(iv) Chunk of knowledge fits somewhere in the set of all knowledge and knowing more or less where that knowledge fits can help an algorithm reason about how to use it.

2.1 Lane Division Scenario

Consider a situation of a traffic movement scenario(figure 2) with Road network defined mathematically as:

Junction Set: J = {A, B, C, D, E, F, G, H, I}
Connectives Set: E = {e1, e2, e3 .e23}
Minor Sublanes Set: M= {m1, m2, m3, m4}

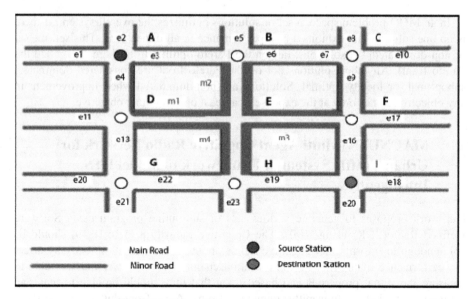

Fig. 2. Multi lane Diagram showing Lane bypass

The flow of traffic flow is determined by the set of connectives that are used from the starting place to the target place. Some connectives function as the bypass connectives. The bypass connectives are used when the other connectives are engaged in the traffic jam. VMS (Variable Message Signboard) technology is employed in this approach. The commuters will establish the signal that will divert the route before one crossing through minor sub lane generation if there exists any traffic jam at the next crossing. The source to destination should be minimized as much as possible. In order to overcome this setback, sensors are placed at each junction to intimate the traffic congestion to the previous junction through VMS. It recognizes the traffic congestion at junction J through VMS when he is on the route e4 at the junction. In order to deal with this condition the control system generates the signal for route diversion using minor lane bypass and the commuter gets the signal to select the route m1 and m4 to reach its destination. The procedure to divert the route should be included with the traffic phase and movement decision approaches. We can develop OWL for the above situation as:

```
<owl:Class rdf:ID='Sub_lane'>
<rdfs:subClassOf>
<owl:Class rdf:ID='Main_Road'/>
</rdfs:subClassOf>
</owl:Class>
```

Knowledge representation and semantics can provide such a cognitive radio community architecture with a shared base of constructs that can enable a more flexible and dynamic environment(figure 3). According to the IEEE 802.11 specification, in order for a station (STA) to use the medium it must first associate with an access point (AP), i.e., STA must transmit a message with certain information to the AP. In particular, the type of the message must be management and its subtype must be association request. (The message must also include address identifiers for the STA and AP, but those are ignored here.)

We use *Message(x)* to encode *x* is an IEEE 802.11 protocol message, *msgTypeIs(x, y)* to encode the type of message *x* is *y*, and *msgSubTypeIs(x, y)* to encode the subtype of message *x* is *y*. The following written axiom encodes this information more compactly:

If
Message(x) /& msgTypeIs(x, MANAGEMENT) /&
msgSubTypeIs(x, ASSOCIATION REQUEST) Then
AssociationRequestMessage(x)

where *MANAGEMENT* and *ASSOCIATION REQUEST* are individual constants that denote the IEEE 802.11 message parameters. In terms of an OWL ontology, it states that if an individual is an instance of the class of IEEE 802.11. Let the predicate *Cognitive Sensor(x)* encode *x* is a sensor, let *SensorInUse(x, t)* encode sensor x is in use at time *t*, let *ControlCentreCheckedAtTime(x, t)* encode a *ControlCentreCheckof Channel x* occurs at time *t*, let TimeDifference(t_1, t_2) be a function that returns the time difference in seconds between the time values of instants t_1 and t_2, let *ControlCentreCheckTimeThreshold* be an individual constant with the value difference between t1-t2, and finally let *RequireControlCentreCheck(x,t)* encode a *ControlCemtreCheck* on sensor x is required at time t. Then consider the following rule:

If
Sensor(x) &
Instant(t1) & SensorInUse(x, t1) &
Instant(t2) & ControlCentreCheckedAtTime(x,   t2) &
TimeDifference(t1, t2)   CONTROLCENTRE CHECK TIME
THRESHOLD Then
RequireControlCentreCheck(x,t1).

The above method shows the semantics between Control Centre and the Cognitive radio network. Depending on the velocity of information (Complexity) we have enhanced our software network and written all the possible sub-routines for software defined radio network for traffic modeling.

2.2   Applying EMO Algorithm Using NSGA-II with Elist Non-dominated Genetic Algorithm

Let MACNUTS be defined as MC ; representing the entire control situation consisting of various variables like(figure 4): Multi agent based Cognitive radio subroutines, Multiagent percepts Ontology Rule base(Traffic Domain).

```
Step I: Mc:= Initialize (Mc)
Step II: while termination condition is not satisfied, do
Step III: Mc:= Selection (Mc)
Step IV: Mc := Genetic Operations(Mc)
Step V: Mc := Replace (MC U MC)
Step VI: end while
Step VII: return (non dominated solutions (Mc))
```

The entire set of Cognitive radio readings are adjusted according to the non-dominated sorting and crowding distance. The XML program which is going to be executed will be dependent on Pareto dominance relation based on the rank of the current population.

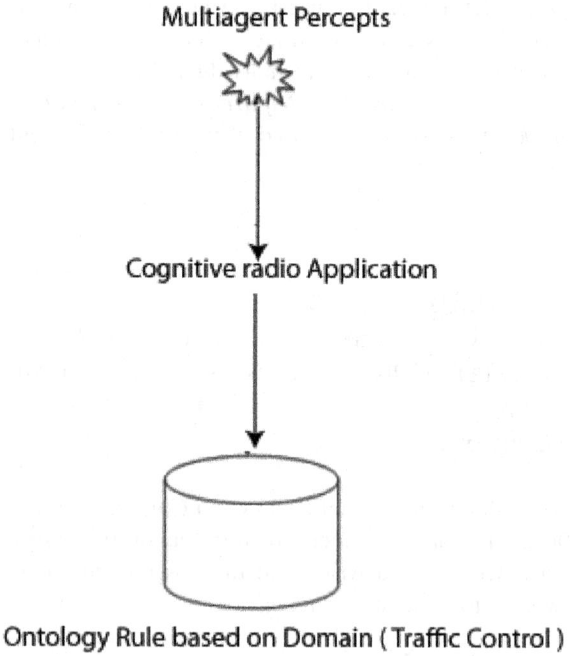

Fig. 3. Architecture of MACNUTS Components

Solutions in the current population are sorted to assign a rank to each solution. The outcome is formulated using crowding distance which is the sum of the calculated distance over all the objectives. On plotting this for two objective(variables) we got large number of non-dominated solution for multi objective optimization (figure 5).

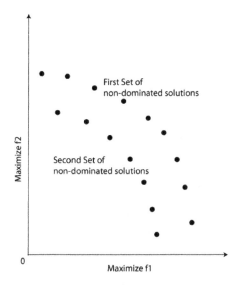

Fig. 4. Multi lane Diagram showing Lane bypass

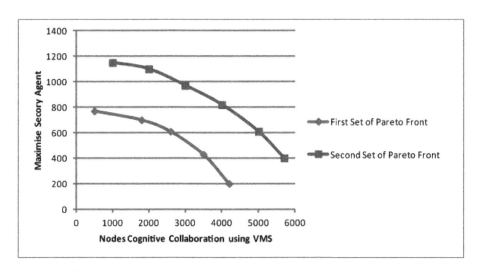

Fig. 5. . Experimental by single sum of Two Objective using NSGA-II

Results and Analysis of the Model

Solutions in the current population are sorted to assign a rank to each solution. The outcome is formulated using crowding distance which is the sum of the calculated distance over all the objectives. On plotting this for two objective(variables) we got large number of non-dominated solution for multi objective optimization (figure 6,7,8).

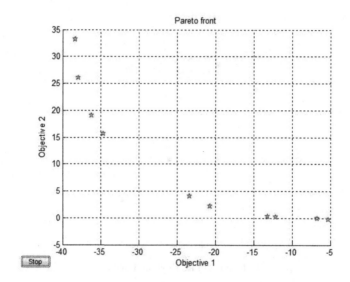

Fig. 6. Plot between rank of the CORSIM

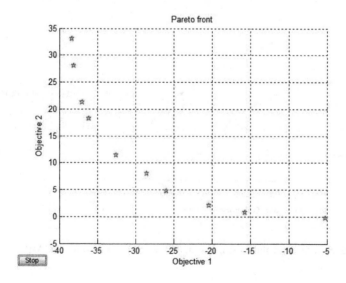

Fig. 7. Plot between rank of the ITEGRATION

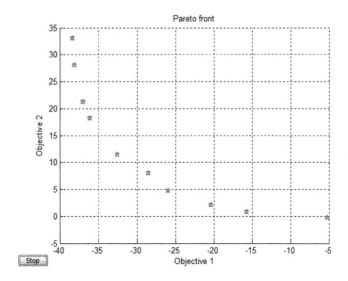

Fig. 8. Plot between rank of MACNUTS

From figure 5.4, 5.5 and 5.6 we are able to compare two other proposed models using multi objective optimization, we found that out of the two our model optimized with perfect non dominated set of variables. As we increase the cognitive radio sensors the accuracy of the model increases leading it high complexity(number of rules set) of the model, these are the factors in which we have used   EMO to maximize.

3   Conclusion

As we have seen, cognitive radio, as an engineering endeavor, bridges community and device architectures. We have tried to bridge linguistic styles of both declarative and imperative semantics using radio device technology, software defined radio (SDR) and finally extending it to object-oriented programming, the semantics of which is essentially imperative in nature.

References

1. Srivastava, A.K., Darbari, M., Ahmed, H., Asthana, R.: Capacity Requirement Planning Using Petri Dynamics. International Review on Computers Software 5(6), 696–700, 4 (2010)
2. Fette, B.: Cognitive Radio Technology. Elsevier, Boston (2006)
3. Yagyasen, D., Darbari, M., Shukla, P.K., Singh, V.K.: Diversity and Convergence issues in Evolutionary Multiobjective Optimization: Application to Agriculture Science. IERI Procedia 5, 81–86, 4 (2013)
4. Cabric, D.: Cognitive radios: System design perspective, Ph.D. dissertation, Univ. of California, Berkeley (2007)

5. Siddiqui, I.A., Darbari, M., Bansal, S.: Application of Activity Theory and Particle Swarm Optimization Technique in Cooperative Software Development. International Review on Computers Software 7(5), 2126–2130, 7 (2012)
6. Peha, J.M.: Sharing Spectrum Through Spectrum Policy Reform and Cognitive Radio. Proceedings of the IEEE (2009)
7. Mitola III, J.: Cognitive radio: An integrated agent architecture for software defined radio. PhD thesis, Royal Institute of Technology (KTH), Stockholm, Sweden (2000)
8. Mango Communications, http://www.mangocomm.com
9. Nekovee, M.: Dynamic spectrum access concepts and future architectures. BT Technology Journal 24, 111–116 (2006)
10. Darbari, M., Dhanda, N.: Applying Constraints in Model Driven Knowledge Representation Framework. International Journal of Hybrid Information Technology 3(3), 15–22,9 (2010)
11. Darbari, M., Medhavi, S., Srivastava, A.K.: Development of effective Urban Road Traffic Management using workflow wechniques for upcoming metro cities like Lucknow (India). International Journal of Hybrid Information Technology 1(3), 99–108, 8 (2008)
12. Darbari, M., Asthana, R., Ahmed, H., Ahuja, N.J.: Enhancing the capability of N-dimension self organizing petrinet using neuro-genetic approach. International Journal of Computer Science Issues (IJCSI) 8(3), 4 (2011)
13. Darbari, M., Singh, V.K., Asthana, R., Prakash, S.: N-Dimensional Self Organizing Petrinets for Urban Traffic Modeling. International Journal of Computer Science Issues (IJCSI) 7(4), 37–40 (2010)
14. Dhanda, N., Darbari, M., Ahuja, N.J.: Development of Multi Agent Activity Theory e-Learning (MATeL) Framework Focusing on Indian Scenario. International Review on Computers Software 7(4), 1624–1628, 4 (2012)
15. Murphy, P., Sabharwal, A., Aazhang, B.: Design of WARP: A Wireless Open-Access Research Platform. In: European Signal Processing Conference, Florence, Italy (2006)
16. Koch, P., Prasad, R.: The Universal Handset. IEEE Spectrum (2009)
17. Shukla, P.K., Darbari, M., Singh, V.K., Tripathi, S.P.: A Survey of Fuzzy Techniques in object oriented databases. International Journal of Scientific and Engineering Research 2(11), 1–11, 6 (2011)
18. WARP: Wireless Open Access Research Platform, http://warp.rice.edu/
19. Jing, X., Raychaudhuri, D.: Global Control Plane Architecture for Cognitive Radio Networks. In: IEEE CogNet 2007 Workshop (with IEEE ICC) (2007)

A Task Management in the Intranet Grid

Petr Lukasik and Martin Sysel

Department of Computer and Communication Systems
Faculty of Applied Informatics
Tomas Bata University in Zlin
nam. T.G. Masaryka 5555, 760 01 Zlin
CZECH REPUBLIC
plukasik@tajmac-zps.cz, sysel@fai.utb.cz
http://www.fai.utb.cz

Abstract. The main purpose of this work is to explain the management and distribution of tasks in a grid middleware. The aim is to propose an environment that enables designing the batch jobs that use the standard software and system resources for communication and data exchange. The article explains JSDL specification for defining and management of one batch jobs. The motivation is to design a tool for an easy job definition. The result is a tool that does not require special programming objects for solving a specific task. This work deals with the mechanisms for monitoring and gathering information about the result of each processing task. The JSDL definitions provide the mechanism for the solution of these problems. Grid Scheduler more easily recognizes the status of a specific task.

Keywords: Grid, JSDL, POSIX, fault tolerance, return code.

1 Introduction

Planning and distribution of tasks are essential elements of a grid services. A tool that allows easy definition of the role and its distribution in the environment is a prerequisite for high-quality and user-acceptable Grid Services. The user should have a freedom as well as resources to easily tracking of their own processing. An important feature is that the Grid service has the least restrictive conditions for a successful job execution. (Type or version of software, operating system and hardware features). The user of the grid should to have a certain freedom. Not to be tied up of restrictive rules, except the rules relating to information security and data processing. The POSIX interface provides extensive options in the use of standard programs for communication and distribution of batch job. It also provides excellent portability of applications in various types of operating systems. This feature is convenient for defining the tasks in a grid.

2 The Job Distribution in a Grid Environment

JSDL (Job Submission Definition Language) is an XML-based computational specification for the management and distribution of batch jobs in a Grid

© Springer International Publishing Switzerland 2015
R. Silhavy et al. (eds.), *Software Engineering in Intelligent Systems*,
Advances in Intelligent Systems and Computing 349, DOI: 10.1007/978-3-319-18473-9_8

environment, developed by OGF JSDL-WG [1][2][3]. A current version 1.0 (released November 7, 2005) has also the definition of the POSIX support. This allows implementing the requirements described above. Important is Open Grid Forum support. JSDL should be considered as a standard [6].

The life cycle of the task instance represents the current state of the program which is currently executed (or pending to be executed). Each instance of the job it has a direct impact on the overall result. Status of the tasks that provides the Grid client is very important for the scheduler. The scheduler is responsible for the successful solution of the assigned task. Whereas it applies that in a number of fault conditions, this information is not delivered to the planners [5].

The same mechanism of the lifecycle also applies to the run of the whole batch job. Failure of a single part means a bad result of the entire job. The grid scheduler has to keep track of each state of all processed instances of the job. Scheduler based only on all available information, can successfully manage the processing tasks and to resolve error conditions. The states that can occur in the life cycle flow of the job are described in the fig.[1]. The task progress can take three states. These states can be categorized into two groups. The first group is the state when the task will return result of the processing (correct or incorrect).In the second group, cannot send a return value. Example is violent interruption of the task or power failure in the some parts of the infrastructure.

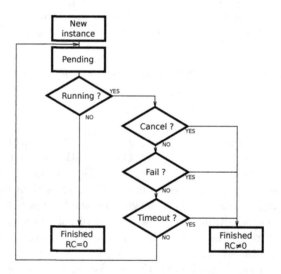

Fig. 1. Task and return code management

Grid scheduler has to solve the state, if the work cannot send a return value. The principle is as follows. Scheduler receives information about the interruption of the communication channel. If this information is not available, the planner has to monitor timeout, which is set as a parameter to the task, which is currently running. After a timeout, the task instance is declared as lost. This means that an instance of task must be rescheduled again and sent back to the processing. To specify a timeout value is relatively difficult. Usually effective processing length, from which the value can be inferred is not known. Correct timeout can be specified based on some experience with specific job. There is, of course, possibility that the problem causes the task, currently running that do not have correctly handled error conditions (memory leak, divide zero). This task can crash the entire grid infrastructure. Therefore, the grid scheduler must solve this situation as well, according to predetermined rules. Job Submission Definition Language includes support that allows you to define ranges of computing resources (length of treatment, number of running threads, disk space). The task does not run if any of the parameters is exceeded. You can also define rules for communication with Grid scheduler while processing the return codes and rules in case of failure of one of the running instances.

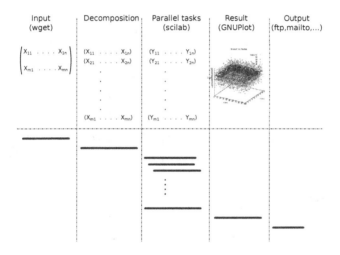

Fig. 2. Lifecycle of the job and task

3   The Standard Programs for Communication and Task Distribution

POSIX interface and its features are used for the distribution of tasks, data transfer and communication between the various components in the grid. This interface provides portability of program objects and commands on the system console. This allows the use of existing software. Grid user therefore does not need to write a special programs or objects to solve their task and enables the use of standard software (Python interpret, Mathworks, Scilab). The advantage is a choice of suitable existing components and using the standard and well-known user environment. See fig.[2].

This solution is a perspective in the local environment, such as on the university or corporate grid, which guarantee a certain standardization of software and system tools. The example below defines this job. Service wget retrieves data from various data repositories (http). The GnuPlot program generates a graph. The result is sent to the user via email. The return value is sent to the grid scheduler using the Web services or the Intranet Grid services fig.[3]. This particular example shows that the task does not require special software, but will allow the user to utilize the standard software. The JSDL has a definition for single-program-multiple-data (SPMD) parallel techniques. This extension supports various MPI environments. (MPI, GridMPI, Torch / MPI, MPICH). Parameter Sweep definition to the JSDL is also available for a parallel tasks. This extension defines rule to explicitly submitting various number of individual tasks of the same base job template. One definition allows generating a large number of parallel executable tasks [4].

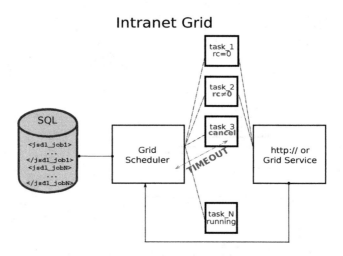

Fig. 3. Management of return values

4 Principles for Resolving of the Error States

The weak part in communication of the grid scheduler and grid agents is a violent termination of the currentl running job. Fatal error of the application causes a loss of information about the job. Grid scheduler does not get back information about this error. Fault Tolerance subsystem and its properties has a major impact on the reliability of the entire grid services. The challenge is to increase tolerance to the system failures, and recognize various error scenarios. Based on the information available, choose the correct recovery procedure. JSDL specification enables the user to define some limit values. This increases the intelligence of the system and also allows to better evaluating error conditions.

The basic limit values of the batch jobs is memory size, file size, number of threads and maximum run time. Exceeding some of the limit values indicates incorrect job status. Type of the limit value that was exceeded, allows solve to better the error status of the job. For example, invoke the rollback or send a job to the processing again.

5   Tools for the JSDL Support

G-Eclipse Plugin extends the Eclipse Project `eclipse.org` and provides support for Grid services. This environment provides services independent of a specific grid middleware. Plugin architecture enables developers to extend the g-Eclipse on a new feature. Example is integration of another middleware. Support for the gLite and GRIA middleware is also completed [6].

5.1   g-Eclipse - Support for Users

- *Grid user:* Has no detailed knowledge of Grid technologies. Allows you to run and monitor the progress of a job.

- *Grid operator:* Has a detailed knowledge of the Grid infrastructure. Operator has support for the management of local resources, and also supports external resources so called virtual organizations, where operators are the members.

- *Development engineer:* Expert on programming in the Grid applications. Developer has tools to develop, debug and deploy the application. Graphical editor for generating JSDL file is also available. Editor supports POSIX and Parameter Sweep extension.

5.2   A Program for the JSDL Parsing

The parser was designed for the Intranet Grid Middleware[7]. We use the Apache XMLBeans library and DOM model. The DOM advantage is parsing of XML document in the memory - no I/O operation is required. The aim was to determine the performance of the parser.

It has been shown the low requirements of the JSDL tools to system resources. Two types of tasks were measured. The simple Sweep-loop in the first task, (see the fig.[4] *sweep(1)* label) and the triple Sweep-loop in the second task, (fig.[4] *sweep(3)* label). There were generated the scripts in the number from one to 10000. Five measurements were performed for each job, and was carried out on two systems with different performance. It was verified by the linear increase of the time and a low system load, inspite of the XML based JSDL. In the fig.[4] is only average value presented. Detail of the measurement is in the 1 presented.

Table 1. Results for the sweep loop (1) and sweep loop (3)

sweep(1) M420

number of the jobs	time[s]
1	0.555 0.555 0.554 0.557 0.558
2	0.559 0.558 0.564 0.559 0.560
5	0.537 0.533 0.536 0.536 0.537
27	0.606 0.606 0.608 0.604 0.606
256	0.842 0.830 0.810 0.828 0.819
2000	2.067 2.070 2.142 2.056 2.030
13824	9.218 9.088 9.209 9.233 9.177

sweep(3) M420

number of the jobs	time[s]
1	0.553 0.559 0.550 0.548 0.548
10	0.559 0.557 0.600 0.556 0.566
100	0.672 0.661 0.679 0.676 0.706
1000	1.445 1.425 1.404 1.454 1.489
10000	7.412 7.524 7.438 7.501 7.625

sweep(1) T1700

number of the jobs	time[s]
1	0.156 0.155 0.158 0.157 0.156
2	0.160 0.160 0.175 0.158 0.156
5	0.149 0.152 0.150 0.151 0.151
27	0.168 0.170 0.170 0.170 0.170
256	0.233 0.230 0.228 0.228 0.226
2000	0.595 0.552 0.571 0.539 0.482
13824	2.613 2.355 2.591 2.546 2.482

sweep(3) T1700

number of the jobs	time[s]
1	0.155 0.152 0.155 0.156 0.151
10	0.158 0.158 0.158 0.157 0.157
100	0.197 0.193 0.193 0.191 0.190
1000	0.457 0.497 0.483 0.506 0.455
10000	2.534 2.607 2.504 2.590 2.493

5.3   The XML Schema of the Sweep(1) Loop

```
<!-- SWEEP(1) - job_NNNNNN.sh -->
<!-- LEVEL1.1 -->
<sweep:Sweep>
  <sweep:Assignment>
    .
  </sweep:Assignment>
</sweep:Sweep>
```

5.4   The XML Schema of the Sweep(3) Loop

```
<!-- SWEEP(3) - job_XXYYXX.sh -->
<!-- LEVEL1.1 -->
<sweep:Sweep>
   <sweep:Assignment>
   .
   </sweep:Assignment>
   <!-- LEVEL2.1 -->
   <sweep:Sweep>
     <sweep:Assignment>
     .
     </sweep:Assignment>
     <!-- LEVEL3.1 -->
     <sweep:Sweep>
        <sweep:Assignment>
        .
        </sweep:Assignment>
     </sweep:Sweep>
   </sweep:Sweep>
</sweep:Sweep>
```

6   Result

The text describes the use of JSDL generator as a batch jobs. The primary objective was to take maximum advantage of the standard and existing software and system services. This allows the user, to be more creative in the design of task. At the first glance, it may also appear that JSDL brings some complications

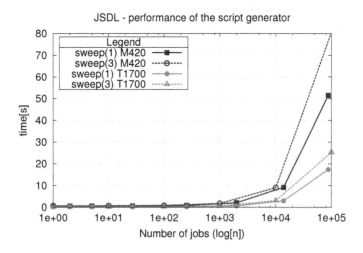

Fig. 4. Sweep parm and system load

Table 2. A list of workstations where measurements were made

A List the hardware where it has been JSDL parser tested

1	**DELL T1700 64 bits**
Memory	16GB RAM
CPU	Intel(R) Xeon(R) CPU E31225 @ 3.10GHz 64 bits
OS	Ubuntu 14.4 LTS
Java	OpenJDK 1.7

2	**Fujitsu Siemens Celsius M420 32 bits**
Memory	1.96 GB RAM
CPU	Intel(R) CPU E 6750 @ 2.66GHz 32 bits
OS	Ubuntu 12.4. LTS
Java	OpenJDK 1.6

(XML parser, job generator). The advantage is obvious when it is necessary to generate and manage thousands of concurrent instances of a single task.

Another direction for development of intranet grid middleware will be design and use of JSDL for applications in the Apache Hadoop cloud. Hadoop is the executive and promising platform. Problem of forcibly terminated task is solved by using redundant processes.

7   Conclusion

This work is focused on the job definition in Grid computing environment. The Grid usually have a large amount of heterogeneous sources that have many different configuration and many different internal rules. JSDL has a set of instructions that unify these rules. JSDL has the tools, for example G-Eclipse Plugin that facilitate definition of the tasks. Users do not to have detailed JSDL knowledge. Measurements that have been made not demonstrated great demands on the computing performance of the parser. The advantage JSDL is easy definition of parallel tasks. Parameter sweep easily defines the number of jobs and their identification.

Comparing JSDL and similar technologies, which are used to define the job in grid computing, for example JDL (Job Definition Language), we came to this conclusion. JSDL is suitable for the commercial applications that require easy design and definition of the task.

JDL is a Python-based language for describing jobs.[9] It requires an experienced user. Provides more possibilities for the definition of the job. Deployment JDL is preferable in the research centres or universities.

References

1. Anjomshoaa, A., Brisard, F., Drescher, M., Fellows, D., Ly, A., McGough, S., Pulsipher, D., Savva, A.: GFD-R.136 Job Submission Description Language (JSDL) Specification (July 28, 2008), `http://forge.gridforum.org/projects/jsdl-wg`, Copyright (C) Open Grid Forum (2003-2005, 2007-2008). All Rights Reserved
2. Humphrey, M., Smith, C., Theimer, M., Wasson, G.: JSDL HPC Profile Application Extension, Version 1.0 (July 14, 2006) (Updated: October 2, 2006) Copyright Open Grid Forum (2006-2007). All Rights Reserved
3. Drescher, M., Anjomshoaa, A., Williams, G., Meredith, D.: JSDL Parameter Sweep Job Extension Copyright © Open Grid Forum 2006–2009. All Rights Reserved (May 12, 2009)
4. Savva, A.: JSDL SPMD Application Extension, Version 1.0 (draft 008) Copyright Open Grid Forum (2006, 2007). All Rights Reserved
5. Theimer, M., Smith, C.: An Extensible Job Submission Design May 5, 2006 Copyright (C) Global Grid Forum (2006). All rights reserved, Copyright (C) 2006 by Microsoft Corporation and Platform Computing Corporation All rights reserved
6. Bylec, K., Mueller, S., Pabis, M., Wojtysiak, M., Wolniewicz, P.: Parametric Jobs – Facilitation of Instrument Elements Usage In Grid Applications. In: Remote Instrumentation Services on the e-Infrastructure. Springer US (2011), `http://dx.doi.org/10.1007/978-1-4419-5574-62`, 978-1-4419-5573-9
7. Lukasik, P., Sysel, M.: An Intranet Grid Computing Tool for Optimizing Server Loads. In: Silhavy, R., Senkerik, R., Oplatkova, Z.K., Silhavy, P., Prokopova, Z. (eds.) Modern Trends and Techniques in Computer Science. AISC, vol. 285, pp. 467–474. Springer, Heidelberg (2014)
8. Rodero, I., Guim, F., Corbalan, J., Labarta, J.: How the JSDL can exploit the parallelism? In: Sixth IEEE International Symposium on Cluster Computing and the Grid, CCGRID 2006, May 16-19, vol. 1, pp. 275–282 (2006), `http://ieeexplore.ieee.org/stamp/stamp.jsp?tp=&arnumber=1630829&isnumber=34197`, doi:10.1109/CCGRID.2006.55
9. Developer Guide and Reference, Novell ® PlateSpin Orchestrate 2.0.2 (July 9, 2009), `https://www.netiq.com/documentation/pso_orchestrate25/`

Experiences with DCI Pattern

Ilona Bluemke and Anna Stepień

Institute of Computer Science, Warsaw University of Technology,
Nowowiejska 15/19, Warsaw, Poland
I.Bluemke@ii.pw.edu.pl

Abstract. The DCI architectural pattern for software, introduced by Reenskaug, contains three parts: Data, Context and Interaction. Data represent domain knowledge while Context and Interaction represent the business logic by implementing communication between objects. Context dynamically injects roles into objects. This design pattern is especially appropriate for agile software development. The goal of our paper is to present some practical experiences with the DCI design pattern. We used the DCI pattern in an exemplary application. We proposed and described the association and cooperation between the DCI and MVC (Model View Controller) pattern. Some remarks on the usage of the DCI design pattern in the software development are also given.

Keywords: design patterns, DCI, MVC.

1 Introduction

In software engineering, a design pattern is a general, reusable solution to a commonly occurring problem within a given context in the software design. A design pattern is not a finished design that can be transformed directly into the source or machine code. It is a description or template for how to solve a problem that can be used in many different situations. Patterns are formalized best practices that the programmer can implement in the application.

Patterns originated as an architectural concept by Christopher Alexander (1977/79) [1]. In 1987, Kent Beck and Ward Cunningham began experimenting with the idea of applying patterns to programming and presented their results at the OOPSLA conference [2]. Design patterns gained popularity in computer science after the book Design Patterns: Elements of Reusable Object-Oriented Software was published in 1994 by the so-called "Gang of Four" [3]. In 1995 the Portland Pattern Repository [4] was set up for the documentation of design patterns.

Object-oriented design patterns typically show relationships and interactions between classes or objects, without specifying the final application artifacts that are involved [5]. Design patterns reside in the domain of modules and interconnections. At a higher level there are architectural patterns that are larger in scope, usually describing an overall pattern followed by an entire system.

© Springer International Publishing Switzerland 2015
R. Silhavy et al. (eds.), *Software Engineering in Intelligent Systems*,
Advances in Intelligent Systems and Computing 349, DOI: 10.1007/978-3-319-18473-9_9

There are many types of design patterns, for instance:

- Algorithm strategy patterns addressing concerns related to high-level strategies describing how to exploit application characteristics on a computing platform.
- Computational design patterns addressing concerns related to key computation identification.
- Execution patterns that address concerns related to supporting application execution, including strategies in executing streams of tasks and building blocks to support task synchronization.
- Implementation strategy patterns addressing concerns related to implementing source code to support program organization, and the common data structures specific to parallel programming.
- Structural design patterns addressing concerns related to high-level structures of applications being developed.

Design patterns can speed up the development process by providing tested, proven development paradigms [3, 4, 5]. Effective software design requires considering issues that may not become visible until later in the implementation. Reusing design patterns helps to prevent subtle issues that can cause major problems, and it also improves code readability for programmers and architects who are familiar with the patterns. Meyer and Arnout were able to provide full or partial componentization of two-thirds of the patterns they attempted [6].

In this paper we concentrate on a design pattern which draws attention and discussion for several years called DCI (Data, Context, Interaction). The DCI pattern was introduced in 2008 by Trygve Reenskaug [7, 8, 9, 10] and is being used, elaborated by James O. Coplien and others e.g. [11, 12, 13, 14, 15, 16]. The main idea of DCI is to separate the code that describes the system state from the code that describes the system behavior. The code is organized in three perspectives each of which focuses on a certain aspect of the code, these perspectives are called:

- **Data** - contains the computer representation of the user's mental model,
- **Context** is the specification of a network of communicating objects that realizes a user command,
- **Interaction** - specifies how objects communicate when executing a user command.

The idea of separating what the system is and what it does seemed very interesting to us. In the literature we were able to find only small examples of the DCI usage e.g. for one simple use case, so we decided to implement a case system. Our contribution is the usage of the DCI paradigm in the system, called DCI –Project, which we designed and implemented. We also present how DCI and MVC (Model View Controller) [17] patterns can be combined in the architecture of the system (section 4.1). In this paper we present some of our experiences gathered during the development of our system.

The organization of our paper is as follows. Firstly, we describe briefly the DCI concept and related work (section 3). In section 4 we introduce our system and give an example. Finally, section 5 contains some concluding remarks.

2   DCI

Applying the DCI pattern we should view the program in three perspectives:

1. Data,
2. Context,
3. Interaction.

In the **Data** perspective the conceptual schema is coded as a set of class definitions. The only methods that are in these classes are data access methods and methods for derived attributes. The data classes are instantiated to form an object structure that corresponds to the conceptual schema of the user's mental model.

The system behavior is the system's response to user commands. A user command starts a method in one of the objects. This method sends further messages to other objects so that several objects communicate to execute the command. In the DCI program, communicating objects are coded as a network of connected roles. The establishment and maintenance of the runtime network structure is centralized in an element called the **Context**. Context is a class specifying a network of communicating objects as a similar structure of roles and connectors. It specifies also methods that bind roles to objects at the runtime.

Interaction specifies how objects interact to realize a system operation expressed in terms of its roles. All objects that play a given role shall process the same interaction messages with the same methods. These methods are called role methods and are properties of the role. The role methods are *injected* into the data classes, and the classes may not override them.

In Figure 1 the DCI paradigm is presented with an injection relation from role to class. The instances of these classes will give priority to the role methods above any methods defined in the class itself.

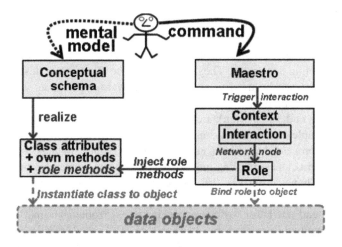

Fig. 1. DCI paradigm –source [8]

The Context object is encapsulated and its inner contents is invisible from the outside so it can be anything. The role abstraction describes how the object is used together with other objects while the class abstraction describes the inner construction of an object.

Data classes constitute objects that interact to implement system operations. Object interaction takes place within a Context where the objects are identified by the Roles they play. Objects are "temporarily equipped" with role methods while they are playing a role. As the role methods are associated with the roles rather than with classes we can reason about system operations without having to study the classes of the role-playing objects.

3   Related Work

The idea of DCI is very interesting and it seems that this pattern could be very useful and convenient in the software development especially with Lean [18] architecture or Agile development [18]. Since its inception the DCI approach has spread in many languages such as Ruby, Python, C++, C#, Java and many more. In strongly dynamic languages like Ruby or Python, a context can dynamically inject roles into an object so the implementation of DCI in these languages is easy and natural. More implementation problems occur in Java and C++. In [19] we presented how to efficiently implement DCI in these languages.

As the examples available in the literature present usually partial code (e.g. [20]), also only very simple examples can be found in blogs on DCI e.g. [21, 22, 23] so several researchers decided to check the real usefulness of this pattern in a case project. We also found the available examples unsatisfactory and designed and developed an exemplary project using the DCI paradigm, presented in section 4. When we started our project (in 2011) the works described below were not published yet.

Hasso and Carlson [24] in 2013 introduced a conceptual framework and an implementation model for software composition using the DCI architecture. Theirs compositional model is based on design patterns by abstracting behavioral model using role modeling constructs. They describe how to transform a design pattern into a role model that can be used to assemble a software application. Theirs approach offers a complete, practical design and implementation strategies, adapted from DCI (Data, Context, and Interaction) architecture. They demonstrate this technique by presenting a simple case study, system RESORT, implemented in C# language. They also created a process that should guide practitioners and students, learning how to use design patterns, in assembling individual components created by different teams.

Qing and Zhong in 2012 [20] proposed a "seamless software development approach using DCI". They use a method based on Problem Frame to capture the requirements and DCI for software architecture. Problem frame is a software requirement method proposed by Jackson [25, 26] and provides a schema for analyzing and decomposing problems. The Problem Frame has five patterns: Demand behavior frame, Command behavior frame, Information display frame, Simple tools frame, and Transformation frame. Qing and Zhong observed that DCI architecture and Problem frame have a similar vision how to construct programs. DCI can capture the end user cognitive model of roles and interactions between them. They integrate Problem Frame and DCI and claim that such approach enables to develop the software seamlessly.

Hayata, Han and Beheshti in 2012 [27] proposed a framework showing how agile practices could be fulfilled by introducing "lean" practices under the DCI paradigm. In the proposed framework, the lean architecture is complementary to agile principles and supports the agile development. The DCI approach is the backbone for lean architecture. Developers separate "what-the-system-is" (domain) from "what-the-system- does" (business-logic features). This separation of rapidly changing features from the stable domain architecture enables developers to focus on coding fast with confidence rather than do several iterations. Lean architecture achieves consistency by applying the DCI approach, allowing developers to separate the long-term structure of domain from its rapidly changing features.

4   System Based on DCI Paradigm

The idea behind DCI looks promising and it seems that this pattern could be very useful and convenient in the software development, especially combined with Lean architecture or Agile development [18]. We decided to check the real usefulness of this pattern on an exemplary project. Similar approach was used by others and theirs systems are mentioned in section 3. We designed and developed an application named *DCI-Project*. The detailed description is given in [28]. The main goal of DCI – Project is the management of projects, tasks and documents. It can be used by a group of people working on several projects. It enables to create, edit, delete, view projects. The actors of this application are: user, collaborator and project owner. The project owner (project manager) can create tasks and assign them to staff members. Users of this system can also monitor some items (tasks or projects) and comment them. The use case diagram for a project owner is given in Fig. 2. DCI-Project was written in *Ruby* [29] with the usage of *Ruby on Rails* [30] framework.

We started with the use case model as a "driving force" to implement the application. The architecture of the DCI-Project comprises of the Data part which describes the core of the system and the Interaction part which describes system's functionality. The third element – Context dynamically connects these parts. We have objects to represent the applications' domain objects; roles and their methods to represent system behavior and contexts to represent use cases – interactions between roles.

The architecture of the application makes a clear distinction between design activities corresponding to each of the artifacts namely the Data, Context, and Interaction. It also makes traceability between what the user wants and where it is implemented in the code clear through the context construct in the architecture. The domain object knows everything about its state and how to maintain it. Coplien refer to these domain objects as dumb objects that know nothing about other objects in the system. The interaction between domain objects is a system functionality captured as the system behavior and assigned to Interaction objects. In a typical use case scenario system entities interact with each other through defined roles. These roles will be assigned to domain objects instantiated at the runtime. At the design time these object roles and the expected behavior should be identified. System functionality i.e. functionality that does not belong to any specific object type at design time is injected into objects at the runtime.

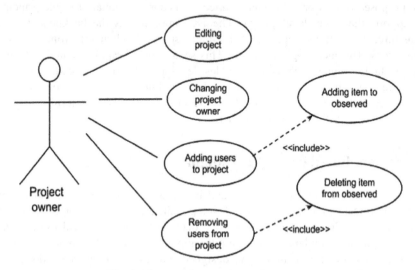

Fig. 2. Use case diagram for Project owner

4.1   DCI and MVC

Analyzing DCI we may have some associations with the MVC (Model View Controller) [17]. MVC was also introduced by Trygve Reenskaug but in the 1970s in Smalltalk. In late eighties this concept became popular in object technology initially in the development of graphical user interfaces, currently is being used in different parts of system.

MVC divides a software application into three interconnected parts so internal representations of information is separated from the ways that information is presented to the user. The central component (Model) consists of application data, logic and functions. A View can be any output representation of information such as a chart or a diagram. Multiple views of the same information are possible e.g. bar chart and a table. The third part - Controller accepts input and converts it to commands for the model or view.

Our exemplary system (the DCI Project) was implemented in *Ruby on Rails* framework [30] which imposed the usage of MVC. We have to combine the MVC and the DCI in the architecture of our system. In Fig. 3 the architecture of the DCI Project is shown. As the Model in MVC is responsible for data it was obvious that it contains Data (from DCI). Controller in MVC is responsible for the logic and is the central part of the application. Similar is the role of Context (in DCI) so these two perspectives are combined. In Controller the context classes are being executed. The difference of this architecture compared to the classical MVC implementation is that the Controller does not have direct access to data. The communication Controller – Data is performed trough Context.

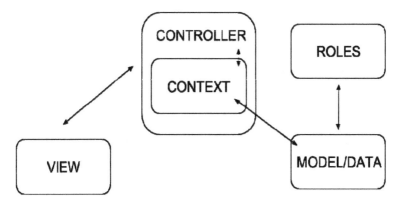

Fig. 3. DCI and MVC in DCI- Project application

MVC (Model View Controller) nowadays widely used pattern and DCI are complementary. MVC transforms the data into a physical form while DCI creates a program that represents the human mental model. The two patterns meet when MVC is used to bridge the gap between the human mind and the model implied by the DCI-based computer system.

4.2   Exemplary Code

In Fig. 2 one of use case diagrams from our application DCI-Project is shown. The implementation of one use case from it i.e. "changing the project owner" is presented in Fig. 4. The code was written using DCI paradigm principles. In the application each use case is represented by a separate context. Such separation allows contexts to be reusable and easy to maintain.

Class `ChangeProjectOwnerContext` implements the use case - changing the ownership of a project in the DCI-project application. `ChangeProjectOwnerContext` defines two roles: `SystemUser` (line 5) and `ProjectOwner` (line 4). Each of them contains one role method, which is responsible for the specific behavior of an actor.

Inside the `initialize` method (lines 3-8) roles are bound to actors. Since then actors are identified by their role names and are able to call the role, it means execute the methods assigned to a role. One of the distinguishing features of DCI is that role bindings are not permanent. Role methods are injected into their actors only during the execution phase . If the execution of the context is finished the actor object returns to its previous state.

```
1.class ChangeProjectOwnerContext
2.    include Context
3.    def initialize(project_owner, system_user, project_id)
4.       role ProjectOwner, project_owner
5.       role SystemUser,  system_user
6.
7.       @project_id = project_id
8.    end
9.
10.   def execute
11.      interaction do
12.         project = ProjectOwn
r.change_ownership_of_project(@project_id)
13.         Response.new(project)
14.      end
15.    end
16.    module SystemUser
17.       extend Role
18.       class << self
19.          def is_assigned_to_project?(project_id)
20.             projects.exists?(project_id)
21.          end
22.       end
23.    end
24.    module ProjectOwner
25.       extend Role
26.       class << self
27.          def change_ownership_of_project(project_id)
28.             unless SystemUser.is_assigned_to_project?(project_id)
29.AssignUserToProjectContext.new(context.roles[ProjectOwner],
context.roles[SystemUser], project_id).execute
30.             end
31.             project = Project.find(project_id)
32.             project.owner_id = SystemUser.id
33.             project.save
34.             project
35.          end
36.       end
37.    end
38.end
```

Fig. 4. Exemplary code

It is worth to mention that any actor object can play multiple roles. On the other hand there is a restriction regarding role uniqueness – a role should be assigned to only one object.

The crucial part of DCI – interaction is implemented by the `execute` method (lines 10-15), which triggers use case execution. Inside `interaction` block (lines 11-14), actors communicate with each other using role methods. Interaction block plays a significant part in DCI implementation as it defines the scope for actors' role methods thus making them not available outside of the interaction.

As contexts are separate units of code they can be reused or even nested within one another. For example `AssignUserToProjectContext` (line 29) which represents a separate use case is used as a part of `change_ownership_of_project` (line 27) role method.

5   Conclusions

In this paper we presented some of our experiences gathered during the design and development of the DCI-Project - system based on the DCI paradigm. We have shown in section 4 how the architecture of such system implemented in *Ruby on Rails* can look like. We described the relations of MVC and DCI elements and demonstrated theirs complementarities. We also presented and explained the implementation of one use case extracted from the DCI-Project.

In our application each use case is represented by a separate context. Such separation allows contexts to be reusable and easy to maintain. It is also easy to add new functionalities (use cases) to an existing system.

DCI based systems are very flexible much more than the traditional ones, this is due to the fact that static (Data) and dynamic (Context, Interaction) parts of the system are separated and separation of concerns is very powerful strategy to master the complexity.

DCI is strictly tied to the system architecture what we shown in section 4.1 so it seems to us that it would be very difficult to use it in legacy systems especially if these systems were designed without the DCI paradigm. Applying DCI pattern in a MVC based system should require less effort because these two techniques are complementary and can be easily used together.

References

1. Alexander, C., Ishikawa, S., Silverstein, M.: A Pattern Language. Oxford University Press, New York (1977)
2. Beck, K., Cunningham, W.: Using Pattern Languages for Object-Oriented Program. In: OOPSLA 1987 Workshop on Specification and Design for Object-Oriented Programming (September 1987)
3. Gamma, E., Helm, R., Johnson, R., Vlissides, J.: Design Patterns: Elements of Reusable Software. Addison-Wesley, Reading (1995)
4. Portland pattern repository, http://c2.com/ppr/ (access 2014)
5. Shalloway, A., Trott, J.R.: Design Patterns Explained: A New Perspective on Object-Oriented Design, 2nd edn. Addison-Wesley (2007)

6. Meyer, B., Arnout, K.: Componentization: The Visitor Example. IEEE Computer (IEEE) 39(7), 23–30, doi:10.1109/MC.2006.227
7. Reenskaug, T.: http://heim.ifi.uio.no/trygver/ (access 2013)
8. Reenskaug, T.: The Common Sense of Object Orientated Programming, version of September 11 (2008)
9. Reenskaug, T.: A DCI Execution Model, v 1.0 (2012)
10. The DCI Architecture: A New Vision of Object-Oriented Programming, http://www.artima.com/articles/dci_vision.html (access 2012)
11. Reenskaug, T., Coplien, J.O.: DCI as a New Foundation for Computer Programming, draft v.1.7 (2014)
12. Coplien, J.O., Reenskaug, T.: The Data, Context and Interaction Paradigm. In: SPLASH 2012, Tucson, Arizona, USA, October 19-26, p. 227, ACM 978-1-4503-1563-0/12/10 (2012)
13. Coplien, J.O.: Reflections on Reflection. In: SPLASH 2012, Tucson, Arizona, USA, October 19-26, pp. 7–9, ACM 978-1-4503-1563-0/12/10 (2012)
14. Coplien, J.O.: Objects of the People, By the People, and For the People. In: AOSD 2012, Potsdam, Germany, March 25-30, pp. 3–4, ACM 978-1-4503-1092-5/12/03 (2012)
15. Reenskaug, T., Coplien, J.O.: The DCI Architecture: A New Vision of Object-Oriented Programming (March 2009)
16. DCI – Data Context Interaction, http://fulloo.info/ (access 2014)
17. MVC: http://en.wikipedia.org/wiki/Model-view-controller (access 2014)
18. Coplien, J.O., Bjørnvig, G.: Lean Architecture for Agile Software Development. Wiley (2010)
19. Bluemke, I., Stepień, A.: DCI implementation in C++ and JAVA - case study. In: Madeyski, L., Ochodek, M. (eds.) Soft. Eng. from Research and Practice Perspectives, ch. 9, Nakom, pp. 154–163, ISBN 978-83-63919-16-0 (2014)
20. Qing, C., Zhong, Y.: A Seamless Software Development Approach Using DCI, pp. 139–142, IEEE 78-1-4673-2008-5/12 (2012)
21. Mahlen, P.: Blog [online]: DCI Architecture – Good, not Great, or Both, http://pettermahlen.com/2010/09/10/dci-architecture-good-not-great-or-both/ (access 2014)
22. Oberg, R.: Blog [online], http://java.dzone.com/articles/implementing-dci-qi4j (access 2014)
23. Parson, C.: Blog [online]: A fresh take on DCI with C++ (with example), http://chrismdp.com/2012/04/a-fresh-take-on-dci-with-c-plus-plus/ (access 2014)
24. Hasso, S., Carlson, C.: Design Patterns as First-Class Connectors. In: RIIT 2013, Orlando, Florida, USA, October 10-12, pp. 36–42, ACM 978-1-4503-2494-6/13/10 (2013)
25. Jackson, M.: The meaning of requirements. Annals of Software Engineering 3, 5–21 (1997)
26. Jackson, M.: Problems & Requirements. In: Proc. IEEE Second Int. Symp. on Requirements Engineering, pp. 2–8. ACM Press (1995)
27. Hayata, T., Han, J., Beheshti, M.: Facilitating Agile Software Development with Lean Architecture in the DCI Paradigm. In: Ninth International Conference on Information Technology-New Generations, pp. 343–348 (2012), doi:10.1109/ITNG.2012.157
28. Stepień, A.: DCI – Data Context Interaction – evaluation of usability. BA thesis, Institute of Computer Science, Warsaw University of Technology (2013) (in Polish)
29. Ruby: http://www.ruby-lang.org/ (access 2013)
30. Ruby on Rails: http://rubyonrails.org/ (access 2013)

Data Modeling for Precision Agriculture

Jan Tyrychtr[1], Václav Vostrovský[2], and Martin Pelikán[2]

[1] Czech University of Life Sciences in Prague
Faculty of Economics and Management
Department of Information Technologies, Prague, The Czech Republic
tyrychtr@pef.czu.cz
[2] Czech University of Life Sciences in Prague
Faculty of Economics and Management
Department of Software Engineering, Prague, The Czech Republic

Abstract. Presented article deals with issue of data modelling for managerial decision making of entrepreneurs with regards to knowledge management principles. The core line of the approach is capturing of explicit knowledge relevant to given business activities into multidimensional databases that would become part of utilized ICT/IT. These issues are demonstrated on the agriculture domain where the need to computer storage of relevant knowledge and provide them on-demand is very up to date. Recently, it has become very necessary in frequently discussed agriculture technique called precision agriculture.

Keywords: multidimensional modelling, knowledge support, multidimensional database, knowledge rules, precision agriculture.

1 Introduction

One of the possible solution for facilitating the transformation of knowledge especially in matters of transfer of tacit knowledge to explicit knowledge represents the knowledge engineering that is part of artificial intelligence and deals with methods of obtaining, formalization, coding and testing the knowledge gained from the human source or inductively derived from other information sources (databases, scholarly texts, etc.) during the creation of the corresponding knowledge base of the knowledge systems. Knowledge engineering was first defined in 1983 Feigenbaum and McCorduck as „scientific discipline concerned with integrating knowledge into computer systems designed to solve complex problems requiring highly skilled expertise" [1].

The computerization of our society currently has penetrated into all areas and could be observed almost everywhere. The decision making processes has been significantly influenced by software tools that improve their efficiency. One of these means are knowledge-based systems. The stormy and uncoordinated development of knowledge-based systems in recent years, however, did not contribute to their

R. Silhavy et al. (eds.), *Software Engineering in Intelligent Systems*,
Advances in Intelligent Systems and Computing 349, DOI: 10.1007/978-3-319-18473-9_10

credibility, and vice versa, lowered the credit to certain extent in the eyes of agricultural users. Now it is necessary to look for new ways and opportunities for further meaningful use in precision agriculture.

1.1   Theoretical Principles of Multidimensional Databases

Currently, the knowledge base of explicit knowledge is often implemented as relational databases [2, 3, 4, 5]. An innovative approach to storing explicit knowledge through multidimensional databases is presented in this paper.

Multidimensional databases are implemented for this reason. *Multidimensional databases* are suitable for storage of large amount of (multidimensional) data that are mostly analysed and summarized due to decision- making. The term *multidimensional data* represents data of aggregated indicators that were created with various groups of relational data aimed for online analytical processing (OLAP). OLAP is an approach to decision support that serves for data warehouse, data mart retrieval [6]. The solely way of data organization in multidimensional database is implemented as data cube.

Data cube is a data structure for storage and analysis of huge amount of multidimensional data [7]. In general, it is interpreted as a basic logical structure that describes multidimensional databases same as a relation describes relational databases. Data cube represents an abstract structure that, in contrary with classical relational structure, has not the unique definition. There are plenty of formal definitions of data cube operators, a compact overview is e.g. in [8]. At large, the data cube consists of dimensions and measures. *Dimension* is a hierarchically organized set of dimensional values that provide category information characterizing certain aspects of data [9]. *Measures* (observed indicators) of cube are mainly quantitative figures that could be analysed.

For physical storage of multidimensional data (and the implementation of OLAP applications) can be used several technologies. The two main ways to store data include so-called *multidimensional OLAP* (MOLAP), *relational OLAP* (ROLAP) [10]. MOLAP physically stores the data in an array like structures that are similar data cube shown on Figure 1. In the approach ROLAP, data is stored in a relational database using a special scheme instead of the traditional relational schema. There are also other approaches, so-called *hybrid OLAP* (HOLAP) that combine the properties of ROLAP and MOLAP [11] and so-called *desktop OLAP* (DOLAP). DOLAP is capable to connect to a central data store and download the required subset of the cube on a local computer [12].

For the design of multidimensional databases it is used the multidimensional modelling. *Multidimensional modelling* nowadays is mostly based on the relational model, or on the multidimensional data cube. *Multidimensional models* categorize the data either as facts with associated numerical measure, or as dimensions that characterize the facts and are mostly text. Facts are objects that represent the subject of the required analysis that has to be analysed for better understanding of its behaviour [13].

The multidimensional data model based on the relational model distinguishes two basic types of sessions that are called dimension tables and tables of the facts. They can create a star structure (*star schema*) [14, 15, 16, 17, 18], various forms of snowflakes (snowflake schema) [15, 16, 18] and constellations (constellation schema) [19]. The issue of choosing an appropriate structure is solved in the paper Levene et al., [20].

2   Methods

For the design of the innovative solution to capture the knowledge of agricultural activities through multidimensional databases, we use the databases of the Farmer's Portal run by Czech Ministry of Agriculture. These databases are basically implemented using standard relational tables containing evidence of treatment, evidence of veterinary medicinal products, evidence of medicaments administration and registry of animals with details of their application and effectiveness (see example Table I).

Table 1. Relational table containing a set of registered medicaments for treatment

Diagnosis	Medicaments	Species of animal	Breed	Dose ml/mg/tablet
vaccination	Probicol	Bovinae	C78 A22	13
prevention	Nutradyl	Bovinae	H100	2
prevention	Nolvasan Otic	Bovinae	C78 A22	1
prevention	Mikros PV	Bovinae	C78 R22	1
vaccination	Dinalgen	Pig	6000	300

For the design of conceptual schema it is used multidimensional Model Entity Relations (MER model) [21] and schema-type star [18]. For creating of the prototype of the multidimensional knowledge base is used ROLAP technology and applications PowerPivot. The reason for this choice is the flexibility of the solution in creating of ad-hoc queries and the ability to work with relational databases instead of creating a data warehouse. For these purposes the design method by [22] is used. The basic assumption of the designed prototype is that of requirements analysis represents only and just demand for analytical application of the knowledge-based rules in agriculture.

Knowledge retrieval in the farm is a very complex and time consuming development stage of an usable knowledge system. If the agricultural business entity stores their explicit knowledge in existing database applications, it can be effectively used such knowledge in managerial decisions of this business entity. These databases can be processed using specialized tools to create the knowledge base (in the form of rules) for applied knowledge system. Creating a knowledge base, however, is a nontrivial problem, because when storing explicit knowledge in database it reverses transformation of knowledge on data. Such an approach of the storing knowledge tends to be incomplete, inconsistent and almost invalid.

Suppose for example the following rule:

IF diagnosis (prevention) AND species of animal (bovinae) AND breed (C78 R22) THEN medicaments (Mikros PV) AND dose (1 tablet)   (1)

The rule given above can be interpreted such as: To treat a heifer (breed C78 R22) preventively, use the medicament Mikros PV with dosing 1 tablet. Such rule would be useful as the indication of the final efficiency of the recommended product and application site (housing). However, from rules conceived this way and stored in the database it is not clear whether the rule is still valid and there is also missing an effective tool for their analysis (analysis of explicit knowledge). The multidimensional approach for storing explicit knowledge allows analyzing knowledge in this context by various dimensions such as the time, the place, the evolution of the weather, etc. These dimensions in terms of the agricultural sector are very useful since it is possible to determine not only the recommended medicaments (quantity and efficiency), but also to find out how this efficiency has changed in time and in space (i.e. location) compared to other monitored medicaments. In this context it should be noted that the knowledge gained in the past may not be in full force at present (e.g. change of technology preparation, changes in breed quality, etc.). And this analytical approach allows to edit / create new explicit knowledge to increase (economic, technical) efficiency in the agricultural enterprise.

Finally, the general scheme for the design of multidimensional databases is specified that leads to the elimination of these problems in order to improve the quality of the databases used for storing explicit knowledge.

3   Results

3.1   Transforming Knowledge Rules

From the above table 1 we can derive some examples of knowledge-based rules similar to rule (1). Such rules under which the farm can effectively decide on the use of suitable equipment are beneficial. However, for such knowledge (stored in a relational database) lacks an effective tool for their analysis. Suppose there is a set of knowledge as IF E THEN H where E and H are predicates in implication (according to [3]) then for multidimensional approach it is needed to limit H predicates to those related with observed indicators. Knowledge rule that complies with this condition can look like this:

IF diagnosis (prevention) AND species of animal (bovinae) AND breed (C78 R22) AND medicaments (Micros PV) AND dose (1 tablet) AND application day (Q1 2014) THEN efficiency % (85) AND cost CZK (2.500)   (2)

Rule (2) might be interpreted such as: If we use Mikros PV medicaments to preventively treat heifer C78 R22 in dose 1 tablet in 1st quarter 2014, then we can reach 85 % efficiency at total costs 2.500 CZK.

This type of explicit knowledge will allow the farm management to decide what product to choose in comparison with other criteria such as efficiency and cost. Sometimes it may be advantageous for agricultural enterprise to use cheaper alternatives with lower efficiency, and sometimes vice versa. This given type of explicit knowledge can be transformed into a multidimensional schema for the application of knowledge-based analysis. Further, default conceptual and logical schema for the design a multidimensional database will be created.

In the first phase of conceptual design of the multidimensional database, must be created a fact table into an empty conceptual schema. Based on the knowledge rule (2) it can be told that E predicates will represent dimensional tables while H predicates are particular measures (indicators) in the fact table. Because there is a predicate on time (Application day) in the knowledge rule (2), therefore the scheme will contain time dimension. Fact table will be associated with roll-up relationship (N:1) with all necessary dimensions. During of the transfer to the logical schema, each dimension will be provided with a numerical primary key and will be associated with the foreign key into the fact table (Figure 1).

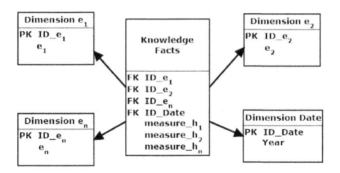

Fig. 1. Logical scheme of the transformation of the knowledge rules (own work)

At this stage was created the analogy of the knowledge rules with the multidimensional paradigm, and it was possible to transform these rules into conceptual and logic multidimensional data model. The authors of this paper believe that the multidimensional database created through such approach allows to design OLAP applications to the support of the knowledge-based decision of the responsible managers of agricultural businesses.

3.2   Creation of Prototype

To get an accurate preview of the future OLAP solution is created through the transformation of knowledge rules the prototype of a multidimensional database.

Creating a prototype allows to obtain the advantages and limitations of the proposed process of transforming knowledge into a multidimensional database in the agricultural farm. For the creating of the prototype is used the knowledge-rule by:

IF diagnosis ∧ species of animal ∧ housing ∧ breed ∧ medicaments ∧ dose ∧ application day *THEN* efficiency % ∧ cost CZK.   (3)

Conceptual Design Prototype

Application of transformation of the knowledge rule (3) leads to the identification of measures and dimensions of the conceptual schema. Fact table for the whole diagram is only one (due to the fact that in the methodological approach scheme is used the type of the star). Identified measures thus represent attributes of the table of the facts.

Table 2. Description of the results of the transformation of knowledge-based rules

Measure	Dimension	Dimension Date
Efficiency (%)	Diagnosis	Date
Cost (CZK)	Species of animal	
	Housing	
	Breed	
	Medicaments	
	Dose	

The result of the transformation of verdicts of knowledge of the rules at the dimensions and measures (Table II) allows the creation of conceptual schema.

Logical Design Prototype

The identified dimensions are associated with the table of facts in the logical design. The basic logic scheme is extended by the additional dimension attributes. For the dimensions Species of animal, Medicaments and Housing, the attributes specifying the names are created. For the dimension Dose, two attributes are created. The attribute Quantity, which will contain size dose of protective agent, and the attribute Measure to indicate the measurement units. For the dimension Medicaments, the attributes Batch, Number of packages, Origin, Veterinarian, Expiration date and Active substance are created. For the dimension Diagnosis, the attributes Identification and Details are created. In Dimension Date are added attributes that will allow analysis of both short term (Application day) and in the long term (Year). For the knowledge analysis is important to implement the analytical processing both the short and long term, since not all knowledge expressions are in a short period unchanged. In this stage of design of the logic scheme prototype is selected the snapshot granularity of data. Data (relating to individual knowledge statements) then will access the database at the same time intervals (e.g. quarterly). The dimension of the time takes into account the year and application period. Just in these intervals is possible to identify changes in each dimension (statements). The resulting prototype logic diagram is shown in Figure 2.

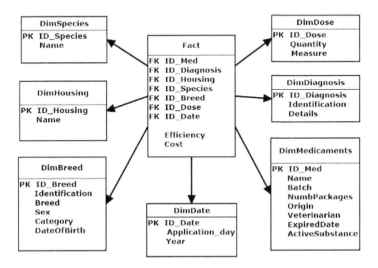

Fig. 2. Logical schema of the prototype (own work)

Physical Design Prototype

The essence of the design phase is complete logical model of the physical characteristics that are typical of the OLAP technology and specific database system. However, during the physical design of the prototype are not important optimum specific settings of proposed database solution, but especially the possibility of the verification the proposed logic model. In our methodological approach is to use for physical design of prototype the software Microsoft ® Excel 2013 and PowerPivot, which is sufficient for the creating of this prototype.

Firstly, relational data are integrated into fact tables and particular dimensions in PowerPivot, then relations are created after proposed logical scheme (Figure 2). Integrated data does not represent specific data of agricultural holding. Data are only theoretical (from Table I), represent a small farm. Integrated tables contain attributes designed in the logical schema. Other attributes, especially those that could be added the hierarchy to individual dimensions are not included in the prototype.

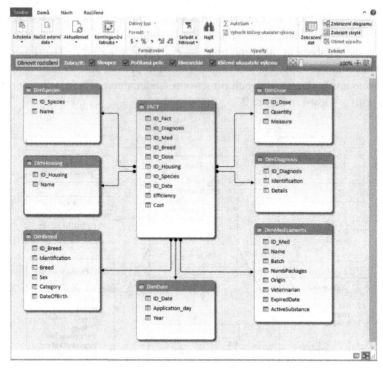

Fig. 3. Multidimensional model in PowerPivot 2013 (own work)

Physical design of prototype enables not only to verify that the logical model of our designed approach is realizable, but also verifies whether the solved transformation of knowledge rules into multidimensional databases for OLAP has practical meaning.

For example, to answer the following two analytical questions:

What was the average efficiency of the Nutradil preparation applied during the spring of 2014?
What were the total costs (in CZK) of medicaments application on individual breeds?

To answer these analytical questions and to test prototypes of multidimensional databases in practice the corresponding pivot tables are created in PowerPivot. This type of output is mostly supported by all client applications for OLAP. In the physical design, the rate is selected so that it will be displayed inside the pivot table and dimensions are placed in rows and columns where their values are calculated within.
By creating pivot tables (or graphs) farm management can increase effectiveness in decision making on the basis of available knowledge in the agricultural farm. Outputs in PivotTables allows to summarize and aggregate the data according to different knowledge statements. Technological approach used to design the prototype can be used for the real application of OLAP in agriculture. This is mainly because the software MS Excel with PowerPivot is readily available to users in agriculture and the

majority of users controls the work in Excel. The authors consider the selected technological solution useful, mainly because it allows to create a clear analytical views and reports with low entry costs compared to a robust Business Intelligence solutions.

Such an approach will find a management of agricultural holding the various combinations of protective agents that lead to approximately the same level of efficiency applications. It will also be possible to identify the maximum cost value in the given period and to deduce the changes of the efficiency of protective agents in time and so on.

4   Discussions

The designed prototype OLAP application verifies that the present conceptual and logical model is feasible, while also confirms that this new approach dealing with the transformation of knowledge-based rules to multidimensional databases for OLAP has the practical meaning.

Given the nature of the used methods of analogy during the transformation of the knowledge rules as the thinking process, it is clear that conclusions analogues are only probable, not conclusive. Therefore, it may be to exist the other permissible transformation of knowledge rules into the conceptual and logical schema. In further research in this area it must be taken into account and extend knowledge support of activities for the whole area of precision agriculture.

Acknowledgments. The results and knowledge included herein have been obtained owing to support from the Internal grant agency of the Faculty of Economics and Management, Czech University of Life Sciences in Prague, grant no. 20141040: "New methods for the support of managers in agriculture".

References

1. Feigenbaum, E.A., McCorduck, P.: The Fifth Generation: Artificial Intelligence and Japan's Challenge to the World (1983)
2. Fonkam, M.: On a Composite Formalism and Approach to Presenting the Knowledge Content of a Relational Database. In: Wainer, J., Carvalho, A. (eds.) SBIA 1995. LNCS, vol. 991, pp. 274–284. Springer, Heidelberg (1995)
3. Vaníček, J., Vostrovský, V.: Knowledge Acquisition from Agricultural Databases. Scientia Agriculturae Bohemica 39 (2008) ISSN 1211-3174
4. Hawryszkiewycz, I.: Knowledge Management: Organizing Knowledge Based Enterprises. Palgrave Macmillan (2009)
5. Pankowski, T.: Using Data-to-Knowledge Exchange for Transforming Relational Databases to Knowledge Bases. In: Bikakis, A., Giurca, A. (eds.) RuleML 2012. LNCS, vol. 7438, pp. 256–263. Springer, Heidelberg (2012)

6. Abelló, A., Romero, O.: On-Line Analytical Processing. In: Liu, L., Özsu, M.T. (eds.) Encyclopedia of Database Systems, pp. 1949–1954. Springer US, USA (1949), doi:10.1007/978-0-387-39940-9_252, ISBN 978-0-387-35544-3

7. Pedersen, T.: Cube, pp. 538–539 (2009), doi:10.1007/978-0-387-39940-9_884, ISSN 978-0-387-35544-3

8. Vassiliadis, P., Sellis, T.: A Survey of Logical Models for OLAP Databases. ACM Sigmod Record 28(4), 64–69 (1999) ISSN 0163-5808

9. Pedersen, T.: Dimension, p. 836 (2009), doi:10.1007/978-0-387-39940-9_886, ISSN 978-0-387-35544-3

10. Datta, A., Thomas, H.: The Cube Data Model: A Conceptual Model and Algebra for on-Line Analytical Processing in Data Warehouses. Decision Support Systems 27(3), 289–301 (1999) ISSN 978-0-387-35544-3

11. Khan, A.: SAP and BW Data Warehousing: How to Plan and Implement. iUniverse (2005) ISBN 9780595340798

12. Novotný, O., Pour, J., Slánský, D.: Business Intelligence: Jak Využít Bohatství Ve Vašich Datech. Grada Publishing, Praha (2005) ISBN 80-247-1094-3

13. Pedersen, T.B.: Multidimensional Modeling, pp. 1777–1784 (2009), doi:10.1007/978-0-387-39940-9_229, ISSN 978-0-387-35544-3

14. Wu, M., Buchmann, A.P.: Research Issues in Data Warehousing. In: Dittrich, K.R., Geppert, A. (eds.) Datenbanksysteme in Büro, Technik und Wissenschaft. Informatik aktuel, pp. 61–82. Springer (1997) ISBN 978-3-540-62569-8

15. Chaudhuri, S., Dayal, U.: An Overview of Data Warehousing and OLAP Technology. ACM Sigmod Record 26(1), 65–74 (1997) ISSN 0163-5808

16. Ballard, C., et al.: Data Modeling Techniques for Data Warehousing. IBM (1998) ISBN 9780738402451

17. McGuff, F.: Designing the Perfect Data Warehouse (1998), http://Members.Aol.Com/ Fmcguff/Dwmodel/Fmcguff/Dwmodel (Abruf: September 30, 1999)

18. Boehnlein, M., Ulbrich-Vom Ende, A.: Deriving Initial Data Warehouse Structures from the Conceptual Data Models of the Underlying Operational Information Systems. ACM, NY (1999) ISBN 1-58113-220-4

19. Abdelhédi, F., Zurfluh, G.: User Support System for Designing Decisional Database. IARIA, Nice (2013) ISBN 2308-4138

20. Levene, M., Loizou, G.: Why is the Snowflake Schema a Good Data Warehouse Design? Information Systems 28(3), 225–240 (2003) ISSN 0306-4379

21. Sapia, C., Blaschka, M., Höfling, G., Dinter, B.: Extending the E/R Model for the Multidimensional Paradigm. In: Kambayashi, Y., Lee, D.-L., Lim, E.-p., Mohania, M., Masunaga, Y. (eds.) ER Workshops 1998. LNCS, vol. 1552, pp. 105–116. Springer, Heidelberg (1999)

22. Rizzi, S., Abelló, A., Lechtenbörger, J., Trujillo, J.: Research in Data Warehouse Modeling and Design: Dead Or Alive? ACM (2006)

Organization of Knowledge Management Based on Hybrid Intelligent Methods

L.A. Gladkov, N.V. Gladkova, and A.A. Legebokov

Southern Federal University, Rostov-on-Don, Russia
leo_gladkov@mail.ru, legebokov@gmail.com

Abstract. The present article is concerned with topical issues of creation and organization of knowledge management systems. The background is briefly run through; current trends of information system development are noted. Different types of knowledge are analyzed; the significance of implicit knowledge types for cognitive activity and new knowledge creation is noted. Advantages and application of artificial intelligence methods and models within creation of knowledge management systems are discussed.

Keywords: Knowledge management systems, Categories of knowledge, Implicit knowledge, Experience, Artificial intelligence, Systems based upon fuzzy rules, Genetic algorithms.

1 Introduction

The process of knowledge accumulation, spread and transfer is a crucial factor of the modern civilization history. The problem of knowledge management became acute in recent times. The reason for this is that information revolution occurred, the development of computing machinery and mechanical media made it possible to reduce the complexity of information receiving, data processing and storage, simplify interpersonal communication and extend accessibility to various data.

All these achievements left a handle for transition to a new social pattern: from industrial to information-oriented society where key factors are information and knowledge rather than wealth [1]. We are eyewitnesses of new economy formation based upon intake and production of knowledge. These processes turn to be effects of globalisation on the one hand and its driving force on the other hand. The reflection of this new economic and social model is soaring knowledge content of goods and services, intellectualization of production technology field, appearing of economy sector specialized in production and providing intellectual services and products (consulting, intellectual property, etc.).

All the mentioned processes and trends led to necessity for classification of actual knowledge management models and technologies, efficient learning methods and development of new ones. We consider key problems associated with organization and knowledge management.

© Springer International Publishing Switzerland 2015
R. Silhavy et al. (eds.), *Software Engineering in Intelligent Systems*,
Advances in Intelligent Systems and Computing 349, DOI: 10.1007/978-3-319-18473-9_11

2   Problems of Knowledge Management Organization

As defined in "Wikipedia", the open on-line encyclopedia, [2] knowledge management comprises a range of strategies and practices used in an organization to identify, create, represent, distribute, and enable adoption of insights and experiences. Such insights and experiences comprise knowledge, either embodied in individuals or embedded in organizations as processes or practices.

The first works given attention to knowledge management problems dated back in the early 1990's. The following years reflected a surge in interest to the knowledge management problem in all the fields of activities, science and education inclusively; a significant number of publications and web-sites on the subject appeared. Today knowledge management methodology and generation of software and engineering tools are forming to provide natural knowledge circulation and knowledge generation.

It is evident that knowledge is not only an independent value but has a direct impact on efficiency of other manufacturing forces. The process of knowledge accumulation and processing results in formation of competences that are themselves the basis for building up the product and service market as well as labour market. That way, knowledge management involves integration of various disciplines, such as Human Resources Management, Marketing, Economics, Psychology, Information Science etc.

Knowledge is a specific individual nonprogrammed concept of environment, its laws and phenomena. Knowledge accumulation can be caused by direct interaction with the present environment (personal knowledge) or human cognitive activity. Knowledge can be nominally divided into two categories. The first category is an explicit knowledge that can be written down, depicted, documented (theory, methods, technology). Another category is an implicit knowledge that is hard or impossible to document (experience, skill, intuition). Implicit knowledge reflects personal experience, human adaptability; they are in minds of professionals, experts, people who have an intimate knowledge of a certain activity area (occupation), great life experience. In this regard it is important for an expert to increase and check constantly on his (her) knowledge, examine it in correspondence with the results of his (her) practice and theoretical activity. Cerebration may be said to be through trial and error when useful knowledge and skills are formed on the ground of observations and their further analysis. Knowledge accumulation, experience and intuition are evolution dynamics of cogitative brain activity; they implement "from simple to complex" motion principle; enable to get answers for intricate question instinctively [3].

When constructing complex information systems including knowledge management systems, scientists try to imitate the nature, employ artificial analogues of natural biosystems, principles and structures taken as the basis of natural systems functioning. Research area dealing with artificial system construction is called artificial intelligence. Let us take a brief look at the range of issues connected with artificial intelligence methods and models to create and organize knowledge management systems.

3   Using of Artificial Intelligence Methods to Construct Knowledge Management Systems

Today, quality improvement and complexity increase problems of information systems developed in different branches of sciences and engineering are associated with their possible intellectualization, in other words adding a number of functions generally performed by a human being to created technical objects and systems. These functions can be considered to be as following: analytical work and decision making within incomplete, fuzzy or controversial inputs, searching and selection of previously unknown, nontrivial but practically useful common factors in input data files, validation and interpretation of these factors. In this regard, one of the problems of great importance is to create efficient tools of processing and data mining, knowledge acquisition and management, as well as tools for common factors search to implement them in decision-making systems [4, 5, 6].

One of the most effective current tools to solve the problems mentioned above is fuzzy hybrid methods, models and algorithms. The methods applied are genetic, evolutional, bionic, adaptive and other searching methods. These methods and approaches can be integrated into independent interdisciplinary field, artificial intelligence, where some aspects of other research areas, namely computational intelligence, soft computing, database theories etc., are implemented.

When applied, generation of mathematically sound clear models and methods is either economically unacceptable or practically unrelizable. While the systems functioning on the base of integrated, fuzzy hybrid mechanisms and models, work well in solving such problems and appear to be the most rational compromise.

In this regard the usage of hybrid methods to solve the problem discussed suggests itself: fuzzy models, evolutional and genetic algorithms, multiagent organizations. They allow performing at their best with fuzzy, poorly formalized information and have a serious mathematical framework on the other hand to provide sufficient safety factor.

Today, the greatest success in system integration, fuzzy logic approaches and genetic algorithms (GA) was achieved in the two spares [7]:

1. Usage of mechanisms of genetic and evolutional algorithms to solve the problems of information search and data mining; usage of systems based on fuzzy rules. Hybrid methods are used for learning and setting the components of fuzzy rule system including automatic generation and knowledge bases check, setting the output function [8].
2. Usage of methods based on fuzzy logic for modeling different components and genetic algorithm operators, as well as adaption and dynamic setting of genetic algorithm control values.

The solution to the first class problems directly associated with the problems of effective organization of knowledge data bases, construction of quality control systems. The problem of setting fuzzy model parameters becomes relevant in this case. The possibility for dynamic change and optimization of model parameters during development and test process is compelling in this regard.

Such a problem can be solved with various methods. Using methods of genetic and evolutional search is one of these methods. Numerous examples of solutions for such problems applying genetic algorithms are known, for instance, configuration and configuring of artificial neural networks with the aid of genetic algorithms. The advantages of using genetic algorithms to set the structure and parameters of fuzzy models are their simplicity, accountability for the whole range of feasible constraints that can be met in such problems [9].

The fuzzy model behavior is characterized by the multitude of linguistically displayed rules based on expert knowledge which can be presented in the following form [10]:

```
If IF (set of conditions fulfilled),
then THEN (set of actions carried out).
```

Besides, the fuzzy model can include more than one inputs and outputs. For every fuzzy rule appears to be fuzzy relation, the behavior of fuzzy control system is characterized by the set of the relations mentioned above [8]. The procedure to obtain fuzzy inference results implementing knowledge base comprises the following steps:

3. Identifying the trigger lever for each rule.
4. Finding the result of fuzzy inference for each rule.
5. Aggregation of special fuzzy inference results into the general result typical for the whole fuzzy inference base.

From the start it is important to identify the set of all possible model rules implementing the preset membership functions. Model rule base may involve elementary rules, as well as generalizing rules. Generalizing rules are essentially logical combinations of elementary rules.

Generalizing rules used in the fuzzy rule make it possible to reduce the total number of the rules applied, in other words to decrease dimension of the model. At the same time the use of generalizing rules has an adverse effect on model accuracy [9].

It is plain that before the genetic algorythm starts functioning, it is necessary to find the set of all the rules applied in the model. As for solution encoding technics, various approaches can be operated. For instance, every element of the rule base can be given a certain position in the numerical sequence (chromosome).

Standard binary encoding can be used as well as more complicated encoding types making it possible to consider the importance and significance of certain rules, fuzzification/defuzzification methods applied, different modifications of genetic operators, etc.

These aspects must be considered when choosing an objective function (fitness function) and valuation of the solutions obtained. When evaluating the quality of each solution and, consequently, structural optimality of a particular rule base, proper allowance must be made for model accuracy, the total number of rules and the number of rules exploited in the structure under estimation. As these characteristics are closely related, to find a rational balance between them is of primary importance.

4   Conclusion

The experience of recent years showed that implementation of similar methods, i.e. methods appropriate for one scientific paradigm, in Information Technology is not always successful. In hybrid architecture connected several paradigms the effectiveness of one approach can compensate the weaknesses of another. One can avoid disadvantages inherent with each of them in particular combining various approaches. Thus, amplification of integrated and hybrid systems became one of the leading tendencies to define the development of contemporary informational systems and knowledge engineering. Such systems include various elements joined in the interests of objects in view. Integration and hybridization of various methods and information processing technologies afford to solve complicated problems impossible to be solved on the ground of separate methods or technologies. In case of integration of heterogeneous information processing technologies, one should expect synergistic effects of higher order than when grouping of various models within specific technology [11].

The choice of knowledge extraction and processing depends upon specific features of the solved problems, their development level, number of quality and quantity parameters. It is therefore necessary to define applicability conditions of every discussed technology and develop methods and algorithms for adapting technologies to problem solution of problem domain.

Acknowledgment. This research is supported by grants of the Ministry of Education and Science of the Russian Federation, the project # 8.823.2014.

References

1. Toffler, A.: The Third Wave. AST, Moscow (2010)
2. Wikipedia, http://ru.wikipedia.org/
3. Tuzovskiy, A.F., Chirikov, S.V., Yampolsky, V.Z.: Knowledge Management Systems. In: Yampolsky, V.Z. (ed.) Methods and Technologies. NTL Publishing House, Tomsk (2005)
4. Gladkov, L.A.: New Approaches to the Design and Creation of Artificial Systems of Hybrid Components. In: Intelligent Systems. Multi-author book, pp. 143–163. Fizmatlit, Moscow (2010)
5. Bova, V.V., Kureichik, V.V., Legebokov, A.A.: The integrated model of representation model of representation oriented knowledge in information systems. In: 8th IEEE International Conference "Application of Information and Communication Technologies – AICT 2014", pp. 111–115. IEEE Press, Astana (2014)
6. Kravchenko, Y.A., Kureichik, V.V.: Knowledge management based on multi-agent simulation in informational systems. In: 8th IEEE International Conference "Application of Information and Communication Technologies – AICT 2014", pp. 264–267. IEEE Press, Astana (2014)

7. Gladkov, L.A., Gladkova, N.V.: Features Using Fuzzy Genetic Algorithms for Optimization and Control. J. Izvestiya SFedU. Engineering and Industrial Technology Sciences. Special issue "Intelligent CAD System" 4, 130–136 (2009)
8. Yarushkina, N.G.: Basic theory of fuzzy and hybrid systems. Finance and Statistics, Moscow (2004)
9. Pegat, A.: Fuzzy modeling and control. BINOM. Knowledge Laboratory, Moscow (2009)
10. Gladkov, L., Gladkova, N., Leiba, S.: Manufactoring scheduling problem based on fuzzy genetic algorithm. In: Proceeding of IEEE East-West Design & Test Symposium – (EWDTS 2014), Kiev, Ukraine, pp. 209–212 (2014)
11. Kureichik, V.M., Pisarenko, V.I.: Synergetics in Education. J. Open Education 4, 33–45 (2010)

Development of Distributed Information Systems: Ontological Approach

V.V. Bova, Y.A. Kravchenko, and V.V. Kureichik

Southern Federal University, Rostov-on-Don, Russia
{vvbova,krav-jura}@yandex.ru, vkur@sfedu.ru

Abstract. The conceptual basics of using ontologies in the process of developing information systems were summarized and analyzed in the paper in the article. Author consider systematic and at the same time cognitive approach to development of distributed information systems. Here the ontology is used for managing, presentation and integration of diverse knowledge in a unified semantic model multifunctional system. As well, the article describes the composition of a single ontology as the basis of the semantic structure of the knowledge base model of distributed information system. The author proposes the principles of system-cognitive analysis of object-oriented domain knowledge and ontological synthesis algorithm and formalization description of models of the semantic structure of objects, processes and classes of problems the problem space of knowledge.

Keywords: Distributed information system, Ontology, Semantic model, Knowledge base, Knowledge representation and processing, Knowledge management system.

1 Introduction

Development of distributed information systems is closely related to the conceptualization of ontological categories, the improvement of the hierarchical structures of knowledge at all levels, building formal systems axioms and constraints that ensure the solution of specific problems in the areas of applied ontology-controlled systems [1]. At the same time, the ontological tool acts as an integration mechanism for semantic-based access to heterogeneous knowledge resources and decision-making model semistructured knowledge management tasks.

As well, the problem of developing methodologies and tools aided design formal (computer) domain ontology distributed information systems is relevant for the direction of ontological engineering [2-4]. It remains an open question of the constructive theory and the development of formal ontologies as a universal semantic model of pre-compilation and processing of heterogeneous knowledge.

This article examines the creation of a unified approach to ontology information systems, based on the methods and technologies of knowledge management, as well as the algorithm of development of domain ontology and the associated problem of objects' ontology, processes and ontology tasks. It is shown that the knowledge base

© Springer International Publishing Switzerland 2015
R. Silhavy et al. (eds.), *Software Engineering in Intelligent Systems*,
Advances in Intelligent Systems and Computing 349, DOI: 10.1007/978-3-319-18473-9_12

is formed on the basis of this ontology allows to accumulate experience in building architectures and system relations of distributed information systems and re-use it to develop new information systems.

2   Approach to Building a Unified Ontology Information Systems: Justification

The purpose of modeling the knowledge base of information systems is a systematic description of knowledge used in the process of knowledge management. The study revealed the following problems of modeling knowledge: knowledge representation as semantic relations between domain objects, modeling of knowledge about the dynamics of the behavior of objects, modeling operations and methods of knowledge processing for forming recommendations for decision-making [3-5]. To solve these problems we propose an approach to the construction of a common ontology information system, which is based on the integration of various models of knowledge management.

Analysis of the well-known approaches to modeling knowledge space of information systems [1,4-6] allows the following types of knowledge for further harmonization and integration in a single model of the ontology:

- semantic meta-knowledge represented in the ontology;
- formal knowledge presented in the form of production rules (rule base), on the basis of which the inference solutions;
- knowledge of the real situation (events) and managerial decisions (precedents);
- factual data obtained as a result of monitoring the state of the information system.

For constructing the ontology of information system we propose a methodology of system-cognitive analysis of the problem space of knowledge, methods of integrating object-oriented, ontological and semantic analysis to the problem of representation of heterogeneous knowledge.

Object-oriented analysis is a method of analyzing, studying the system requirements of knowledge in terms of identifying the cognitive elements (classes and objects), based on the dictionary domain [2].

Ontological analysis is the level of analysis of knowledge based on the description of the subject area in terms of rules and relationships with objects, characterizing the state of the system. The interpretive model area of expertise of the information system is the result of such analysis [3].

Semantic analysis is the analysis of the subject area, aimed at the identification and description of the basic elements of the subject area, the establishment of relationships between them and the characterization of the relationship [4].

The authors have developed the following principles of system-cognitive analysis:

- the principle of hierarchical decomposition of knowledge based on the use of various forms of abstract knowledge representation and polymorphism describes the relationship between objects and semantic network polymorphism inference engine;

- the principle of inheritance of properties, based on the ratio of "class - subclass" in the description of concepts in various forms of abstraction of knowledge representation;
- the principle of integration of ontological analysis and semantic modeling domain based on a hierarchy of concepts of cognitive elements;
- the principle of introducing concepts in ontology-based object-oriented modeling of heterogeneous knowledge management process [5];
- the principle of interpretation of the concept of similarity in the semantic networks based on cognitive analysis of the terms in the domain and the development of a common the ontology.

Initially, in accordance with the methodology of object-oriented analysis, we define the set of significant cognitive elements (the set of base classes and domain objects). Then, meaningful relationships that exist between the selected elements have been identified. The next step defines, which operations of interaction of objects are important. At this stage the behavior of objects is modeled. Subject-oriented ontology of the relevant information and logical and functional aspects of the system under study provides a knowledge of information systems based on the results of modeling based on ontological analysis. In conclusion, significant relationships are created syntactically and with usage of axioms in semantic web.

In the semantic web attitudes needed for representation and retrieval of knowledge in the system are allocated, such as: logical relations, causal relations, the relations of synonymy, depending on the implementation. These relations are characterized by the compatibility of the individual events and facts in the problem domain allow us to construct procedures verifying the integrity and consistency of knowledge. For modeling of these relationships, we need a special presentation model intentional knowledge, similar models used in knowledge bases [1].

Processing knowledge with this approach is based on the use of interpreters semantic networks based on inferential (deductive) capabilities of semantic networks and implement the inference engine, search engines, analyze, identify objects that make up the network of semantic knowledge space [7].

3   Building an Ontology of Information Systems Based on System-Cognitive Analysis

The central idea of systemic cognitive analysis is the development of support tools ontological representation, knowledge organization and data in the system, taking into account its functionality [1, 4] - multifunctional ontological system.

Depending on the complexity of the system under construction and elaboration of its domain, ontology knowledge of the system can be based either on the ontology knowledge representation or by completion and development of previously established basic or applied ontologies (Figure 1).

As we consider three basic points of the ontology: ontology of activity that constitutes the basis of the ontology problem domain of information system; ontology of knowledge, based on which to build the ontology field of knowledge of information system; and the ontology of basic tasks of information system, which is used for building the ontology of problems of information system.

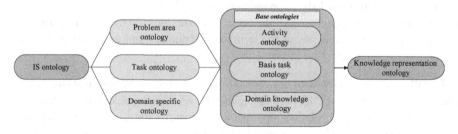

Fig. 1. Ontology scheme

First basic ontology describes the problem area of the system and, in fact, a top-level ontology. As such the ontology can be, for example, an ontology of scientific and educational activities, which includes the following cognitive elements (classes of concepts): person, organization, events, activities, and class information resource that is used to describe information resources available on the Internet [3].

The second basic ontology is ontology of knowledge sets of meta-concepts and it describes the cognitive elements of possible areas of expertise. These ontologies can be, for example, an ontology of scientific knowledge, captures the main content structure that can be used to build ontologies specific areas of knowledge. In particular, this ontology contains meta-notions such as a branch of science, research methods, object of study, the subject of research, scientific results. Using these meta-notions we can identify and describe important sections and subsections for the field of study, ask typing methods and objects of study, describe the results of operations.

Ontology basic tasks describes the basic functionality of information system, so it can be regarded as a specification of requirements for the user interface of information system. This ontology includes such basic concepts as search, navigation, browsing, filtering, etc., which may be specified in the design of the ontology problems of specific information system.

Notion of basic ontologies linked associative relationships, which are selected not only based on the completeness of the problem and the subject area of information systems, but also taking into account the ease of navigation on its information space and information retrieval.

According to the proposed approach, a single ontology representation space of knowledge can be seen as a universal model of semantic knowledge space information system that is designed for semantically-based access to heterogeneous knowledge resources of distributed information system.

Let us consider the methodology for constructing ontology model of information system. In general, as is it shown in Figure 1, it includes the problem domain ontology, domain ontology (the knowledge) and ontology tasks.

Conceptual part of such a system is described by two subparts, including ontology domain (*SD*), which consists of the ontology of objects, processes and of the ontology of tasks.

$$O^{IS}=<O^{SD} (O^O,O^P) O^Z>,$$ (1)

where O^O – ontology is a set of objects (concepts) *SD*, which is seen as a hierarchical structure of classes, subclasses and classes of elements.

O^P – ontology is a set of processes *SD*, which is seen as a hierarchical structure of processes, sub-processes, activities and operations.

O^Z – ontology is a set of objectives (standard sets) that can be formulated and solved to *SD*. Considered as a hierarchical structure of tasks, sub-tasks, procedures and operators. Fig. 2 is a schematic diagram ontologies element of cognitive domain and the problem space (*PA*).

PA – is a model of aspects of expert activity and component *SD*, which are linked (directly or indirectly) with the required knowledge for solving various problems in this *SD* [2]. *PA* consists of two units: the invariant (relatively constant) portion and a plurality of variable parts, corresponding to the individual tasks.

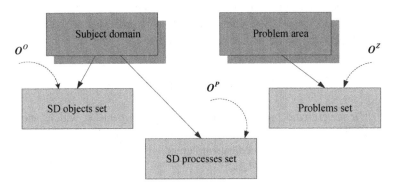

Fig. 2. Diagram of components of ontology cognitive elements

Under the ontology objects *SD* we understood the following:

$$O^O=<X, R, F, A(D,Rs)>,$$ (2)

where $X = \{X_1, X_2,...,X_i,...,X_n\}$, $i= \overline{1,n}$, $n = Card\ X$ - a finite set of concepts (concepts of objects) given *SD*.

$R = \{R_1, R_2,...,R_k,...,R_m\}$, $R \subseteq X_1 \times X_2 \times ... \times X_n$, $k = \overline{1,m}$, $m = Card\ R$ - a finite set of semantically meaningful relationship between the concepts of objects *SD*. Type of relationship between the concepts of general interest to subsection (of which distinguish partial order relation) and specific relationship given *SD*.

$F: X \times R$ - a finite set of functions the interpretation given to the concepts of objects and / or relations. $\{F\} = \{f_h\}$, where $h=\overline{1,H}$, where H - the set of functions on the interpretation of the tuple $<X, R>$, that is optimal for SD.

A – a finite set of axioms, which consists of a set of definitions D_i^l and a set of constraints Rs_i^t for the notion X_i. Definitions are written in the form of identically true statements that can be taken from a thesaurus SD. They may be given as an additional relationship concepts X_i and X_j. Limitations on the interpretation of the relevant concepts X_i can be set in a variety of restrictions Rs_i .

The subject and the problem of knowledge in general stand concepts and processes SD, concepts and classes of problems PA, problem-solving methods and algorithms that implement the appropriate methods. These types of divided into static and dynamic and by the ontological scheme on [1]: ontology objects PA (thesaurus); ontology processes SD; application ontology; ontology class of PA; ontology of problem-solving methods

There are various schemes design (synthesis) of the ontological structure of subject knowledge and problem [1,7,8]. In this paper, ontology SD included objects and processes (statistical knowledge), which are grouped into separate ontologies and ontology tasks included classes of problems, their solutions and related algorithms. In this scheme, group cognitive elements of knowledge, ontology basic tasks divided into static and dynamic views. If you change the class of problems for solving, then the ontology of objects and processes are ready for re-use.

Tasks ontology model is described as the triple

$$O^Z = <Z^P, M, Q >, \tag{3}$$

where Z^P is a generalized problem space consisting of p tasks, which, accordingly, contain $w = \overline{1,W}$ of fragments (each of them) in Fig. 3. At the same time, each fragment is represented by the procedure implemented on the set $v = \overline{1,V}$ operations (each of them). Furthermore, the task

$$Z^P = <D_{in}^P, R^P, C^P, D_{out}^P> \tag{4}$$

is defined by a set of input data D_{in}^P, requirements (conditions, limitations) R^P, task context C^P and output data D_{out}^P; M – a variety of methods problem solving, which is defined as a mapping

$$M^P: (D_{in}^P, R^P, C^P) \rightarrow D_{out}^P, \tag{5}$$

whose components are as defined above; Q – is problem solvers.

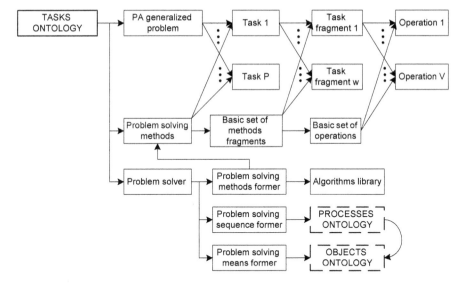

Fig. 3. Ontology tasks scheme

4   Designing of Ontology: Algorithm

We consider an algorithm of designing ontology *SD* and ontology tasks shown in Fig. 4. Ontology *SD* is based on ontology and ontology objects processes for which there are separate branch circuit. It is assumed that the set of functions in the interpretation of the ontology model is identical to the set of axioms [9].

Description of the algorithm of designing ontology of objects, processes and ontology tasks with reference to paragraph (block numbers) are shown in Fig. 4.

In the figure the abbreviations are:

- *PA* – the problem space.
- T-O, T-P – terms of objects and processes terms.
- T-Z– task lists, procedures, and methods for their solution.
- XO, RO, FO – finite sets of concepts, relations and functions interpretation ontology objects.
- XP, RP, FP – finite sets of concepts, relations and functions, the interpretation of the ontology of processes.
- XZ, RZ, FZ – finite sets of concepts, relations and functions, the interpretation of ontology problems.
- OGO OGP, OGZ – ontograf objects, processes and tasks.

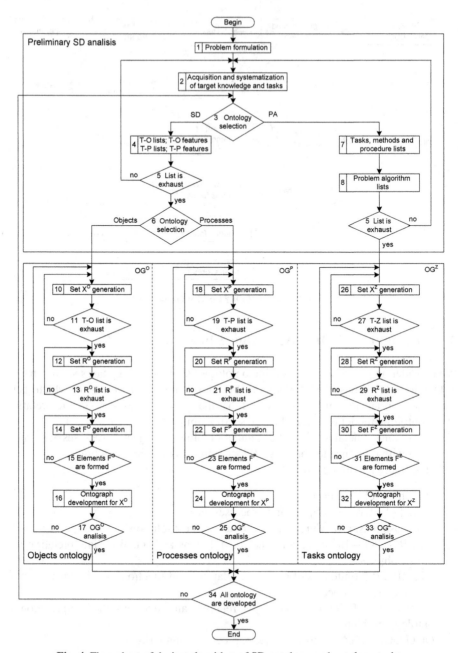

Fig. 4. Flow chart of design algorithm of SD ontology and ontology tasks

When the formulation of the problem for the design of a given ontology *SD* and associated *PA* of systematized methods and means of research subject and problem domain knowledge, and identifies the sources of their descriptions. Then, the extraction and systematization of knowledge given *SD* and multiple tasks *PA*. The algorithm provides a consistent building ontology *SD* (step 4), and task ontology (step 7).

As a result, we have the lists *T–O, T–P* and their properties with regard synonymy (step 4) and lists of tasks, procedures, and methods for their solution (step 7). It should be noted, that the steps 7 and 8 perform a predetermined space forming problematic *SD*, including a space for the information is specified solvers (step 8).

Next, we analyze the completeness of the generated list (in case of incompleteness of any of the lists you return to step 2) and the transition to the design of ontology of objects, processes and ontology for ontology-cottages, which consists of three blocks for each of them (O, P, Z):

- construct the set X (steps 10, 18, 26);
- construct the set R (steps 12, 20, 28);
- construct the set F (steps 14, 22, 30).

In the construction of the set R we get a set of pairs: (x_i^o, x_j^o), (x_i^p, x_j^p), (x_i^Z, x_j^Z) and, in accordance with the ratio of "above-below", analyzes and establishes the relationship $(x_i^o, R_k^o x_j^o)$, $(x_i^p, R_k^p x_j^p)$, $(x_i^Z, R_k^Z x_j^Z)$, accordingly, and built pre-ontographs X^O, X^P, X^Z. In the construction of the set *F* properties are attributed to each X^O, X^P and X^Z from the list of properties *T–O, T–P* and *T–Z* accordingly, as well as interpretation function are formed $\{f_l^O\}$ for each X^O, $\{f_l^P\}$ for each X^P, $\{f_l^Z\}$ for each X^Z of the definitions of a glossary description of a typical set of tasks and procedures and methods for their solution. Next, we build the finite ontograph OG^O (step 16), OG^P (step 24) and OG^Z (step 32) by elements of the set of concepts X^O, X^P and X^Z accordingly; analyzed for completeness and consistency ontograph OG^O (step 17), OG^P (step 25) и OG^Z (step 33) and their formal descriptions [10]. If the specified criteria are not met, then we will return to steps 10, 18 and 26 for the reengineering of the ontology of objects, processes, or ontology tasks accordingly. Algorithm of designing of the ontology *SD* and tasks is a sequence of steps of analysis and synthesis of heterogeneous knowledge of the ontological structure of objects, processes and classes of problems. Blocks 1-9 of algorithm describes the process of analyzing and blocks 10-17, blocks 18-25 и blocks 26-33 – processes of semantic synthesis of the ontology objects, processes and of ontology tasks.

5   Conclusion

Thus, the article proves approach to development a unified ontology based on a previously performed analysis and study of the problems of modeling and methodologies for developing computer domain ontology knowledge of distributed information systems [1, 3-5]. This approach is called systematic ontological. The author proposes the principles of system-cognitive analysis of object-oriented domain knowledge and ontology allows formalizing the semantic structure of the knowledge base model at the stage of studying the problem domain and requirements analysis to develop an information system. Components of the semantic structure of the ontological knowledge base are the ontology of objects and processes ontology tasks for which developed model of representation and algorithm design. Appropriate technology implementation require *SD* analysis, extraction, processing and presentation of subject knowledge. Automate acquisition and processing of knowledge issues continue to remain relevant and requires improvement of intelligent technologies, methods and tools for automated creation ontologies.

Acknowledgment. The study was performed by the grant from the Russian Science Foundation (project # 14-11-00242) in the Southern Federal University.

References

1. Bova, V.V., Kureichik, V.V., Legebokov, A.A.: The integrated model of representation model of representation oriented knowledge in information systems. In: 8th IEEE International Conference "Application of Information and Communication Technologies – AICT 2014", pp. 111–115. IEEE Press, Astana (2014)
2. Rodzin, S., Rodzina, L.: Theory of bioinspired search for optimal solutions and its application for the processing of problem-oriented knowledge. In: 8th IEEE International Conference "Application of Information and Communication Technologies – AICT 2014", pp. 142–147. IEEE Press, Astana (2014)
3. Bova, V.V.: Conceptual model of knowledge representation in the construction of intelligent information systems. In: Proceedings of SFU, vol. 156, pp. 109–117. TTI SFU, Taganrog (2014)
4. Kravchenko, Y.A., Kureichik, V.V.: Knowledge management based on multi-agent simulation in informational systems. In: 8th IEEE International Conference "Application of Information and Communication Technologies – AICT 2014", pp. 264–267. IEEE Press, Astana (2014)
5. Bova, V.V., Legebokov, A.A., Gladkov, L.A.: Problem-Oriented Algorithms of Solutions Search Based on the Methods of Swarm Intelligence. World Applied Sciences Journal 27(9), 1201–1205 (2013)
6. Kravchenko, Y.A., Bova, V.V.: Fuzzy modeling of heterogeneous knowledge in intelligent tutoring systems. Open Education (4), 70–74 (2013)
7. Bova, V.V., Kureichik, V.V.: Simulation of the process of knowledge representation in intelligent tutoring systems based on competence approach. Open Education (3), 42–49 (2014)
8. Kravchenko, Y.A., Markov, V.V.: Decision-making in integrated information model based on the analytic hierarchy process. In: Proceedings of SFU. Technical Sciences, vol. 136, pp. 212–216. Publishing House TTI SFU, Taganrog (2012)
9. Zaporozhets, D.Y., Zaruba, D.V., Kureichik, V.V.: Hybrid bionic algorithms for solving problems of parametric optimization. World Applied Sciences Journal 23, 1032–1036 (2013)
10. Gladkov, L.A., Kravchenko, Y.A., Kureichik, V.V.: Evolutionary Algorithm for Extremal Subsets Comprehension in Graphs. World Applied Sciences Journal 27, 1212–1217 (2013)

Decision Support Systems for Knowledge Management

V.V. Kureichik, Y.A. Kravchenko, and V.V. Bova

Southern Federal University, Rostov-on-Don, Russia
vkur@sfedu.ru, {krav-jura,vvbova}@yandex.ru

Abstract. This article deal with the development of imitative and mathematical decision making models in knowledge management training systems based on fuzzy modeling and multi-agent technologies. These approaches enable the authors to solve a lot of problems with the help of classifier learning agents and classifier combinational agents connected by informative exchange procedures. Information management is conceived as a combination of systematic attainment, synthesis, knowledge exchange and use.

Keywords: Knowledge management, Intelligent agents, Simulation, Decision support, Mathematical models, Information flows.

1 Introduction

Information management in intelligent systems is connected with the solving of problems difficult to formalize under conditions of uncertainly. Consequently, this requires design of real-time algorithms for the determination and adjustment of external and internal parameters. Such systems are characterized by the following factors:

1. advanced communicative functions;
2. solving of problems difficult to formalize;
3. self-organization and self-learning;
4. adaptability.

All these characteristics are integrated into knowledge management systems. The main function of knowledge management training systems (KMTS) is informational management as a combination of systematic attainment, synthesis, knowledge exchange and use. In terms of knowledge management it is necessary to involve both internal and external information resources.

Knowledge management training systems are a set of elements such as methods of acquisition, synthesis, exchange and the knowledge use. They also include experts, tutors, students as a subjects of knowledge management and computer models of different subject areas.

Relationships between these system elements provide the integration of heterogeneous knowledge sources into a unified information space within the learning environment. A distinctive feature of such knowledge management systems is the integration of knowledge from different subject areas to provide the necessary adaptability in the term of training management.

© Springer International Publishing Switzerland 2015
R. Silhavy et al. (eds.), *Software Engineering in Intelligent Systems,*
Advances in Intelligent Systems and Computing 349, DOI: 10.1007/978-3-319-18473-9_13

An important task is the extraction of implicit knowledge with the help of a variety of analytical tools. These tools enable the researchers to detect patterns within the learning environment and identify the most rational individual learning trajectory under the current conditions. In this case experts, tutors and lecturers using KMES become researchers who put forward various hypotheses regarding information management in the learning process to achieve the desired results.

2   Problem Definition

A hierarchical model of the learning process allows considering the future professional activity at the level of target competencies of a specialist that characterize his/her knowledge, skills and mastery. The sequence of competencies specifies relationships between the learning elements of the educational process and underlies a set of alternatives for building learning trajectories. A decision-making model for evaluating the effectiveness of these alternatives cannot be designed on the basis of mathematical analysis methods due to the uncertainty of the problem. The problem of estimating the learning trajectory can be described as follows. Let there be a set of n alternatives $T = \{t_1, t_2, t_3, ..., t_n\}$, the preference of each alternative on one of the criteria Q_i, $i = 1 ... k$, is given by using fuzzy set membership function μQ_i, defining the extent to which alternative t_j, $j = 1 ... n$ the concept matches determined criteria Q_i. The alternative is the best if it meets each of the selected criteria. A fuzzy set of the degree of alternatives t_n to criterion Q_i can be described as follows:

$$Q_i = \left\{ \mu_{Q_i}(t_1)\Big/_{t_1}, \ \mu_{Q_i}(t_2)\Big/_{t_2}, ..., \ \mu_{Q_i}(t_n)\Big/_{t_n} \right\}. \tag{1}$$

The rule of the best alternative selection shall be given as intersection of the respective fuzzy sets:

$$S = Q_1 \cap Q_2 \cap ... \cap Q_k. \tag{2}$$

The operation of the fuzzy sets intersection corresponds to the operation of the minimum membership functions:

$$\mu_S(t_j) = \min \mu_{Q_i}(t_j), i = 1, ..., k; j = 1, ..., n. \tag{3}$$

We assume that the best alternative to t* with the highest value of the membership function

$$\mu_S(t^*) = \max \mu_S(t_j), j = 1, ..., n. \tag{4}$$

An implementation of the proposed algorithm requires the specification values of the membership function, which can be accomplished, for example, based on integrated assessments.

If considered as a basis for constructing learning trajectory the ordering training modules, then it is possible to formulate a multiobjective problem [2], in which each of the selected criteria should get the value closer to the optimum one. Lets describe a possible set of criteria for the task.

The first criterion is called the total competency significance. This criterion means at the each step of the trajectory selection of the training module with the highest competency significance. Assume that we have P training modules preliminary distributed on S training units, so it is necessary to arrange training modules in the block for C specializations with the maximization the competency significance criterion.

$$Q_1 = \sum_{i=1}^{P} \sum_{k=1}^{S} \varepsilon_{ij}\, \delta_{ik} \to max, \tag{5}$$

where ε_{ij} - the coefficient of the significance competency training module i in the specialization j, (j = 1 ... C), δ_{ik} - binary value that is 1 if unit i is included in the block k, and 0 - otherwise.

In the second criteria an interdisciplinary should be considered, considering the training modules in the specialization as the elements of a variety of relationships (links) between the subject areas. It will give the opportunity to develop systematic knowledge of future specialists.

$$Q_2 = \sum_{i=1}^{P} \sum_{y=1}^{P} \sum_{k=1}^{S} \gamma_{iy}\delta_{ik} \to max, \tag{6}$$

where γ_{iy} - the coefficient matrix of interdisciplinary connections that define the significance of a training module i while the module y is learning. These coefficients are determined on the basis of expertise and take values in the range [0, 1].

The third criteria take into account the level of residual knowledge, i.e. we will solve the problem of minimizing the loss of the gained knowledge. On the basis of this criterion specific disciplines will be located as close as possible to the end of the learning process by increasing their competency significance.

$$Q_3 = \sum_{k=1}^{S} \sum_{i=1}^{P} \varepsilon_{ij}\, \delta_{ik}(S - k)\varphi_{ij}, \tag{7}$$

where φ_{ij} – a coefficient involves consideration of only the most significant in terms of professional competence of disciplines, taking the value 1 if $\varepsilon_{ij} \geq 0{,}8$, and 0 - otherwise.

The generalized optimization criterion presupposes the construction of a learning trajectory, all the described criteria taking extremum values. Thus, the generalized criterion can have the following form:

$$Q_{int} = \tau_1 Q_1 + \tau_2 Q_2 + \tau_3 Q_3 \to max, \tag{8}$$

where τ_i is the weight of each of the considered criteria, defined on the basis of expert assessments.

The disadvantage of the generalized criterion is its inability of taking into account the learner's individual characteristics of in terms of the construction of his/her individual learning trajectory. The use of the generalized criterion allows to generate the optimal learning trajectory for a selected group of students with similar personality characteristics. The construction of an individual trajectory requires the consistent application of optimization criteria and regular changes of the subject features, the uncertainty of behavior leads to the nonlinearity of the researched process of acquiring the target professional competencies.

Let the learning process be an x-step process of the transition from the initial state Z_{begin}, corresponding to the beginning of the learning process, to the final state Z_{end} [2], which means getting specialization C_j with the Z_{end} equal to 1 in this model

Let us consider a current system state refer to as Z_x; the set of admissible solutions will be a number of unexplored, but mandatory training modules P_x. Then each of the following intermediate system state on the basis of the previous ones as follows:

$$Z_{x+1} = Z_x \cup P_x. \tag{9}$$

It is obvious that at each step there will be several options for destination modules. The current state of the system is determined by the set of solutions obtained within last step, the maximum corresponding to the set of the target competency requirements.

$$Z_{x+1} \rightarrow max. \tag{10}$$

In this case the optimal solution satisfies the following condition:

$$Z_{x+1}(Z_x, P_x) = max_{P_x}(Z_x + f(Z_x, P_x)), \tag{11}$$

where $f(Z_x, P_x) = \Delta Z_{x+1}$.

The objective function based on the accumulation of the performance indicators of each step is additive. So the optimal sequence of training modules constituting the learning process shall be:

$$Z_{optimum} = \sum_{x=1}^{S} Z_x. \tag{12}$$

We can conclude that set P_x includes training modules corresponding to all the constraints. The modules, do not have any ancestors or their ancestors have already been studied. Also, these modules are a set of feasible solutions for the current state of the system Z_x. At each step of the learning process development one can select training modules for subsequent blocks on the basis of the assessment of the previous modules.

3   Decision Support via Hyper Surface Classifiers

For a Hyper Surface Classifier (HSC) combination to be implemented, a multi-agent technology is used. Agents can simulate the decision making within the student's reference group to solve problems or the implementation of educational projects, such as the expert group consultations [2]. It is known that the hyper-surface model is obtained during the learning process; then it is used to classify a large database depending on the parity or the oddness of the number of turns based on the Jordan Curve Theorem [3, 4]. The experiments show that the HSC method can effectively and precisely classify large volumes of data in the two- and the three-dimensional space.

Although the HSC method can theoretically classify data with a higher dimension in accordance with the Jordan Curve Theorem, its implementation in the space of more than three-dimensional is complicated enough. So, an algorithm, which can process both large-volume and high-dimension is required.

Considering the problem of the distributed computing environment and the fact that agents can simulate epy decision-making by a group of people (e.g. a group expert counseling) we introduce methods of multi-agent systems for understanding the combination of hypersurface classifiers. This combination is a set of classifiers, in which individual classification decisions are combined in a certain way, as a rule, by weighted or equal voting. This is done to classify new examples. There are two combination methods - horizontal and vertical combinations. In this paper, we consider the vertical combination. The HSC combination for high dimensions data is constructed by dividing the dimension, rather than by its reduction. We will use two types of agents: classifier training agents and classifier combinational agents [4]. Each classifier training agent reads the vertical portion of the samples and teaches the local classifier, while the classifier combinational agent is designed to combine the results of the classification of all classifier combinational agents [5].

The HSC is a universal classification method based on the topological Jordan Curve Theorem. The main difference from the well-known technique "Support Vector Machines" is the fact that HSC allows to solve the nonlinear classification problem within initial space. Consequently, there is no need to display data in a higher dimensional space. The core function is not needed to be used, either. [4]

We will use a multi-agent technology as a combination of hyper surface classifiers. Agents can simulate decision-making by the reference group during the educational design, the brainstorming or other situation. Each classifier is developed as an agent with the given conditional attributes. There are two types of agents - classifier training agents (CTA) and classifier combinational agents (CCA) [4]. After the agent-coordinator (AC) determines the correspondence between the student model, reference groups and learning processes, learning and classification model are formed. Classifier learning agents examine the classification model using the specified algorithm on a data array and predict the emergence of instances with the use of a learning model. Classifier combinational agents put together the results of the predictions by multiple classifier training agents and find unlabeled sample instances. Thus adjustments are made to the researched learning and classification models (Fig. 1).

When agents work together to accomplish the classification task each classifier learning agent should first finish their studies independently and establish an independent classification model with a variety of given conventional signs. Further on, when the problem of predicting the classification is complete, classifier combinational agents send the samples predicted to each classifier training agent. These agents use the classification model to predict the sample label, while assessing its properties and sending the assessment complete results with the class label to the classifier combinational agents. The latter allow the final class label based on the results of all the classifier learning agents in terms of the control logic, such as voting or weighted voting. Thus, the classifier combinational agents may exceed the limits of individual classifiers abilities.

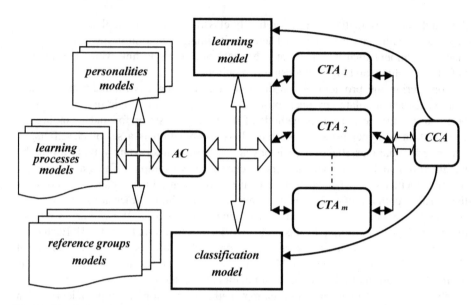

Fig. 1. Simulation model of decision-making based on a combination of learning agents and classifiers

Consequently, the HSC system of heterogeneous classifiers based on agents can achieve high classification accuracy without additional training of a group of classifiers. There are two combining methods - horizontal and vertical combination. The most significant difference lies in obtaining subsets by dividing features, but not the initial set of samples. So if there is no incompatibility, the size of each subset is equal to the size of the original set of samples, taking only a little more space in the memory than the initial set. Therefore, the vertical combination is preferred.

The HSC combination classifies high dimensional data set by analyzing multiple sections of training and verification samples. Furthermore, the combination of classifiers is much less error-prone than a single HSC classifier [5, 6].

Among the methods of the HSC combination the majority decision is the easiest to implement, as it requires no prior learning [7]. The use of this method is especially suitable in situations where other forms of the quantitative output cannot be easily obtained by a classifier agent, or when other precise combination methods could be too difficult. This combination method was very effective [7,8, 9, 10, 11, 12]. By the simple majority voting, when the decision of each classifier has equal weight, a various data processing can be performed. One can assign different weights for each agent when the work of the combined agent classifier is optimized within the training set. For the initial weight processing values can be assigned using the genetic algorithm and exhibited to the vote to determine the optimal value for the target function. This function may include conditions of the recognition and frequency of errors. In this case the majority vote method is applied. Classifier combinational agents make a decision about the belonging of the test sample to the desired class C_j, if the number of classifiers, which support this class, is substantially greater than the number of classifiers, which support any other class.

Multi-agent environments are very important for Combination Hyper Surface Classifiers Systems (CHCS). For the functioning of such systems it is necessary to develop a number of agents specializing in the autonomous partitioning problems. Such agents will be coordinated through a standardized language of agent interaction. The language is supported by means of communication protocols within a multi-agent environment.

4   Conclusion

The article discusses intelligent agents simulating decision-making by a group of people to solve problems in various subject areas (e.g. group experts counseling. Two types of agents are described - classifier training agents and classifier combinational agents. Each classifier learning agent is developed as an agent with given conditional attributes instead of general ones, which is important for reading the vertical section of samples and training local classifiers. The classifier combinational agent is designed to put together the results of the classification of all classifier learning agents by voting. The most important difference between the HSC combination and the traditional combination is the fact that data subsets are obtained by dividing feature but not the actual set of samples. Thus in case of inconsistency, the size of each subset of data samples is equal to the initial set, while the size of memory is not considerably greater than the volume required to store the initial samples. The time complexity of the HSC combination algorithm is a $o((nd + n^2d)/3)$, where n is the size of a plurality of samples and d is a number of dimensions.

Acknowledgment. The study was performed by the grant from the Russian Science Foundation (project #14-11-00242) in the Southern Federal University.

References

1. Tel'nov, Y.F.: Intelligent Information Systems, p. 26. Moscow State University of Economics, Statistics and Informatics (2003)
2. Klebanov, B.I.: Simulation technology socio-economic dynamics of the municipal education based on multi-agent approach. In: Simulation. Theory and practice. Proceedings, vol. II, St. Petersburg, JSC "SSTC", pp. 113–118 (2009)
3. Gorohov, A.V.: Approach to quality management of educational activities based on the simulation. In: Simulation. Theory and Practice. Proceedings., St. Petersburg, JSC "SSTC", vol. II, pp. 57–59 (2009)
4. He, Q., Zhao, X.-R., Luo, P., Shi, Z.-Z.: Combination methodologies of multi-agent hyper surface classifiers: Design and implementation issues. In: Gorodetsky, V., Zhang, C., Skormin, V.A., Cao, L. (eds.) AIS-ADM 2007. LNCS (LNAI), vol. 4476, pp. 100–113. Springer, Heidelberg (2007)
5. Bova, V.V., Kureichik, V.V., Legebokov, A.A.: The integrated model of representation model of representation oriented knowledge in information systems. In: 8th IEEE International Conference "Application of Information and Communication Technologies – AICT 2014", pp. 111–115. IEEE Press, Astana (2014)

6. Kureichik, V.V., Kravchenko, Y.A.: Bioinspired algorithm applied to solve the travelling salesman problem. World Applied Sciences Journal 22(12), 1789–1797 (2013)
7. Shi, Z.Z., Zhang, H., Cheng, Y., Jiang, Y., Sheng, Q., Zhao, Z.: MAGE: An Agent-Oriented Programming Environment. In: Proceedings of the IEEE International Conference on Cognitive Informatics, New York, pp. 250–257 (2004)
8. Petrovsky, A.B.: Decision theory. Textbook for university students. Publishing Center "Academy", Moscow (2009)
9. Kureichik, V.V., Kravchenko, Y.A.: Knowledge management on multi-agent simulation in informational systems. In: 8th IEEE International Conference "Application of Information and Communication Technologies – AICT 2014", pp. 264–268. IEEE Press, Astana (2014)
10. Vagin, V.N., Golovina, E.J., Zagoryanskaya, A.A., Fomina, M.V.: Reliable and believable conclusion in intelligent systems. Fizmatlit, Moscow (2004)
11. Larichev, O.I.: Theory and methods of decision-making, and chronicle in a fairyland. Logos Moscow (2000)
12. Kureichik, V.V., Kureichik, V.M., Rodzin, S.I.: The concept of evolutionary computation inspired by natural systems. In: Proceedings of the SFU. Engineering. The Thematic issue "Intelligent CAD", vol. 4(93), pp. 16–25. Taganrog Publishing House of the TTI SFEDU (2009)
13. Kureichik, V.M., Pisarenko, V.I., Kravchenko, Y.A.: Innovative educational technologies in building support systems group decision. In: Proceedings of the SFU. Engineering. The Thematic Issue "Intelligent CAD", vol. 4(81), pp. 216–221. Taganrog Publishing House of the TTI SFEDU (2008)

Innovation in Services: Implications
for Software Maintenance

Mircea Prodan[*], Adriana Prodan, Andreea Ogrezeanu, and Anca Alexandra Purcarea

University Politehnica Bucharest, Faculty of Entrepreneurship,
Business Engineering and Management, Str. Splaiul Independentei 313, Sector 6
Corp BN, Bucharest, Romania
{pro100mir,burduja_adriana}@yahoo.com,
{ogrezeanu.a,apurcarea}@gmail.com

Abstract. Software maintenance is often perceived as an activity without innovation by the personnel involved in the process, especially due to corrective maintenance. Although it's an important part of the software lifecycle, especially from length in time and costs perspectives, this phase has been less researched and emphasized as others, for example software development. It is our intent to put together some of the researches on innovation, especially in services, and use the common themes as guidelines in application maintenance. The goal is to advance some theoretical fundaments to foster innovation in software maintenance, which could be used and tested in practice.

Keywords: innovation, services, software maintenance, corrective.

1 Introduction

Software maintenance is an important phase in the software lifecycle. It is the phase which accompanies the use of the software, in which previously undetected errors are being solved, as well as adapting the software to new changes in environment, new business requirements, and legislation changes. As we can imagine, a significant portion of the total cost with a software is spent on maintaining it.

In its essence, software maintenance (SM) it's a service, delivered by developers, also called maintainers, to the users of the application. Most of the time, the end users are not the same persons with the ones contracting the service, moreover, not the same with the ones ensuring governance and monitoring service quality. To make things even more challenging, the people maintainers are usually different than the people who have developed the software.

Due to all these and due to corrective maintenance, the component activity concerned with removing defects, SM has long been perceived as problem solving, fixing bugs type of activity, where innovation cannot have place. It is considered a boring activity by the developers, most of them being more interested in starting on a project to develop a new software or functionalities than to do corrective maintenance.

---

[*] Corresponding author.

© Springer International Publishing Switzerland 2015   131
R. Silhavy et al. (eds.), *Software Engineering in Intelligent Systems*,
Advances in Intelligent Systems and Computing 349, DOI: 10.1007/978-3-319-18473-9_14

Making corrective maintenance a more innovative activity will have an impact not only on making SM more appealing for professionals, but could be delivered in a more efficient way and with lower costs.

2   Methods

Through this paper, we propose to review existing literature on innovation in services, identify common themes which can be adopted and used in software maintenance. This could be translated even further into a better service for clients and an improved performance for a company delivering this type of service. The fundamental principle we are using is the proximate genus and specific difference. We are seeking to understand how innovation is working in a similar environment, and use the learnings into the specific case of software maintenance. It is a literature review study, compiled with author's practical experience in the field.

3   Research Results

3.1   Software Maintenance

Maintenance is one of the phases of a software lifecycle, the longest in time and the most costly. All software, regardless the size, complexity, code language or supplier are going through this stage, since this phase is associated with operating the application.

Software maintenance is defined by Institute of Electrical and Electronics Engineers as "the modification of a software product after delivery to correct faults, to improve performance or other attributes, or to adapt the product to a modified environment." [1]. It is not just about modifying the code, but also modifying the documentation associated with the application.

To maintain the software operational and adapt it to new changes is an expensive task. The cost of the maintenance phase is 50-80% of the total cost of the software across its lifecycle [2],[3].

Reactive modification of a software product performed after delivery to correct discovered faults it is named Corrective Maintenance (CM) [4]. Estimated cost for CM is about 20% of the total maintenance cost [5].

Spending a lot with maintenance leave companies with less to invest in new software, less than they would want. Hence, any improvement on the cost side of maintenance may translate in an increased volume of development of new applications. This can be achieved by making SM, and CM in particular, more innovative.

3.2   Innovation in Services

The topic of innovation in services is not new, significant research has been conducted by researchers.

The definition for innovation is [6]: *"the act or process of introducing new ideas, devices, or methods"*. In his extensive study of innovation, Fuglsang [7] finds that innovation composed by two areas: creativity and innovation. First area is focused on generating innovative ideas, while the second one is concerned with exploiting the ideas. His model propose a three stage approach: generating idea, development and implementation.

Gallouj and Djellal [8] look at the innovation from the output to the customer point of view. Innovation has to have a result the introduction of something new in the processes which are related with customers.

Anthony [9] has the following definition: "Something different that has impact", highlighting that words like "technology" and "never been done before" are not part of his definition, and

Gallouj [10] classifies innovation in services as:

a. Radical – creation of a new set of characteristics of the service
b. Ameliorative – Improving one or more characteristics of a service
c. Incremental – additional characteristics
d. As hoc – interactive (social) construction of a solution to a particular problem posed by a given client
e. Re-combinative – combining or splitting of groups of characteristics
f. Formalization – formatting and standardizing characteristics

Leifer [11] speaks about incremental and radical innovation and each of them require different approaches. While incremental innovation builds on existing products or services, radical innovation is concerned with development of new business. Since the two are so different, the process in the early stages is significantly different [11],[12].

Sundbo [13] shows that innovations are technological or depend on technology (46%) while the rest are non-technological. Technology plays a significant role in the innovation process. Gallouj and Djellal [8] also sustain that information and communication technologies have an important role, being also a precondition for innovation.

Even in these conditions, more than half of the innovations in services are not related with technology. Innovations in services are related to behaviors. [7].

In services, innovation is very often small improvements, rare radical innovations [7],[10],[13]. Gallouj and Djellal also support the idea that innovation in services is rather incremental, having some disadvantages being easy to imitate and cannot be patented. [8]

Ettlie [12] states that services are extremely difficult to standardize compared with products because services are often tailored for each unique customer.

In these conditions, one idea seems to get agreement from most of researchers: the personnel involved in the process [14] and the customers are playing a key role in innovating the service. In fact, they are considering customers as co-producers of the service [7], [15]. Sundbo[14] is basing his general innovation theory on involving employees.

Howells[16] support the idea of customers involved in the innovation process, but he is showing concerns that they shouldn't develop the ideas themselves. Their involvement is through interaction with the employees.

Ulwick[17] insist that the customer needs to be asked only about the outcome, and not to come up with solutions. Tushman and O'Reilly III [18] propose having a structured process that allows customers to contribute to innovation.

Gallouj and Djellal [8], in their extensive study, notice the evolution from a bipolar (producer and user) view to a "co-production" of innovation, emphasizing the collaboration between suppliers and users.

Gustafsson and Johnson [19] assimilates co-involvement of customers in innovation process with using customer information as input for product innovation, and such the customers are a source of valuable ideas.

Ettlie [12] find difficult to deliver well service innovation because most services are coproduced by providers and clients, and this is changing all the rules for quality delivery.

An interesting topic is the source of innovative ideas. Innovations in services are not laboratory or science based (Sundbo [13], Gallouj [10]) as they typically are in manufacturing. Often innovations are ad hoc, based on ideas from employees or managers.

Guo et all. [20] suggest using Root Cause Analysis in the innovative process. Edvardsson et all. 2000 [21] consider new solutions based on observations of service quality problems.

John E. Ettlie [22] suggest that innovations based on ideas from customers are more likely to lead to market success. Also Gallouj and Djellal [8] promote that a solution needs to be adapted to a specific problem and/or a particular type of client.

Tushman and O'Reilly III's [18] study on airline services shows clearly that ordinary users are the most important source for innovation, surpassing industry experts and Lead Users.

Baldwin et al.[23] finds that information from external agents is a key factor in the innovation process, especially for in technology business services. Customers and competitors are deemed to be among the most valuable of all sources of information.

Gallouj [10]'s study shows that clients, sales force and contact personnel are by far the most valuable sources if innovation.

Ettlie [22], on the other hand, shows that significant number of ideas comes from outside the company.

3.3   Models for Innovation

Researching the literature several models have been designed to enhance innovation in services:

1. Tushman and O'Reilly III model [18] is presented in Figure 1. In their vision the first step is "Focusing the Energy", which assume that the organization realize that need to switch focus towards innovation. Another characteristic is that the ideas need to go through three different gates: Strategy Gates, Cultural ate and Organizational Change Gate. Only the ideas aligned with the three gates will advance forward.

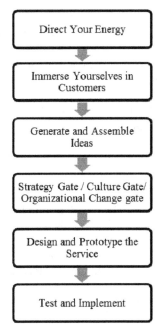

Fig. 1. Tushman and O'Reilly III service innovation process

2. Gallouj and Djellal [8] model sees innovation as having four distinct layers but interactive, with most problematic being Layer 2: The service as a result defined and required by the service provider.
    Layer 1: Service-based innovation: Why and for Who?
    Layer 2: The concept of service: What?
    Layer 3: Organizational innovation: How?
    Layer 4: The methods and resources implemented: With What and with Whom?
3. L. Zhang et al. [24] are proposing a three dimensions Innovation Process Capability Assessment Tool (IPCAT) – 3 domains (Process, Technology and Culture), 3 main processes of innovation (problem solving, internal diffusion, project management) and the CMM stages. The tool objective is to gauge innovation process capability and to contribute a guideline and instruction of innovation process performance improvement. It's still in early age, but it's useful to understand the innovation dimension.
4. Van Zyl [25]   speak about innovation as a competence which will become the differentiator in the software world. In his view, innovation has four areas:
    -    Organizing for innovation - organizations classify themselves around market positioning and design their businesses appropriately.
    -    Designing for innovation - complexity analysis methods are required to handle the many challenges when designing products in the software producing organization.

-      Planning for innovation - once architectural views have been created and form the basis of product development, strategic product development is used to materialize innovations.

-      Human innovation - people is the cornerstone of all innovation abilities in an organization. Focus on using alternate methods for dealing with the increased innovation abilities.

5. Pikkarainen et all. [26] identified 8 areas which provide software companies with structure to use to organize innovation :

    a.   The Art of Focusing

    b.   The Art of Idea Harvesting

    c.   The Art of Idea Valuation

    d.   The Art of Openness

    e.   The Art of Optimizing the Impact of Critical Experts

    f.   The Art of Crafting Smart Products

    g.   The Art of Innovation Stimulation

    h.   The Art of Innovation Incubation

6. Gallouj [10] identifies three different models in which organizations are innovating:

a. Professionals in partnership – innovation is informal, pragmatic, individual, no defined structures for innovation

b. Managerial – there is an innovation process, however there are no department in place. Project groups are formed to carry on innovation.

c. Traditional Industrial – formal structure in place with defined processes.

7. Prahalad and Krishnan [27], in their book "The new age of innovation" state that "Value is based on unique, personalized experiences of consumers. Firms have to learn to focus on one consumer and her experience at a time (N = 1), even if they serve 100 million consumers." Their vision about how companies should innovate themselves in the global market is completed by access (not ownership) to resources from a variety of other firms, creating a true global ecosystem(R = G).

8. Anthony [9]'s model of innovation is based on 4 major phases, and promotes an 28 days program on innovation for large companies.

    a. discovering opportunities

    b. blueprinting ideas

    c. assessing and testing ideas

    d. moving forward

Nonaka and Takeuchi [28] suggest that improvements in innovation ability need to become a matter of organizational learning.

The five critical practices for innovation identified by Ettlie [12] are:

- Commitment not Contribution of Senior Management

- Clear and Stable Vision

- Improvisation – team needs to be flexible to try different ideas and iterate rapidly

- Information Exchange – sharing information formally and informally
- Collaboration Under Pressure – focus on goals and objectives. Building coherent teams

3.4   Related Work on Innovation in Software Maintenance

Innovation in SM is a particular case for innovation in services, however not so much research exists specific for this area.

Adler [29] introduce the concept of interpreter, who's role is to translate customer requirements in the language known by the developers. The interpreter has a complex role, needing to be knowledgeable about customer requirements and developers capabilities.

Tushman and O'Reilly III [18] describe SM as being defensive because the nature of this service is to remove things gone wrong and build on things gone right in order to satisfy and retain existing customers. However, through innovation, this can be turned around into offensive, and in this way attracting new customers. Even so, they partially accept that software maintenance is lacking innovation through its defensive outlook.

According with Gallouj [10], software maintenance may fall in the "The Professionals in Partnership Model" or "The Managerial Model of Innovation Organization", or at least this is what the customers would think. The first is the model that supply services with a high 'grey matter' content and sell not service 'products' exactly but rather competences and problem- solving capabilities. The research is not formalized, being individual and pragmatic. In the second model, the search for ideas is part of everybody job, while for development ad hoc project teams are dedicated.

Authors practical observations is that innovation is perceived as something which is not present in SM.

4   Common Themes to Foster Innovation in SM

The research helped us to understand some common themes in relation with innovation, applicable in software maintenance, especially for the case of corrective maintenance:

- An innovation needs to address a customer problem, the customer needs to see value in the innovative service in order to generate more revenue. This is the opportunity. To do this you must know who your customers are, including software users, IT department and client's customers. An idea will be proposed to the client and might result in a new project to enhance the current software.
- Innovation can be incremental, which rather small as impact, and radical. Many of the innovation in services are behavioral
- Identify sources of ideas and exploit them. There are many sources to generate ideas, from focus groups with the software users to surfing the internet.

- Starting from the opportunity, an innovation, not only in service, but also in manufacturing, has four main stages:

  ▪ The spark – the "de-click", the thing that is going to differentiate the new service from the old service. This cannot appear without understanding the customer.

  ▪ Idea development – the initial spark is being developed into a comprehensive idea

  ▪ Ideas evaluation – ideas are evaluated from many aspects, including impact/cost analysis, and ideas are prioritized for implementation

  ▪ Idea implementation – last stage after which a new service will be produced

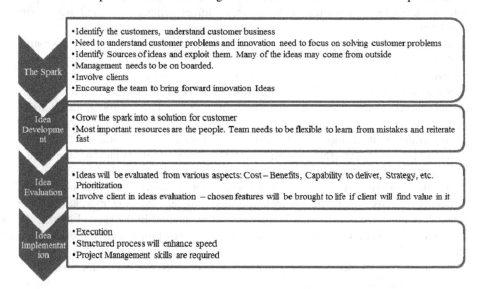

Fig. 2. Principles mapped on the four stages of innovation

- Structured process for innovation will most likely enhance the speed of innovation.
- Management needs to be on boarded on innovation journey. The organization need to embrace innovation.
- Clients are co-producers of the service, therefore they need to be involved in the innovation process. Involve software users in making decisions on what the software should be doing next.
- Many ideas may come from outside the organization. Often, in the IT field, people lacking technical skills don't get much attention. They might be just a fresh eye which can bring new ideas. Don't ignore discussions, articles on forums, blogs or various magazines. Most likely, somebody else had a similar problem that wanted to resolve, why not explore instead of reinventing the wheel?
- Most important resources are the people. The employees need to be encouraged to bring innovative ideas. Many of the innovations are related to human behavior.
- Flexibility for the team, being able to learn from mistakes and reiterate rapidly.

- Stay focused on innovation. This is not one time activity, it is a continuous process. Every team member should innovate. This doesn't mean that structured processes shouldn't be followed. On the contrary. Following structured processes make it easy to innovate.

5   Conclusions

In our paper we proposed to research on how innovation is adopted in services, and more particular in software maintenance, and identify the characteristics that can improve innovation on the corrective type of activities. This is a theoretical approach, as next steps we have proposed to take these concepts and test them in a real SM environment.

From our own experience we have witness many instances when:

- Maintenance was perceived by personnel as mere bug fixing activity, not a service provided for clients
- Clients are not known or their expectations and needs are not known
- Low motivation because software maintenance seemed boring and not challenging
- High attrition due to low motivation
- It is just problem solving, no room for innovation
- Team size is not balanced with demand
- Data not fully utilized to manage the service.

By adopting proposed suggestions, three problems, often encountered in SM, could be resolved:

- Increase client satisfaction by offering an innovative service. Many clients ask their suppliers not only to ensure applications are up and running, but also to innovate and improve the applications.
- Increase peoples morale, by realizing that corrective maintenance is not just another repetitive activity, but rather an innovative one
- Improve business performance by increasing revenue and client retention.

Acknowledgments. Research has been conducted with financial support from "Sectorial Operational Programme Human Resources Development 2007-2013", contract number POSDRU/159/1.5/S/132395 and 132397.

References

1. Abran, A., Moore, J.W., Bourque, P., Dupuis, R., Tripp, L.L.: Guide to the Software Engineering Body of Knowledge. CS Press, California (2004)
2. Schach, S.: Software Engineering, 4th edn. McGraw-Hill (1999)

3. Banker, R.D., Datar, S.M., Kemerer, C.F.: Factors affecting software maintenance productivity: An exploratory study. In: Proceedings of the 8th International Conference on Information Systems, Pittsburgh, pp. 160–175 (1987)
4. IEEE Standard for Software Maintenance, IEEE Std 14764-2006, The Institute of Electrical and Electronics Engineers, Inc.
5. Penny, G., Armstrong, T.: Software Maintenance - Concepts and practice, 2nd edn. World Scientific (2003)
6. Merriam Webster, http://www.merriam-webster.com/dictionary/innovation (accesed on August 2, 2014)
7. Fuglsang, L.: Innovation and the Creative Process: Towards Innovation With Care. Edward Elgar Pub., Northampton (2008)
8. Gallouj, F., Djellal, F.: The handbook of innovation and services. Edward Elgar Publishing Limited, USA (2010)
9. Anthony, S.: The Little Black Book of Innovation: How it Works, how to Do it. Harvard Business Press (2012)
10. Gallouj, F.: Innovation in the Service Economy. Edward Elgar, Northampton (2002)
11. Leifer, R.: An information processing approach for facilitating the fuzzy front end of breakthrough innovations. In: Proceedings of the International Conference on IEMC 1998, pp. 130–135 (1998)
12. Ettlie, J.E.: Managing Innovation, 2nd edn. Elsevier Butterworth-Heinemann (2006)
13. Sundbo, J.: Management of innovation in services. The Service Industries Journal 17, 432–455 (1997)
14. Sundbo, J.: Innovation and Learning in Services – The Involvement of Employees. In: Advances in Services Innovation. Springer (2007)
15. Van Looy, B., Dierdonck, R., Gemmel, P.: Services Management. Pitman, London (1998)
16. Howells, J.: Innovation, consumption and services: encapsulation and the combinatorial role of services. The Service Industries Journal 24(1), 19–36 (2004)
17. Anthony, W.U.: Turn Customer Input into Innovation. Harvard Business Review, 91–97 (2012)
18. Tushman, L.M., O'Reilly III, C.A.: Winning Through Innovation, 2nd edn. Harvard Business School Press, Boston (2002)
19. Gustafsson, A., Johnson, M.D.: Competing in a Service Economy. How to Create a Competitive Advantage Through Service Development and Innovation. John Wiley & Sons, San Francisco (2003)
20. Guo, J.W., Jiang, P., Guo, J., Tan, R.H.: Research and Application of RCA in the Process of Product Innovation. In: International Conference on Mechanical and Automation Engineering, Jiujang, pp. 180–183 (2013)
21. Edvardsson, B., Gustafsson, A., Johnson, M.D., Sandén, B.: New Service Development and Innovation in the New Economy. Studentlitteratur, Lund (2000)
22. Ettlie, J.E.: Managing Technological Innovation. Wiley, New York (2000)
23. Baldwin, J.R., Gellatly, G., Johnson, J., Peters, V.: Innovation in Dynamic Service Industries, Catalogue no. 88-516-XIE, Ottawa, Minister of Industry (1998)
24. Zhang, L., Seidel, R., Shahbazpour, M., Haemmerle, E.: Three-Dimensional Innovation Process Capability Assessment Tool. In: 6th International Conference on Information Management, Innovation Management and Industrial Engineering (2013)

25. Van Zyl, J.: Process innovation imperative [software product development organisation]. In: Change Management and the New Industrial Revolution, IEMC 2001 Proceedings, New York, pp. 454–459 (2001)
26. Pikkarainen, M., Codenie, W., Boucart, N., Alvaro, J.A.H.: The Art of Software Innovation. Springer (2011)
27. Prahalad, C.K., Krishnan, M.S.: The new age of innovation. McGraw Hill, USA (2008)
28. Nonaka, I., Takeuchi, H.: The Knowledge-Creating Company. Oxford University Press (1995)
29. Adler, T.R.: The Innovation Process: Interpreting Customer Requirements. Aerospace and Electronic Systems Magazine 9(6), 17–25 (1994)

A Comparative Study of Educational Laboratories from Cost & Learning Effectiveness Perspective

Krishnashree Achuthan[1,2] and Smitha S. Murali[2]

[1] Amrita Center for Cybersecurity Systems & Networks,
[2] VALUE Virtual Labs,
Amrita Vishwa Vidyapeetham, Kollam-690525, India
Krishna@amrita.edu, smithasm@am.amrita.edu

Abstract. Laboratory education plays a prominent role in the education of engineers and scientists for the practical knowledge and research skill sets they impart. The cost of laboratory set up can vary in magnitudes depending on the level of sophisticated analysis they deliver. Added are the costs of maintenance and facility requirements. Recent advancements in technology enhanced laboratories such as virtual simulation based laboratory and remotely triggerable laboratories, both of which allow e-learning for laboratory education has shown to impact the laboratory education in a significant way. This paper focuses on comparing the tangible and intangible benefits of the various types of laboratories available to students today. Evaluating the differences in infrastructural set up, safety elements, equipment costs, the experimentation time and the effective learning from these laboratories are examined closely. Using pair wise correlation, the advantages and difficulties observed in chemistry based experiments are presented. The compromises, if any, using cost effective laboratories is investigated using N=141 students that underwent undergraduate education. Although most students use physical labs today, the growing trends in online and distance education makes this work significant to teachers and administrators alike. Although not an alternative, the scalability and functional proficiency these virtual laboratories allow bridging the gaps found in traditional laboratories.

Keywords: Cost-Benefit Analysis, Attributes, Virtual laboratory, Physical Laboratory, Analytical Hierarchy Process, Remotely Triggerable Laboratory.

1   Introduction

Scientific laboratories have been an integral part of school and higher secondary education [1]. The importance given to practical hands-on learning is perceived from the separate course allotments and assessments given as part of well established curricula. The quality of laboratory education varies dramatically between institutions. Some of the challenges include poor maintenance of the equipment, periodic up gradation of experiments, increasing cost of accessories and consumables and inadequately trained teachers. Thus the overhead costs of laboratories can be staggering especially when sophisticated experiments are involved. Another problem associated with a typical physical laboratory are the congested labs, students doing experiments in groups rather

© Springer International Publishing Switzerland 2015
R. Silhavy et al. (eds.), *Software Engineering in Intelligent Systems*,
Advances in Intelligent Systems and Computing 349, DOI: 10.1007/978-3-319-18473-9_15

than individually, use of hazardous substances in certain cases, limited time for experiments and in a few cases inadequate safety measures. The advent of virtual laboratories has provided a seminal change in the way laboratory education can be addressed. The introduction of virtual labs as an e-learning tool provides a vast opportunity to students for individualized learning with uninterrupted access all over the world [2,3,4]. In the case of science laboratory instruction, technology has reached a threshold where virtual or simulated approaches can formidably meet or exceed the learning outcomes of traditional approaches. And research suggests that simulated laboratories can dramatically impact learning in positive ways [5,6,7]. Much of the problems stated earlier are circumvented with use of virtual laboratories.

VALUE Virtual labs [8,9] were built with the goal of catering to students, teachers and institutions that have faced regression in student interests with science experimentation. Built on a platform for open access, a consortium of twelve partners were involved in the development of over 1500 experiments as part of 150 laboratories in various thematic areas. Some of the key advantages of virtual labs include an ability to learn about the physical instruments, their operations in an interactive way. By providing features to change various parameters and observe the resulting impact, students are able to visualize the experiment in a far superior way. This is in addition to providing ability to repetitively practice an experiment without a time pressure to complete within a stipulated time and experimenting with a diverse set of experimental conditions can be easily simulated in a virtual environment and those cannot be done in a physical laboratory. Manipulating materials and equipments, recording data, performing the experiment in a collaborative manner and preparing experiment reports [10] and just a few more ingredients that enable constructivist learning in laboratory education. Remotely triggerable labs have real equipments that can be operated from any distance via the internet. The visual changes and/or data values displayed on the equipment are displayed in the remote user's desktop also, making them feel close physical proximity to the equipments. With helpful tips on safety precautions, an awareness of what to expect and how to work in a potentially toxic environment can be well ingrained amongst students. The will reduce significant amount of time spent in describing and demonstrating safety measure as in the real lab. One of the most significant attributes of virtual laboratory is its high degree of flexibility which makes self learning of students easier [8,11,12]. Although there are advantages, a holistic representation from both cost effectiveness point of view and learning is absent. This paper takes a look at the relative degrees of impact from both tangible and intangible factors governing the choice of an experiment or lab.

2   Comparative Analysis of Educational Laboratories

One of the methodologies to compare effectiveness and economic feasibility of experiments in laboratories is by looking at the cost-benefit analysis (CBA) [13,14]. In this work, an analysis on CBA is performed by listing all the direct and indirect contributors. Factors impacting the choice of an experiment or within a lab which are direct factors will involve looking into the equipment cost, the software cost, cost for other network devices, and employing people costs during the project period etc. Indirect costs on the other hand are those associated with time consumed in developing the experiment, its validation, cost of training and maintaining teachers by way of workshops etc. Benefits obtained are of two types – tangible and intangible. In spite of these, many of the costs and benefits are assumed to be realized in the future. This work details four types of labs for comparisons. They are: Physical Laboratory (PL) that contain relatively inexpensive equipments; Virtual Laboratory (VL) that are online simulation based labs; Expensive Physical Laboratory (EPL) that behold expensive equipments and/or complex processes and only accessible physically inside a lab and finally Remote Triggered Virtual Laboratory (RVL) that are connected to the internet and accessible online. RVL is a hybrid lab which may have either PL or EPL instrumentation at one location and VL in others. The next sections will define the cost and benefits of the above labs followed by an elucidation of their attributes. We then look to create a cost benefit decision matrix and do pair wise comparisons to present the cost effective methodologies to adopt for experiments.

2.1   Factors Affecting Deployment of PL, VL, EPL and RL Labs

To deploy and establish laboratories, several key factors need to be rigorously analyzed. A few of the key criteria governing the choice of type of labs are tabulated in Table 1. The factors that are marked as Major or Minor indicate the relative importance of the cost factor with respect to its deployment. When considering the factors for the start up and maintenance of these labs, costs are directly proportional to the infrastructure, maintenance, for PLs and EPLs. However in terms of building VLs integrating with the content etc., there is an appreciable one-time investment involved. The choice in terms of offering alternate types of labs is only possible if the consequential learning is comparable to one another. The next section looks at a case study conducted in this regard.

Table 1. Cost factors in virtual and remote triggered virtual laboratories

Cost Factor	Description	PL & EPL	VL & RL
V-Lab Project Development	Software costs, equipment costs, cost for student material development, instructor costs and subject matter expert costs.	Minor	Major
Delivery	Computers, Internet connection and other Network devices.	Minor	Major
Content	Tracking, Technical support, Content updates and Technology updates.	Minor	Major
Laboratory setup	Building and furnishing cost, equipment cost, reagents cost, energy cost (power and water), safety equipments cost	Major	Minor
Administration and Maintenance	Instructor cost, lab employees cost, maintenance cost, cost for the safe disposal of hazardous substances, cost for the treatment of lab accidents	Major	Minor

3   Educational Outcomes and Learning

In order to study the educational outcomes of online versus physical labs, 141 undergraduate students were exposed to three chemistry laboratory experiments namely - Cryoscopy, Viscosity and Hardness of water [Figure 1, 2, 3]. The experiments had several procedural learning elements. The students were divided into three groups. The first group did these experiments in PL mode, the second did through VL and the third group did both PL and VL. The experiment was designed in such a way that continuous assessment and their knowledge of these topics could be done. They were assessed through a written exam on the knowledge and application of the experiment at the end of lab sessions.

3.1   Cryoscopy

The objectives of this experiment were to determine colligative properties of solutes, freezing point depression of a solution, Van't Hoff factor of solutes and molar mass of unknown solutes. The experiment allows varying the solvents, solutes and the amount of solute that can be added to the solvent as well.

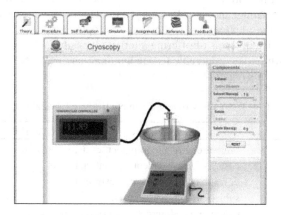

Fig. 1. Virtual Labs for Cryoscopy

3.2   Viscometry

In this experiment, viscosities of polymers are computed by using an Ostwald viscometer. The experiment allows using various polymers and solvents as well as preparing different concentrations of polymer and solvent mixtures. By determining the time of flow of solvents viscosity and the molecular weight of the polymers are determined.

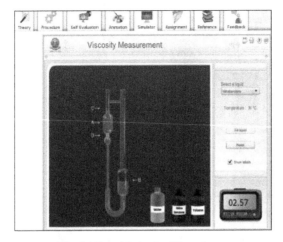

Fig. 2. Virtual Labs for Viscosity

3.3 Hardness of Water

In this experiment, various properties of water namely its hardness, chemical oxygen demand, alkalinity etc. are determined. In the experiment, the titrant, titrate, the speed of titrant and the indicator may be varied.

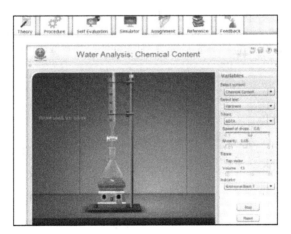

Fig. 3. Virtual Labs for Hardness of water

The results from the assessment are shown in Figure 4. Students that went through PL had an average score of 65% with approximately 42% of them scoring above 80%. Students that went through VL had a higher average score of 78% with over 75% of them scoring about 80%. When students did both PL and VL sequentially, the results were just as good as seen with VL alone. The results show the effectiveness of VL in teaching chemistry laboratory experiments. In spite of VL lab sessions being 50% shorter than PL labs and the individualized learning students received contributed to their improved performance.

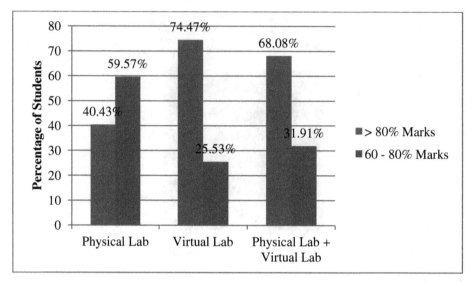

Fig. 4. Student Performance in Three Chemistry Experiments

4   Cumulative Cost Comparisons

A cumulative cost comparison between PL and VL of three experiments described in the previous section is done (Figure 5) based on actual building costs of both PL and VL along with the projected costs of their maintenance over a ten year period.

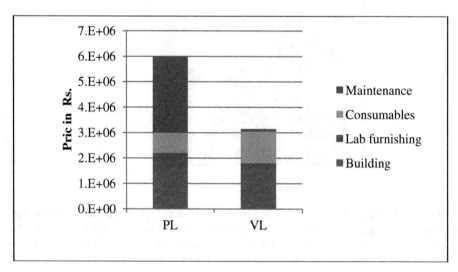

Fig. 5. Cost Comparisons of Three Chemistry Experiments

Cost correlation was based on a standard laboratory classroom size of 800 sq.ft that could accommodate 50 students at a time in both cases. Physical laboratory equipments included the work benches, cupboards, side shelves, exhaust fans etc. The consumables included chemicals, water and other disposable items. Meanwhile virtual lab equipments were primarily computers and networking in addition to basic equipments such as air conditioning devices and furniture. The consumables were non-existent and the maintenance was negligible. The comparisons show 48% savings in cost from VL compared to PL.

5   Economic Attributes of Traditional and E-learning Laboratories

The key contributors towards deciding if an experiment should be built for laboratory education can be summarized by five attributes. They are: 1) Learning - A good and efficient laboratory furnishes an appropriate platform for students to develop constructivist learning; visual & verbal learning; active and reflective learning that enriches their conceptual understanding. The mode of learning becomes less important when concepts are easy to grasp. Hence, a low-cost scalable option is preferable, if in fact it is effective. 2) Time - Time is a great resource and is limited. Options that maximize learning while minimizing the time taken to learn and the flexibility to learn prior to arriving in class are preferable as they allow more interactive time between the teachers and students respectively. 3) Safety - PLs and EPLs are plagued with three types of safety namely personal safety, equipment safety and environmental safety from toxic reagents or products can create highly undesirable repercussions and this will be important to consider. The 4) Set up Cost - The laboratory set up cost can be the single most important attribute that could favor or reject the choice of specific experiments or labs. A low cost yet highly resourceful laboratory that assists students that promotes critical thinking, research skills, and scientific aptitude would be preferable. 5) *Facilities* - This refers to the maintenance and sustenance costs of labs. Commonly, low maintenance yet scalable labs are far more cost effective. A multi-level hierarchical structure is designed based on the above objectives and resolved as criteria, sub criteria and alternatives as shown in Figure 6.

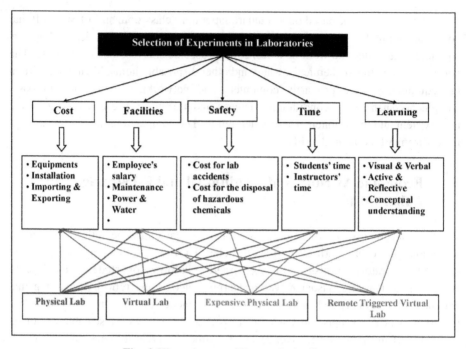

Fig. 6. Hierarchy tree of Economic Attributes

5.1   Optimization Based on Cross Lab Comparisons

Different alternatives from the hierarchical structure are compared on the basis of attributes in a pair wise manner using a one-nine point fundamental scale developed by Saaty [15]. The weightings for the five categories of factors derived based on the observations made in Table 1 and Figure 2 are shown in Table 2. The cells on the diagonal correspond to self-comparisons and weighted the same. As a general rule of thumb, if there is a cost advantage of an option for a specific factor, it is weighted more.

Table 2. Sample Factor Weighting in Pairwise comparison Matrix for Laboratories

Cost	PL	VL	EPL	RL	Learning	PL	VL	EPL	RL	Facilities	PL	VL	EPL	RL
PL	1	0.2	3	0.2	PL	1	0.25	1	0.25	PL	1	0.2	4	0.2
VL	5	1	7	3	VL	4	1	5	1	VL	5	1	5	1
EPL	0.33	0.14	1	0.125	EPL	1	0.2	1	0.14	EPL	0.25	0.2	1	0.2
RL	5	0.33	8	1	RL	1	1	7	1	RL	5	1	5	1
Safety	PL	VL	EPL	RL	Time	PL	VL	EPL	RL					
PL	1	0.14	2	0.14	PL	1	0.5	1	0.5					
VL	7	1	8	1	VL	2	1	3	1					
EPL	0.5	0.125	1	0.125	EPL	1	0.33	1	0.33					
RL	7	1	8	1	RL	2	1	3	1					

The comparative score for each option is calculated and normalized to yield the normalized pairwise comparison matrix. A consistency analysis is done on each pairwise comparison by discerning consistency index and consistency ratio. Table 3 represents a combined pairwise comparison matrix based on the priorities obtained for various criteria and alternatives calculated through the Analytic Hierarchy Process [16]. Higher the priority, the more advantageous is the option with respect to the cost. The data indicated VL is the most economic laboratory option, followed by RVL. Both PL and EPL are not so advantageous from a cost perspective, especially when large number of students are involved.

Table 3. Combined pairwise comparison matrix based on priorities

Criteria	Cost	Facility	Safety	Time	Learning	Priority
Calc. Wt.	0.08	0.0706	0.319	0.1573	0.373	
PL	0.1029	0.1306	0.0739	0.1612	0.3333	0.8019
VL	0.5249	0.4208	0.4386	0.3536	0.1667	**1.9046**
EPL	0.049	0.0639	0.0488	0.1317	0.25	0.5434
RL	0.3231	0.4028	0.4386	0.3536	0.25	**1.7681**

6   Conclusion

Providing high quality laboratory education is a challenge to most institutions due to the exorbitant cost in building them. Needless to say there has to be an alternative to providing laboratory education in a scalable, cost-effective fashion. This paper has looked at the economic factors and learning components of four types of laboratories and cross-compared those to allow teachers and administrators strategize effectively on their choice of labs and experiments for their respective institutions without compromising on the learning outcomes. Five attributes namely learning outcomes, times, set up cost, safety and facility requirements with eleven sub-attributes contributing to them were analyzed. The pairwise comparisons showed significant differences in the overall cost savings from wider adoption of VL and RL. The advantages of using them are investment from one time investment, low maintenance and with high ratings for safety, flexibility and usage. Especially in the area of chemistry, toxic reagents and wastes can be hazardous in nature. Using VL and RL, significant savings in such intangible costs can be realized. From a learning effectiveness perspective, the case study in this work showed no compromise in the learning outcomes when students did VL alone. The conceptual knowledge was equal or more than that gained from PL. This is attributed to the focused attention and individualized learning gained through VL. Today most traditional laboratories are either PL or EPL. Results from this case study strongly indicate that emerging labs such as VL and RL are effective supplemental labs that enrich scientific learning.

Acknowledgment. Our work derives direction and ideas from the Chancellor of Amrita University, Sri Mata Amritanandamayi Devi. The authors would like to acknowledge the contributions of faculty and staff at Amrita University whose feedback and guidance was invaluable. We also like to thank the entire CREATE team for their continuing work and encouragement for the development of virtual laboratory.

References

1. Feisel, L.D., Albert, J.R.: The role of the laboratory in undergraduate engineering education. Journal of Engineering Education 94(1), 121–130 (2005)
2. Ma, J., Nickerson, J.V.: Hands-on, simulated, and remote laboratories: A comparative literature review. ACM Computing Surveys (CSUR) 38(3), 7 (2006)
3. Benetazzo, L., Bertocco, M., Ferraris, F., Ferrero, A., Offelli, C., Parvis, M., Piuri, V.: A Web-based distributed virtual educational laboratory. IEEE Transactions on Instrumentation and Measurement 49(2), 349–356 (2000)
4. Abdulwahed, M., Nagy, Z.K.: The TriLab, a novel ICT based triple access mode laboratory education model. Computers & Education 56(1), 262–274 (2011)
5. Tatli, Z., Ayas, A.: Virtual Chemistry Laboratory: Effect of Constructivist Learning Environment. Turkish Online Journal of Distance Education (TOJDE) 13(1) (2012)
6. Tüysüz, C.: The Effect of the Virtual Laboratory on Students' Achievement and Attitude in Chemistry. International Online Journal of Educational Sciences 2(1) (2010)
7. Patterson, D.A.: Impact of a multimedia laboratory manual: Investigating the influence of student learning styles on laboratory preparation and performance over one semester. Education for Chemical Engineers 6(1), e10–e30 (2011)
8. Achuthan, K., Sreelatha, K.S., Surendran, S., Diwakar, S., Nedungadi, P., Humphreys, S., ... Mahesh, S.: The VALUE@ Amrita Virtual Labs Project: Using Web Technology to Provide Virtual Laboratory Access to Students. In: 2011 IEEE Global Humanitarian Technology Conference (GHTC), pp. 117–121. IEEE (October 2011)
9. Diwakar, S., Achuthan, K., Nedungadi, P., Nair, B.: Enhanced facilitation of biotechnology education in developing nations via virtual labs: analysis, implementation and case-studies. International Journal of Computer Theory and Engineering 3(1), 1–8 (2011)
10. Raman, R., Nedungadi, P., Achuthan, K., Diwakar, S.: Integrating collaboration and accessibility for deploying virtual labs using vlcap. International Transaction Journal of Engineering, Management, & Applied Sciences & Technologies 2(5) (2011)
11. Merchant, Z., Goetz, E.T., Cifuentes, L., Keeney-Kennicutt, W., Davis, T.J.: Effectiveness of Virtual Reality-based Instruction on Students' Learning Outcomes in K-12 and Higher Education: A Meta-Analysis. Computer Education 70, 29–40 (2014)
12. Tatli, Z., Ayas, A.: Effect of a Virtual Chemistry Laboratory on Students' Achievement. Journal of Educational Technology & Society 16(1) (2013)

13. Hanes, N., Lundberg, S.: E-learning as a Regional Policy Tool: Principles for a Cost-benefit Analysis. RUSC. Universities and Knowledge Society Journal 5(1) (2008)
14. Cohen, A., Nachmias, R.: Implementing a Cost Effectiveness Analyzer for Web-Supported Academic Instruction: A Campus Wide Analysis. European Journal of Open, Distance and E-Learning (2009)
15. Saaty, T.L.: How to make a decision: the analytic hierarchy process. European Journal of Operational Research 48(1), 9–26 (1990)
16. Zimmer, S., Klumpp, M., Abidi, H.: Industry Project Evaluation with the Analytic Hierarchy Process. In: 10th Annual Industrial Simulation Conference, pp. 4–6 (June 2012)

Definition of Attack in Context of High Level Interaction Honeypots

Pavol Sokol[1], Matej Zuzčák[2], and Tomáš Sochor[2]

[1] Institute of Computer Science, Faculty of Science
Pavol Jozef Šafárik University in Košice, Jesenná 5, 040 01 Kosice, Slovakia
pavol.sokol@upjs.sk
[2] Department of Informatics and Computers, Faculty of Science,
University of Ostrava, 30. dubna 22, 701 03 Ostrava, Czech Republic
{matej.zuzcak,tomas.sochor}@osu.cz

Abstract. The concept of attack in the context of honeypots plays an important role. Based on the definition of the attack, honeypots obtain information about attackers, their targets, methods, and tools. This paper focuses on the definition of attack in context of high-interaction server honeypots. Paper proposes the definition of attack from the perspective of information security and network forensics analysis.

Keywords: honeypot, computer attack, high-interaction honeypot, network security, server honeypot.

1   Introduction

Network security is an important part of security policies in each organization. Traditional tools (e.g. iptables, snort, surricata) and approaches applied in protection are currently becoming increasingly ineffective. It is due to the fact that the "bad" community (e.g. hackers) is always several steps ahead of defensive mechanisms (firewalls, IDS, IPS etc.). Therefore it is necessary to learn as much information as possible about the attackers. The specialized environment, in which vulnerabilities have been deliberately introduced in order to observe attacks and intrusions, has been introduced 20 years ago and it is called a **honeypot.** It can be defined as "a computing resource, whose value is in being attacked". The honeypot computer system is "a system that has been deployed on a network for the purpose of logging and studying attacks on the honeypot." [1].

Honeypots can be classified into several types. Two classifications are used in this paper. The former classification is **based on the level of interaction**. From this point of view, it is distinguished between low-interaction and high-interaction honeypots. The difference between these types lies in an extent, to which the attacker is allowed to interact with the system. **Low-interaction honeypots** implement targets to attract or detect attackers using software to emulate the characteristics of a

R. Silhavy et al. (eds.), *Software Engineering in Intelligent Systems,*
Advances in Intelligent Systems and Computing 349, DOI: 10.1007/978-3-319-18473-9_16

particular operating system and its network services on a host operating system. Examples of this type of honeypot are Dionaea [2] and HoneyD [3].

In order to get more accurate information about attackers, their methods and attacks, a complete operating system with all services must be used. This type of honeypot is called **high-interaction honeypot.** The goal of this type of honeypot is to provide an attacker access to the complete operating system, where nothing is emulated, nor restricted [1].

Another classification of honeypots is based on the **role** of honeypot. According to this classification, honeypots are divided in server-side honeypots and client-side honeypots. **Server side honeypots** are useful in detecting new exploits, collecting malware, and enriching research of the threat analysis. On the other hand, **client-side honeypots** collect information about client side attacks. They detect attacks directed against vulnerable client applications, when a client interacts with malicious servers. The aim of these honeypots is to search and detect these malicious servers. This paper focuses on server-side high-interaction honeypots.

2   Motivation

The primary motivation for elaborating this paper fact that the definition of an attack against high-interaction honeypots remained unchanged since 2004 [4]-[6]. Honeynet's community [1] considered each connection to be an attack against the honeypot. Such definition is based on the fact that legitimate connections should not be normally directed at honeypots since it does not provide any productive value or service. Therefore any traffic going to the sensor is considered as an attack.

According to the previous paper, where attacks against low-interaction honeypots were defined [7], the above-mentioned approach is inconsistent with techniques used by administrators of sensors to increase the attractiveness of honeypots for research purposes. In order to make a sensor sufficiently sensitive to attacks, it is necessary to raise the awareness of honeypots or pretend already mentioned productive value.

2.1   Presented Contributions

There are two **main contributions** of this paper. The first one is the analysis of the notion of the attack against high-interaction server honeypots from several perspectives in details. This analysis is based on the results of works   [8] and [9]. The second contribution represents a comparison of definition of attack against low-interaction honeypots and high-interaction honeypots. The comparison is based on the authors' paper [7]. The paper focuses on the definition of attack against low-interaction server honeypots for two cases, namely emulation of Windows services and emulation of linux SSH services.

3   Related Works

No systematic research study aimed at definition of attack in the context of low- or high-interaction honeypots has been found. The paper [10] by Oumtanaga et. al. discusses the issue related to the definition of attack. In that paper the authors focus on specification of a model of honeypot attack based on data collected from Leurré.com.

In related fields of network security several relevant papers exist that are aimed to issues of attacks. In [11] the authors focus on attack graphs and their analysis since security analysts use the attack graphs for attack detection, defence and forensic. On the other hand, the paper [12] discusses the issue of ontology for network security attacks that show the relationship many commonly used standard classifications.

4   High-Interaction Honeypot and Definition of Attack

4.1   Comparison to Low-Interaction Honeypot

In the **case of low-interaction honeypots** the emulated protocols and services can be easily and effectively monitored. An example of low-interaction honeypot is honeypot-emulated SMB protocol (e.g. in Dionaea honeypot). Attacker can perform only certain operations (dissemination of malware, shellcode etc.). As [7] shows, only those network connections, in which certain actions occurred after the acceptance of the connection, are considered to be an attack. Another example is Kippo honeypot. Here, when the attacker identifies the SSH shell and its port in the server, logs in using correct combination of name and password, and performs a certain action, the connection can be identified as an **attack** [7].

In the **case of high-interaction honeypot**, the whole system is available for attackers; therefore an attacker's actions are not limited. They can download and install any program or script. They can also use the vulnerability of network protocols or exploit other servers.

```
2014-06-09 13:13:10 - Incoming connection from: 70.27.34.20:54616 -
SSH-2.0-PuTTY_Release_0.63
2014-06-09 13:13:10 - Successful login - Username:administrator
Password:jean-baptiste
2014-06-09 13:13:11 - Session Type: term
2014-06-09 13:13:15 - Entered command: sudo -i
2014-06-09 13:13:16 - Entered command: jean-baptiste
2014-06-09 13:13:17 - Entered command: ls
2014-06-09 13:13:26 - Entered command: pwd
2014-06-09 13:13:29 - Entered command: git clone
git://github.com/pooler/cpuminer.git
2014-06-09 13:14:15 - Entered command: ^C
2014-06-09 13:14:34 - Entered command: passwd administrator
2014-06-09 13:14:39 - Entered command: node34x
2014-06-09 13:14:43 - Entered command: node34x
2014-06-09 13:14:48 - Entered command: logout
2014-06-09 13:14:53 - Lost connection with the attacker: 70.27.34.20
```

Fig. 1. An example of an attack against

The results presented here are based on our research on high interaction honeypots. HonSSH honeypot has been used for this study, where the SSH shell was a primary subject of monitoring. Data collection period was from June 2nd to July 2nd, 2014. An example of attack is shown in Fig. 1. It is a real attack by the human attacker, not bot.

Unlike the case of low-interaction honeypots, it is not possible to monitor the attacker's activities here. In principle, it is not possible to create a table or statistical data as easily as in the former case. Therefore it is necessary to deal with every single attack individually and combine data captured from several monitoring sources (e.g. TCP dump, syslog etc.) When a honeypot in equipped with e.g. a website with a real content in order to increase its attractiveness, it is inevitable to separate possible legitimate traffic from attacks.

If a high-interaction honeypot is operated as a lure, e.g. when its primary web page looks like a legitimate web page offering a certain type of services, the fact that legitimate users visit the server as well, should not be ignored. An attacker should have learnt about the honeypot existence. The owner of the honeypot is not able to distinguish "human" (namely more sophisticated) attackers from legitimate users when advertising it. The honeypot can be advertised e.g. as being indexed in web search services, resulting in possible visits of the web site by real visitors. Certain type and amount of legitimate traffic towards a honeypot imitating real servers in highly realistic way should be expected, accepted and the analysis of captured data must take this fact into account.

The above fact motivate the authors to formulate of specific rules, according to which the concept of the "attack" could be defined and classified in the context of high—interaction honeypots (sensors). Regarding the fact that an attacker may use the system without (visible) limitations, it is necessary to take into account the type of system (Linux, Windows as a model case in this paper) used, available access channels (SSH , RDP, WEB) and actions, allowed in the system for an attacker (limited rights, full rights) by its administrator. The necessary rules are discussed in following sections in more details.

4.2   Model Situation of High-Interaction Honeypot

In the **case of low-interaction honeypots** the emulated protocols and services can be easily and effectively monitored (like in Dionaea honeypot).

Before focusing on the definition of attack, the model situation of high-interaction honeypot should be set up. The purpose of this model situation is to show which system will be deployed as a high-interaction honeypot and how it works. In other words, the system that is directly offered attackers to be attacked and compromised is discussed here. Although the model situation is an abstract example, real honeypots based on Windows and Linux are used as illustrations.

In terms of the definition, running services are elements that need to be considered in the definition of an attack. According to [13], the most often running services on high-interaction honeypots are web application, MSSQL server, SSH server, etc.

The honeypot administrator selects services that are to be provided to attackers, and focuses on their monitoring. Furthermore, there is a strong need for ensuring the security of honeypots to prevent attacks from a honeypot to other systems and networks. This is carried out by the Honeywall [14] in general (e.g. it reduces the network flow to prevent DDoS attack).

Windows OS are remotely managed by the RDP protocol. RDP does not represent a useful solution for attackers due to the high level of interaction with graphical environment. On the other hand, in Linux the SSH protocol is used to manage all system's resources. When an attacker desires to gain control of honeypots, they focus their attention on the SSH protocol therefore. An example of high-interaction honeypot that monitors the SSH protocol is HonSSH [15].

Successful exploitation of honeypot may also result in opening a shell at a non-standard port. An attacker would thus be able to avoid the monitoring of the standard shell. In this situation it is necessary to take appropriate countermeasures (e.g. in Honeywall).

In the past, Sebek [16] was used. Unfortunately, it is already obsoleted and it has no support anymore. Moreover, the "bad" community is aware of it. Also special system daemons (e.g. Auditd [17]) can be used for this purpose. Consequently, it is possible to log all activities in the system in cooperation with syslog. However, sophisticated attackers can detect these measures. Therefore methods based on monitoring of the network flow are used more frequently using Honeywall (e.g. Tcpdump), or using NAT or port forwarding. Possible opening of a shell at another firewall's port can be resolved by using a proper configuration of firewall in the Honeywall.

If the above-described example is accepted as a model for data gathering, the following actions can be considered the attacks:

- unknown network connections into the monitored system,
- modification of files,
- running new processes or services.

5 Approaches to the Attack Definition

Various views on the definition of attack are described in the following sections in more details in order to find the specific definition of attack. The following approaches are further analysed:

- **approach of information security** – value of honeypots and real systems,
- **approach of network forensics** - value for further analysis.

5.1 Approach of Information Security

The **Information security** "encompasses the use of physical and logical data access controls to ensure the proper use of data and to prohibit unauthorized or accidental modification, destruction, disclosure, loss or access to automated or manual records and files as well as loss, damage or misuse of information assets" [18].

L. Spitzner defined the honeypot as "a resource whose value is being in attacked or compromised. ... It means that whatever we designate as a honeypot, it is our expectation and goal to have the system probed, attacked, and exploited" [19].

On the other hand the attack against real system is defined in several standards. The standard ISO/IEC 27001:2009 defines **the attack** as „any attempt to destroy, expose, alter, disable, steal or gain unauthorized access to or make unauthorized use of an asset" [20], while according to RFC 4949 the attack is "an intentional act by which an entity attempts to evade security services and violate the security policy of a system. That is, an actual assault on system security that derives from an intelligent threat". The attack can be also defined as "a method or technique used in an assault" [21].

According to the above-mentioned definition, the attack against a honeypot can be defined as any attempt to gain access to it (e.g. attempts to enter a password via SSH, to change data in the database, DDoS attack, etc.).

An important aspect of any security system is ability to ensure confidentiality, integrity, and availability (CIA triad) [22]. Standard attacks are directed against these basic security features of information systems.

Based on the value and purpose of information system, honeypots solution can be divided into the honeypot itself (a lure), and the system that collects, stores, analyses data and sends them to other place (support honeypot's system). It is desirable that attackers attack against the honeypot itself. Administrators of honeypots want the attackers to compromise authenticity, system integrity, or availability. On the other hand, attacks against the support system are undesirable. In the case of capturing system, it is necessary to ensure the CIA triad. Otherwise, the collected data cannot be used for further network forensics analysis. From this perspective, the attack against the honeypot is desirable and necessary. Conversely, the attack against support honeypot's system is exactly the same as an attack against a real (production) system and thus should be avoided.

5.2   Approach of Network Forensics

An important issue for administrators of honeypots is ensuring that an attack aims against the honeypot but not against support honeypot's system. From this perspective, it is necessary to take into account the value of attack for further network forensics analysis. **Network forensics** is "the field of applying forensic science to the computer networks in order to discover the source of network crimes. The objective is to identify malicious activities from the traffic logs, discover their details, and to assess the damage" [23]. Palmer defines network forensics as "the use of scientifically proven techniques to collect, fuse, identify, examine, correlate, analyse, and document digital evidence from multiple, actively processing and transmitting digital sources for the purpose of uncovering facts,, related to the planned intent, or measured success of unauthorized activities,, meant to disrupt, corrupt, and/or compromise system components,, as well as providing information to assist in response to/ or recovery from these activities" [24].

As mentioned before, monitoring of high-interaction honeypot cannot produce statistical data as easily as in low-interaction honeypots. The reason is that the whole system is available to an attacker and data collection mechanisms are difficult to hide.

In this type of honeypot two major aspects should be taken into account: firstly the path that the attacker used to penetrate the system, secondly an event that will be considered the attack itself.

Monitoring of high-interaction honeypot's activities is related to sophisticated strategies of monitoring of system resources. An attacker has the full access to the system. Therefore it is necessary to take into account the type of operating system and vectors (paths) that are available for attackers and that are used by them. These vectors depend on the operating system of honeypots.

Each attack (or group of attacks) is **evaluated individually**. It is necessary to evaluate data from different resources (Tcpdump, IDS system, operating system of honeypots) and then compose a "mosaic of the attack". It is a complicated process that cannot be fully automated. On the other hand, high-interaction honeypots are designed primarily to detect sophisticated attacks. Thus, it is not expected that the analysis will focus on automated attacks like in low-interaction honeypot. On the other hand, the amount of attacks against high-interaction honeypot is much lower as compared to the low-interaction honeypots, thus making the manual analysis possible. This is confirmed by comparison of number of all connections in the study [7] and results from authors' research on high-interaction honeypots. The number of all connection and the attackers´ countries from high-interaction honeypot is shown in Fig. 2, and combinations of login and password in Fig. 3. Data collection period was one month as mentioned above.

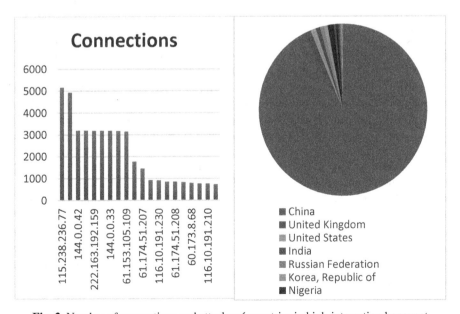

Fig. 2. Number of connections and attackers´ countries in high-interaction honeypot

The evaluation of each attack consists of manual analysis of network traffic, the analysis of IDS logs and a separate output of tools used for monitoring the honeypots. These activities can also be automated (e.g. methods practiced in Honeywall Roo). The definite decision whether the analysed data should be considered as an attack to be investigated is always based on subjective appraisal of honeynet's administrator.

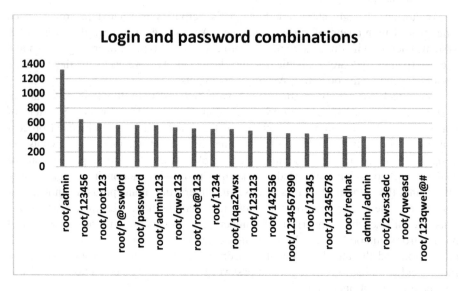

Fig. 3. Combinations of login and password in high-interaction honeypot

In order to eliminate automated attacks, numerous practices are applied. Among others the usage of stronger passwords, usage of operating systems without well-known vulnerabilities and the application of filtering robots' IP addresses using blacklists are considered the most efficient. The administrator of honeypot should be informed (e.g. by a notification) about the fact that an SSH session is opened. In other words, they should be informed about the fact that an attacker penetrated into the honeypot overcoming login. A similar mechanism is also used in HonSSH. The definition of attack also helps administrators in similar situations. An administrator of honeypot should be able to determine the amount of time that would be given to the attacker in the system, and the point when the intervention is necessary. It is useful since there are no 100% implemented countermeasures against possible attacks from the honeypot.

Determination of time of attack (start and end) is closely related to two aspects. The first aspect is security. Start of attack causes an activation of honeypot's security measures (e.g. control of outgoing traffic). The second aspect lies in restoration of honeypots. The system should be renewed from scratch after certain period of being attacked.

6   Proposed Definition

Based on the above outlined approaches, it is necessary to emphasize that high-interaction honeypot needs to be divided into two parts. An attack against the system for the collection, storage, analysis and sending data must be regarded as a real (production) system. Therefore, attack against this system is regarded as attack against the real system.

On the other hand, any attack against the honeypot has to be viewed in the context of further forensic analysis. From this perspective, an attempt to destroy, expose, alter, disable, steal or gain unauthorized access (e.g. scan the ports) or gaining access to the honeypot as such cannot be regarded the attack because it has no value for subsequent analysis. On the contrary, gaining access to the honeypot followed by certain action can be regarded as attack against high-interaction honeypot. An example is unauthorized connect to MySQL database and unauthorized modification of data.

7   Conclusion

The paper focuses on the definition of an attack against to high-interaction server honeypots. As we showed, currently used definitions of attacks are insufficient. Therefore the main focus is given to attacks against this type of honeypot. Based on two approaches the definition of attack was proposed.

In the future research, the definitions of attack should be elaborated. The authors will focus primarily on related problems, such as vector of attack and data collection.

Acknowledgments. Authors would like to express gratitude to their colleagues from Czech chapter of The Honeynet Project for their comments and valuable inputs. The publication was supported by the project "Fuzzy modelling tools for assuring the security of intelligent systems and adaptive searching subject to uncertainty" of the Student Grant Competition Programme of the University of Ostrava (SGS15/PrF/2014).

References

1. Spitzner, L.: The Honeynet Project: Trapping the Hackers. IEEE Security & Privacy, 15–23 (March/April 2004)
2. Dionaea project (2014), http://dionaea.carnivore.it/ (accessed December 1, 2014)
3. HoneyD project (2008), http://www.honeyd.org/ (accessed December 1, 2014)
4. Provos, N., Holz, T.: Virtual Honeypots: From Botnet Tracking to Intrusion Detection. Addison Wesley (2007)
5. Joshi, R.C., Sardana, A.: Honeypots: A New Paradigm to Information Security. Science Publishers, USA (2011)

6. HiHAT project (2007), `http://hihat.sourceforge.net/index.html` (accessed December 1, 2014)

7. Sokol, P., Zuzčák, M., Sochor, T.: Definition of Attack in the Context of Low-Level Interaction Server Honeypots. In: Park, J.J(J.H.), Stojmenovic, I., Jeong, H.Y., Yi, G. (eds.) Computer Science and Its Applications. LNEE, vol. 330, pp. 499–504. Springer, Heidelberg (2015)

8. Rowe, N., Goh, H.: Thwarting cyber-attack reconaissance with inconsistency and deception. In: Proceedings of the 8th IEEE Workshop on Information Assurance, West Point, NY (2007)

9. Briffaut, J., Lalande, J.-F., Toinard, C.: Security and results of a large-scale high-interaction honeypot. Journal of Computers, Special Issue on Security and High Performance Computer Systems 4(5), 395–404 (2009)

10. Oumtanaga, S., Kimou, P., Kevin, K.G.: Specification of a model of honeypot attack based on raised data. World Acad. Sci. Eng. Technol. 23, 59–63 (2006)

11. Jha, S., Sheyner, O., Wing, J.M.: Two Formal Analysis of Attack Graphs. In: Proc. 15th IEEE Computer Security Foundations Workshop, CSFW 2002 (2002)

12. Simmonds, A., Sandilands, P., van Ekert, L.: An ontology for network security attacks. In: Manandhar, S., Austin, J., Desai, U., Oyanagi, Y., Talukder, A.K. (eds.) AACC 2004. LNCS, vol. 3285, pp. 317–323. Springer, Heidelberg (2004)

13. Grudziecki, T., et al.: Proactive detection of security incidents II – Honeypots. European Network and Information Security Agency (2012), `http://www.enisa.europa.eu/activities/cert/support/proactive-detection/proactive-detection-of-security-incidents-II-honeypots/at_download/fullReport` (accessed December 15, 2014)

14. Chamales, G.: The honeywall cd-rom. IEEE Secur. Privacy 2, 77–79 (2004)

15. HonSSH project (2014), `https://code.google.com/p/honssh/` (accessed December 1, 2014)

16. Know Your Enemy: Sebek. A kernel based data capture tool (2003), `http://old.honeynet.org/papers/sebek.pdf` (accessed December 1, 2014)

17. Auditd deamon (2004), `http://linux.die.net/man/8/auditd` (accessed December 1, 2014)

18. Peltier, T.R.: Information Security Risk Analysis, pp. 1–21. Auerbach Publications, CRC Press LLC (2001)

19. Spitzner, L.: The value of honeypots, part one: Definitions and values of honeypots. Security Focus (2001), `http://www.symantec.com/connect/articles/value-honeypots-part-one-definitions-and-values-honeypots` (accessed December 15, 2014)

20. ISO/IEC 27001:2013 Information technology— Security techniques — Information security management systems — Requirements

21. Shirey, R.: Internet Security Glossary, version 2, RFC 4949 (2007), `http://tools.ietf.org/html/rfc4949` (accessed December 1, 2014)

22. Perrin, C.: The CIA Triad (2008), `http://www.techrepublic.com/blog/security/the-cia-triad/488` (accessed December 1, 2014)

23. Stallings, W.: Network Security Essentials: Applications and Standards. Prentice Hall, Upper Saddle River (2000)

24. Chnadran, R.: Network Forensics. In: Spitzner, L. (ed.) Know Your Enemy: Learning about Security Threats, 2nd edn., pp. 281–325. Addison Wesley Professional (2004)

25. Palmer, G.: A Road Map for Digital Forensic Research. In: First Digital Forensic Research Workshop, Utica, New York, pp. 27–30 (2001)

Integrating Constraints in Schematics of Systems

Sabah Al-Fedaghi

Computer Engineering Department, Kuwait University
P.O. Box 5969, Safat, 13060, Kuwait

Abstract. In computer science, a *constraint* represents a restriction related to some element (e.g., a UML class), where it is usually specified by a Boolean expression. Several languages have been developed for constraint specification at the requirements level; still, problems exist, such as the introduction of new notions applied specifically to constraints instead of an attempt to integrate them into the requirements description, e.g., use of Object Constraint Language as an extension of Unified Modeling Language. This paper introduces a new method that integrates constraints into a flow-based diagrammatic conceptual model. Various features of this modeling are applied to several examples from the literature and to a case study that utilizes Role-Based Access Control (RBAC). The results suggest a promising approach to constraints specification as an integral part of the conceptual model of any system.

Keywords: Conceptual model, constraints specification, Diagrams, UML, Object Constraint Language, Unified Modeling Language.

1 Introduction

Over the years, engineering systems have witnessed increasing complexity. To address this issue, a model-based approach has been used to produce coherent system specifications. In model-based engineering, the intended system is depicted graphically and textually at various levels of granularity and complexity [1]. Development of such systems is becoming more and more multidisciplinary, with many stakeholders, including client, architect, engineers, designers, contractors, and consultants, all dependent on each other for information.

In such an approach, Systems Modeling Language (SysML) was created to implement this process [1-4]. It is based on Unified Modeling Language (UML) and, like UML, includes a variety of diagrams and tools with heterogeneous notations and terms, e.g., use cases, blocks, activities, components, parameters, sequences, and so forth. This multiplicity of fragmented representations lacks a central structural notion around which different specification phases can evolve.

Take, for example, the notion of constraints. A *constraint* represents a restriction related to some element (e.g., a UML class or an association path) and is usually specified by a Boolean expression [5]. According to [6], "The reality is that a graphical model, such as a UML class diagram, *isn't sufficient for a precise and unambiguous specification.* You must describe additional constraints about the

© Springer International Publishing Switzerland 2015   165
R. Silhavy et al. (eds.), *Software Engineering in Intelligent Systems*,
Advances in Intelligent Systems and Computing 349, DOI: 10.1007/978-3-319-18473-9_17

objects in the model, constraints that are defined in your supplementary specification... it's complex" (italics added).

In UML modeling, constraints are described via Object Constraint Language (OCL) [7], a textual specification language designed to describe constraints by extending UML. "A fundamental principle of OCL is that a constraint is interpreted on a 'snapshot' of an object configuration. So OCL does not contain any 'commands' to change the system state, but just operations to observe the system state" [8].

The *problem* discussed in this paper is whether a specification language for diagramming is available that would allow us to dispense with these recurrent conceptual add-on expressions at this high level of modeling. The paper proposes a *solution* in an alternative approach called the Flowthing Model (FM) [9-12]. Without loss of generality, the focus here is on the aspect of *constraints specification* in a case study using a diagrammatic language that includes constraints as an integral notion in system descriptions. The *advantage* of this language is the resulting *unified schematic depiction* that reduces mixing of diagrams and English text, as is the case with UML/OCL.

2   Constraints

"Constraints can be a little confusing because of their overlap with business rules and technical requirements" [6]. In engineering, *requirements* are statements in natural language and diagrams that express the functional and nonfunctional predetermined tasks required of the system. *Functionality* here means that the system performs the required work (e.g., sensing and reporting an obstacle), while *nonfunctionality* refers to additional requirements beyond that (e.g., specific minimum distance to obstacle, maximum time needed to reach it, etc.) Accordingly, functional requirements deal with features needed to solve a problem, and nonfunctional requirements refer to additional *constraints* on a solution. In certain situations, it is difficult to distinguish between these types of requirements; for example, a possible functional requirement is that of *Storing data about a task that include its Start and End*. This requirement for storage can be accomplished in an obvious way; however, the value requirement that *end* > *start* seems to be an implicit functional requirement of a consistent system; nevertheless, it appears to be an added constraint on the basic function of storing values.

The notion of differentiating types of constraints has been recognized in many research areas, as in the case of mathematical, physical, logical, space, time, and design constraints. In information systems modeling, *domain constraints* restrict the values that can be represented in terms of length, range, precision, and so forth. In database systems, *referential integrity constraints* require that a value referred to actually exists. This paper focuses on this latter type of constraint, leaving other types for further study. In the paper, a model is demonstrated that integrates all system constraints into one diagram in which all requirements are treated in a uniform way. The next section reviews the modeling tool that will be applied to describe constraints and other requirements. The example developed there is a new contribution.

3   Flowthing Model

The Flowthing Model (FM) is a map of conceptual movement (analogous to the movement of blood through the heart and blood vessels) of things (*flowthings*) existing in a particular state in the flow, for example, in a state of transfer (circulating in blood vessels), or processing (digestion), or creation (oxygenated blood). Goods, people, ideas, data, information, money, and capital move among *spheres* (e.g., places, organizations, machines, cultures) and are flowthings. The movement of flowthings in a regular course of a system is called a *flowsystem*. Fig. 1 shows the general plan of a flowsystem, with five "stages": Create, Release, Transfer, Receive, and Process. Each *stage* has a wide vocabulary of synonyms and concepts:

Create: generate, appear (in the scene), produce, make, …
Transfer: transport, communicate, send, transmit, …
Process: millions of English verbs that change the form of a flowthing without creating a new one, e.g., color, translate, massage, read, …

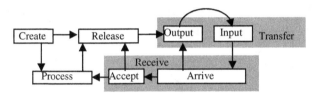

Fig. 1. Flowsystem

FM represents a web of interrelated *flows* that cross boundaries of intersecting and nested *spheres*. A flowthing is defined as "what is created, released, and transferred, arrives, and is accepted and processed" while flowing within and among *spheres*. It has a permanent identity but impermanent form, e.g., the same news translated into different languages. A flowsystem constrains the trajectory of flow of flowthings. A particular flowsystem is the space/time context for *happenings* (e.g., received, released) and existence of flowthings. To simplify the FM diagram, the interior of stages Transfer (input and output) and Receive (arrival and acceptance) are usually not shown.

A flowthing can be in *one and only one* of the FM stages at any one time: Transfer, Receive, Process, Create, or Release – analogous to water being in one of three states in Earth's atmosphere: solid, liquid, or gas. A *state* here is a "transmigration field" of the flowthing that is created, processed, released, transferred, and arrives and is received. It is assumed the flow is "natural" and irreversible, e.g., released flowthings flow only to Transfer, but in a special case, a flow could run "backward." For example, because of a natural disaster, the carrier DHL is not able to transfer parcels from an airport after releasing them for shipment and must truck them back to a warehouse for return to customers.

The exclusiveness of FM stages (i.e., a flowthing cannot be in two stages simultaneously) indicates synchronized change of the flowthing, e.g., a flowthing *cannot be changed in form and sphere simultaneously*. This is a basic systematic property of flowthings. Note the generality of the notion of flow in FM. For example, *creation of* a flowthing is a *flow* (from *nonexistence*, i.e., not currently existing in the system, to *existence*, i.e., appearance in the system).

Initialization, stopping, and continuing of flows occur through *triggering*, a control mechanism. It is the only linkage among elements in FM description besides flow and is indicated by **dashed arrows**. Synchronizations (e.g., join/fork) and logic notions (e.g., and/or) can be superimposed over the basic FM depiction. Note that these mechanisms can be modeled as flowthings.

Example (from [13]): This example includes three entities: Customer, Waiter, and Chef (see Fig. 2). "It can be noted that the responses of the synchronous messages are shown in hashed lines. A synchronous message means that the sender waits until the receiver has finished handling the message, only then does the caller continue. Wherever there is a gap in time line, it shows that there was no real interaction in that time period from the concerned entity" [13]. The sequence diagram of Fig. 2 is "a clear and simple way of depicting to the users, stakeholders, and technical team how the processing of messages will happen and an assessment of this will go a long way in clearing up any gap or misunderstanding at the requirement level" [13].

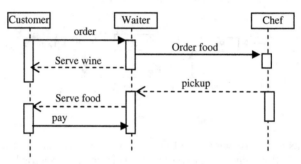

Fig. 2. Sequence diagram (from [13])

Fig. 3 shows the corresponding FM representation. At circle 1 in the figure, the customer generates an order that flows to the waiter (2), then to the chef (3). Delivery of the order to the chef also triggers the waiter to retrieve a bottle of wine from storage (4) and carry it to be received by the customer (6).

The chef (mentally) processes the order (7), which triggers (8) creating the food (9). When the food is prepared and ready to serve, the chef releases it (10) by, say, setting it in a certain location and then ringing a small bell, creating a signal (11) that is heard by the waiter (12).

Accordingly, two events – delivering the wine to the customer (6) and hearing the chef's signal (12) – are modeled by the classical join bar (13), and these trigger (14) the waiter to pick up the food (15) and deliver it to the customer (16). When the customer finishes eating (17), he or she is triggered (18) to produce (create) payment that flows to the waiter.

As can be seen, we had to add several entities (e.g., the chef's signal) to fill the gaps in a model of the sequence of events. Note that Fig. 3 is completely conceptual and uses no technical terms, making it useful as a descriptive map of the modeled situation that can be shared by users, stakeholders, and a technical team.

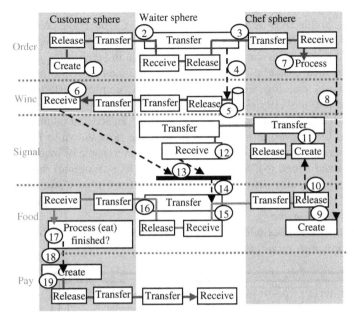

Fig. 3. FM representation

While the figure might look more complicated than the sequence diagram, it is more complete (e.g., includes the chef's signaling process) and contains more details (opens the black box within the processing rectangles). It is actually a simple diagram since the same flowsystems stages are utilized repeatedly.

4   Sample FM-Based Constraints

4.1   Class Ownership

Fakhroutdinov [5] discusses a constraint that applies to two elements, where the constraint is shown as a dashed line between the elements, labeled by the constraint string in curly braces (Fig. 4). Typically, a constraint is formulated at the level of UML classes, but its semantics are applied at the level of objects [7].

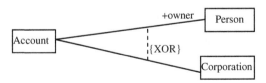

Fig. 4. Constraint: Account owner is either Person or Corporation

FM conceptualizes the involved relationship (edges) in terms of flow. Semantically, in UML the involved constraint is not that *Account owner is either Person or Corporation*; rather it is that *Class Account owner is either Class Person or Class Corporation* – but how can one class own another class?

In FM, the semantics of the constraint are *(Each) instance (a flowthing) of Owner in Account is a flowthing either of Person or of Corporation.* We assume that flowthings of Owner or Corporation are names of that Owner and Corporation. Accordingly, this constraint can be expressed as shown in Fig. 5.

To illustrate such a description, picture the information system of a bank. A new customer requests opening an account. The Bank employee starts by *creating* information about the customer: name, address, telephone, age, etc. Then, the employee constructs a new account, includes creating an account number, an account owner, etc. To fill *account owner*, he/she triggers the flow shown in Fig. 5: *Release* name of that person (circle 1) to flow to XOR (2), then to name of owner (4) where it is *received* (4) and stored (5). XOR is a flowsystem (e.g., program) that ensures that values of the name of owner come from either Person or Corporation. It also *processes* that value to ensure it is a valid value for the name of the owner, e.g., format, length of string, etc.

Note that in such a description the constraint needs no special treatment; it can be modeled like any other requirement.

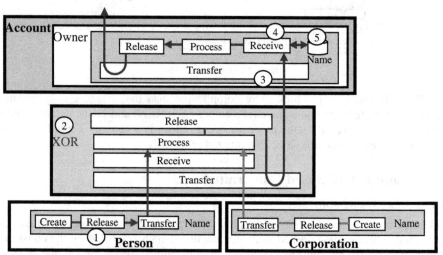

Fig. 5. FM representation of Constraint: Account owner is either Person or Corporation

4.2   Across Attributes Constraints

According to Hußmann [8], invariants constraint is a "condition which always holds" such as, using the UML class diagram: *Meeting inv: end > start.* Fig. 6 shows the FM representation of the constraint, with three spheres: *start, end,* and the process "greater than". A value such as *start* either is created (circle 1) or flows from outside the sphere (2). If it is created (e.g., input from keyboard) then it is immediately released to > (3 and 4), simultaneously triggering the release of current value of *end* to > (5 and 6). In >, the two flowthings are processed (7); if *end* is not greater than *start*, appropriate action is triggered. Similar activities are carried out when *end* is created or received.

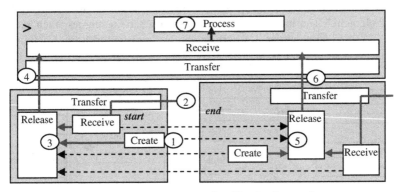

Fig. 6. FM representation of the constraint: *Meeting inv: end > start*

5   Applying FM to Access Control Constraints

Many types of access control methodologies are available, including Discretionary Access Control, Mandatory Access Control, and Role-Based Access Control (RBAC). RBAC introduced the notion of *role* as an intermediate level between users and permissions, where permissions are assigned to specific roles. A subject (e.g., user, business application) is the thing that attempts to access a resource (object). For specifying RBAC policies, several security modeling methodologies have been proposed such as SecureUML [14] and UMLsec [15]. SecureUML "is basically the language of RBAC extended with authorization constraints that are expressed in Object Constraint Language (OCL)" [16].

Cenys et al. [16] utilize SecureUML with the aim of designing RBAC policies. To illustrate their approach, we use a modified version of their example, simplified from the original, of a brokerage corporation application. The system RBAC policies include the following [16]:

A. Roles: {broker, accountant, consultant, customer service representative, internal auditor, broker department manager, customer service department manager};
B. Permissions assigned to roles:
    B. 1 The broker can modify trade data, execute orders, and confirm deals;
    B.2 The accountant can generate reports;
    B.3 The customer service representative can read information related to customers;
    B.4 The consultant can perform select, insert, update, and delete operations on the customer database;
    B.5 The internal auditor has read-only access to all resources.

The RBAC-based language SecureUML is used to model this system, as partially shown in Fig. 7. The figure includes a heterogeneous mix of notations: boxes (module and moduleGroup), submodules of relationships (directed edges), relationships (undirected edges), decisions (diamond), and constraints (dashed arrows). This figure will be compared alongside a maplike representation constructed by the principles of FM (Fig 12).

According to Cenys et al. [13], *module* and *moduleGroup* in the figure "extend *resource* element." The module is a unit "which uses the resources, such as database objects, files, and applications." A corresponding simplified FM representation of the same process is shown in Fig. 8.

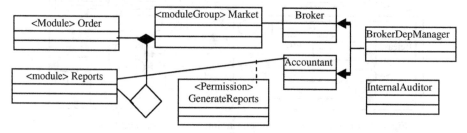

Fig. 7. Specific system modeled with SecureUML (partial view, from [13])

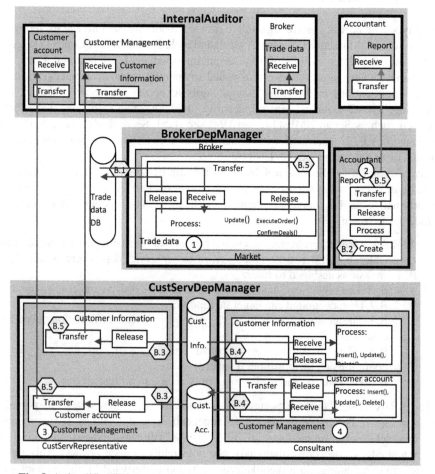

Fig. 8. A simplified FM description corresponding to SecureUML modeling in Fig. 7

Fig. 8 comprises three spheres: BrokerDepManager, InternalAuditor, and CustServDepManager. These correspond to the three nodes on the right in the original SecureUML diagram in Fig. 7. We compare the way permissions are assigned to roles in SecureUML to the way they are assigned in FM. Note that the FM specification enforces the constraints as an integral part of its design.

In Fig. 8, the labels in hexagons refer to the following constraints:

Constraint B.1: *The broker can modify trade data, execute orders, and confirm deals*. For the sake of simplicity, we draw only the flow of *trade data*. In the trade data flowsystem (circle 1), the broker can transfer, receive, and process trade data. Hexagon B.1 represents a guard that controls communication between the trade data process and the trade data database.

Note that B.1 is located at the conjunction between the broker and the database (DB), a natural way of representing permission, in contrast to the boxes of "permissions" in Fig. 7. Thus, FM looks like a city traffic map with constraints such as speed limits superimposed over it. The constraint can be implemented as an independent flowsystem or as part of the flows of two communicating parties.

Constraint B.2: *The accountant can generate reports*. In the Report flowsystem (circle 2), reports are generated. Of course, the BrokerDepManager can access these reports because it contains the subsphere *Consultant*. The assumption here is that Report can access some database (not shown). The link between sources of relevant information for reports and creation of reports is not specified in the SecureUML diagram. In FM it is possible to add a database flowsystem connecting the consultant and, say, broker spheres.

Constraint B.3: *The customer service representative can read information related to customers*. The direction of the rows coming from the involved database indicate reading from the database. Of course, Customer Information and Customer account are in the subspheres of CustServDepManager (circle 3).

Constraint B.4: *The internal auditor has read-only access to all resources*. The internal auditor receives (but does not process) flows of Customer account, Customer Information, Reports, and Orders.

6   Conclusion

This paper explores a way to integrate constraints in a new flow-based diagrammatic representation. Various features of constraints modeling are applied to several examples. The results suggest a promising approach to constraints specification as an integral part of the conceptual model of any system. Future work will investigate representing different types of constraints, such as timing.

References

1. Hansen, F.O.: SysML – a modeling language for systems engineering [slides] (2010), http://staff.iha.dk/foh/Foredrag/SysML-SystemEngineering-DSFD-15-03-2010.pdf
2. Friedenthal, S., Moore, A., Steiner, R.: A Practical Guide to SysML: The Systems Modeling Language, 2nd edn. Elsevier (2011) ISBN 0123852064, 9780123852069
3. Friedenthal, S., Moore, A., Steiner, R.: A Practical Guide to SysML: The Systems Modeling Language. Elsevier (2008)
4. Weilkiens, T.: Systems Engineering with SysML/UML: Modeling, Analysis, Design. Elsevier (2007)
5. Fakhroutdinov, K.: UML Constraint, Blo, @014 (Access), http://www.uml-diagrams.org/constraint.html
6. Ambler, S.W.: Agile Modeling (2003-2014), http://www.agilemodeling.com/artifacts/constraint.htm
7. Warmer, J., Kleppe, A.: The Object Constraint Language – Precise Modeling with UML. Addison-Wesley (1999)
8. Hußmann, H.: Technische Universität Dresden, Lecture Notes (2014) (access), http://www-st.inf.tu-dresden.de/fs/lectnotes/fss5a.pdf
9. Al-Fedaghi, S.: Pure Conceptualization of Computer Programming Instructions. International Journal of Advancements in Computing Technology 3(9), 302–313 (2011)
10. Al-Fedaghi, S.: Schematizing Proofs based on Flow of Truth Values in Logic. In: IEEE International Conference on Systems, Man, and Cybernetics (IEEE SMC 2013), Manchester, UK, October 13-16 (2013)
11. Al-Fedaghi, S.: Flow-based Enterprise Process Modeling. International Journal of Database Theory and Application 6(3), 59–70 (2013)
12. Al-Fedaghi, S.: Toward Flow-Based Semantics of Activities. International Journal of Software Engineering and Its Applications 7(2), 171–182 (2013)
13. BusinessAnalystFaq,com. What is sequence diagram? http://www.businessanalystfaq.com/whatissequencediagram.htm
14. Lodderstedt, T., Basin, D., Doser, J.: SecureUML: A UML-Based Modeling Language for Model-Driven Security. In: Jézéquel, J.-M., Hussmann, H., Cook, S. (eds.) UML 2002. LNCS, vol. 2460, pp. 426–441. Springer, Heidelberg (2002)
15. Jürjens, J., Shabalin, P.: Tools for Secure Systems Development with UML. International Journal on Software Tools for Technology Transfer (STTT) 9(5-6), 527–544 (2004), doi:10.1007/s10009-007-0048-8
16. Cenys, A., Normantas, Radvilavicius, L.: Designing Role-based Access Control Policies with UML. Journal of Engineering Science and Technology Review 2(1), 48–50 (2009)

S-DILS: Framework for Secured Digital Interaction Learning System Using *keccak*

D. Pratiba and G. Shobha

R V College of Engineering, VTU, Bangalore, India
{pratibad,shobhag}@rvce.edu.in

Abstract. Authentication plays a critical role in network security. After reviewing the prior studies, it is found that authentication system focusing on Learning Management System (LMS) based on cloud environment is very few. Majority of the prior security techniques applies conventional cryptographic hash function where enough robustness and resiliency towards various vulnerable attacks on cloud environment is substantially missing. Therefore, this paper presents a novel, simple, and yet cost effective authentication technique that uses the potential characteristics of recently launched keccak protocol by NIST for the first time in securing the accessibility of users in interactive digital library system. Hence, we coined a term S-DILS i.e. secured-Digital Interactive Learning System. The outcome of the system is compared with existing hash protocols to find that S-DILS excels far better in every respect of security.

Keywords: Authentication, Keccak Protocol, Learning Management System.

1 Introduction

E-learning is one of the major component of Learning Management System (LMS), which demands quality education system to be shared using various ICT medium [1]. In e-learning system of the present era, there are various service providers as well as web-based learning portals that offer various interactive learning resources by sharing various valuable medium. Certain learning systems are paid and others are free [2]. One of the important facts, apart from theoretical discussion of e-learning, is the intellectual property as well as the resources being shared between the instructors and students, students and students etc. It is important to understand why to value much in security mechanism of modern LMS. In modern LMS system [3], it is seen that an instructor creates a standard course-materials for all of their online clients (students) and in many cases; such materials are sometime customized depending on the grasping capability of the student. Usually the existences of such course materials are almost nil in market and hence such resources are highly valuable. Not only this, the database of such LMS system consist of personal information of both students as well as instructors, which are highly vulnerable for online attacks as usually such database are stored in normal server or server with less efficient security protocol. Hence, one way to resist such illegitimate access is to have strong and intelligent authentication

© Springer International Publishing Switzerland 2015
R. Silhavy et al. (eds.), *Software Engineering in Intelligent Systems*,
Advances in Intelligent Systems and Computing 349, DOI: 10.1007/978-3-319-18473-9_18

mechanism. In existing system, such authentication mechanism is done by providing standard user ID and static password to users, which are highly vulnerable and susceptible to cryptanalysis. Hence, a strong cryptographic mechanism is required to strength the authentication system in existing LMS. With the evolution of cloud computing, the problems of storage have drastically reduced. Hence, there is higher chances that with the increasing problems of traffic of such LMS application, service providers may migrate their web-based application entirely into cloud as they can save unnecessary expenditure as well as it is possible to provide effective accessibility of the user. However, a closer look into various cloud based authentication system is still found to use conventional cryptographic hash function (e.g. SHA-1 and SHA-2). Although, there are less reported and less documented study on vulnerabilities of SHA-1 and SHA-2, but still various studies have already proved that such algorithm have potential security loopholes [3]. There are various reported cases of illegitimate access or intrusion even in cloud based applications and services. Hence, it can be said that hosting the LMS applications on cloud can just address the storage issues but cannot address the security breaches that are quite common.

Various studies have also evolved out where researchers have introduced sophisticated cryptographic technique to safeguard the authentication policies. However, it is strongly felt that none of the techniques till date have addressed the authentication problems of LMS system in cloud environment. Hence, the prime contribution of this paper is to propose a novel, robust, and highly cost effective authentication mechanism that ensure optimal and fail-proof security. In section 2 we give an overview of related work which identifies all the major research work being done in this area. Section 3 highlights about the identified problem of the proposed study followed by proposed study in Section 4. Design methodology of the proposed authentication system is discussed in Section 5 and in-depth discussion of proposed algorithm is discussed in Section 6. Section 7 discusses about the benchmarked outcomes of the proposed authentication mechanism followed by concluding remarks in Section 8.

2   Related Work

After collecting the consolidated data pertaining to authentication mechanism being adopted in enterprise application, it was found that majority of the systems uses multi factor authentication system in various forms. In this system, the user gets authenticated themselves multiple number of times, where the next level of authentication surfaces only when the first level of authentication succeeds. Significant studies in this direction was first seen in the work carried out by Long and Blumenthal [4] who have introduced the password generator technique to add the manageability to volatile password based authentication while retaining the computation and communication efficiency. However, the discussed work of [4] lacks potential benchmarking.

Usage of mobile device was seen in the work of Hallsteinsen and Jorstad [5] for performing authentication. Certain studies [6] have also introduced Smart cards for security checks for user. The work was mainly focused on mitigating replay attack and uses the one-time password technique and one-way hash functions to reduce the processing overhead. Unlike the weak-password approach, strong password authentication is mostly based on the one-way hash function [6] and exclusive-OR operations (XOR). Enhancement on the one-time password scheme using chaos theory was also seen in literature [7]. Chaos, as characterized by the properties of ergodicity, unpredictability, and sensitivity to parameter and initial condition, is analogous to the requirements of authentication. The chaotic sequence is taken as identity mark sequence, and it is impossible for anyone else to pretend others' identity mark sequence, above which just matches the ideas of one-time password. However, the problem of this framework is similar to that of [4].

Alghathbar and Mahmoud [8], has attempted to alleviate the problem of shoulder surfing or eves dropping by making the replay of a password useless by designing a noisy password scheme.

In majority of above discussed schemes, a user uses different passwords generated by token or software every time they login onto the server, and it is hard to calculate the next valid password from the previous passwords. Once a password is used, it will not be used for a second time. One such attempt was made by Liao et al. [9] to mitigate this problem.

Davaanaym et al, [10] proposed a web/mobile authentication system with one time password that is both secure and highly usable, based on multifactor authentication approach called as Ping Pong based one-time based authentication. However, the study and its associated problems are more or less related to work done by [4] and [5]. Studies to overcome such similar approach by new and unique approach was found in the work carried out by Qing et al. [11] who have proposed a new self-updating hash chain based linear partition-combination algorithm which can share data into m-parts and any part can't leak any information of data.

Li et al. [12] re-evaluate the security of the predicate based authentication service (PAS). The work shows that PAS is insecure against both brute force attack and a probabilistic attack. Similar pattern of studies was again found in [13], which applies almost similar approach like the work in [10][4][5][14]. Among all the approach, uniqueness is found using Single Sign On based techniques. Kim et al. [15] proposed a framework for single sign on for next generation networks. The author has significant raised the compatibility issues of conventional one-time password. A similar direction of studies was again found to be carried out for mitigating one-time password using time and location based technique [16].

Srivastava et al. [17] have addressed the problem of port knocking using advanced cryptography. It was also seen that many organizations generate the OTP and send it to the user through SMS [4][5][16]. However, for most existing two factor authentication schemes, apart from the overhead, the user has to bring an extra token or mobile phone for the authentication. And many two-factor authentication schemes are still vulnerable to the Perfect-Man-In-The-Middle attack. Ren and Wu [18] present a secure dynamic authentication scheme which solves the problems above. Usage of Biometrics based approach was seen in the work of Moon et al. [19]. Recent study of authentication on cloud environment using cellular automata in cloud computing was witnessed in literature of Shin et al. [20]. Moreover with the upcoming trend of pervasive computing, various security challenges rises up owing to multiple access platforms that will be required to be mitigated.

3   Problem Identification

With the increasing adoption of cloud as storage, the interactive learning management system is boosted for its utility and performance to give better pervasive services to its users. However, owing to pervasive nature of such services, authentication is always a bigger challenge. The set of problems that was identified for proposed study are as follows:

- Location of the data is always the critical factor for cloud security.
- The trust factor in LMS is exactly not a technical problem but it is the most critical soft factor that is governed by cloud security where still the encryptions are carried out using conventional SHA protocol.
- Existing security as well as authentication mechanism do control the access rights but do not implicate any security policies for the transactional data. In the area of LMS, there exists various transactional data in forms of course materials that are required to be transmitted from one to another end without any delay.
- The existing cloud service provider adopts various cryptographic protocols as well as various security patches to ensure the optimal authentication procedure. Hence, it can be seen that majority of the problems surfaces out from the fragile authentication system which is developed using existing security protocols or cryptographic hash function. The proposed system, therefore, identifies this problem and introduces a novel cryptographic hash function that was never attempted before in the area of secure authentication system in cloud environment by addressing the above mentioned issues. The paper also considers various actors to ensure better applicability of security authentication.

4   Proposed System

The prime purpose of the proposed system is to introduce a Secured Digital Interactive Learning System (S-DILS). The novelty in the secure authentication system is done using newly launched keccak protocol. Keccak is chosen as conventional SHA algorithms (like SHA1 or SHA2) use the same technique (Merkle-Damgard) [21]. The design of the S-DILS using keccak should therefore have following requirements e.g. i) the candidate hash function should use computational resources as less as possible when hashing massive amount of data, and ii) the keccak protocol should maintain extensive safety factor by ensuring the usage of 4 hash size of prior version (SHA-2) i.e. 224-bits, 256-bits, 384-bits, and 512-bits along with extensive hash sizes depending on security needs of applications. The hash function of keccak protocol considers hash size (n), message text (M), its size (N), and hash variable (H). The simple schema of keccak protocol is shown in Fig.1

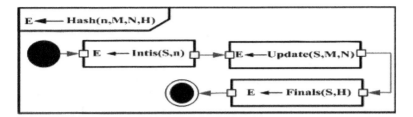

Fig. 1. Keccak Working Principle

S-DILS considers adoption of cloud environment where the knowledge repositories resides and access from the users are granted. Hence, the proposed system targets to provide fail-proof security to the intellectual properties of LMS by safeguarding data residing on cloud servers.

5   Design Methodology

The system considers developing an interface specific to the adopted digital interactive learning system, where it consist of three important modules (or users) e.g. students, instructors, and policy-makers. A generalized interface of S-DILS is designed which will be accessed by all the three users. With an aid of Fig.1, the proposed design methodology for S-DILS is illustrated. The discussions of the design pattern of the individual modules are as follows:

1. Student's Interface:
An interface is developed for meeting the requirements of the student in our proposed S-DILS. In order to start using student's specific interface in the generalized interface of the S-DILS, the student will be required to enroll themselves and the student will be provided with unique user-id and static password. The details of these user id and static password are stored in cloud servers that are assumed to be protected by existing firewall. The multi-stage authentication system will consist of origination of multiple interfaces that will provoke student to re-verify themselves until the last defined stage of verification process. Once verified, a specific interface has been created that enables the student to have privilege of intelligent searching, sharing and resource utilization accessibility of LMS.

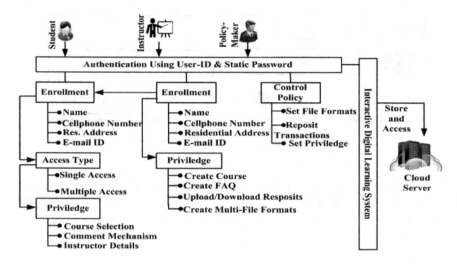

Fig. 2. Proposed Design Methodologies

2. Instructor's Interface:

The instructor is an individual in the proposed S-DILS who impart, share, and assess students using the latest ICT approaches. The prime privilege of instructor is that they can create their own course subject and necessary materials required to be pursued by the student. Hence, an instructor have the rights to upload syllabus or course materials in the form of Microsoft Word, Excel, PowerPoint presentation, PDF, etc. The system also provides flexibility for the instructor to upload Frequently Asked Question (FAQ). The intelligent soft computing approach of analytics and data mining of the LMS is beyond the scope of this paper as here authentication mechanism is emphasized.

3. Policy Maker Interface

As the entire system runs on cloud environment and posse's higher degree of heterogeneity in the users as well as the resources, hence it is quite imperative to implicate superior and skillful administration to the applications. To be very specific, the design of policy makers are endowed with special rights and privilege that will assist certain controls over the users.

6   S-DILS Authentication Mechanism

The authentication mechanism process for the proposed system is applicable for only student and instructor and policy maker is free from getting them verified. As shown in Fig.3, the study considers static password of the user as seed, which will be subjected to the most recently launch keccak protocol in cryptographic hash function in 2014 [22]. Keccak uses the principle of sponge construction where the block of message is basically XORed into a subset of states which is then subjected to mathematical transformation as a whole. Usually the state will consist of multi-dimensional array of 64 bit of words. A sponge function is a class of algorithm with finite internal state considering input of bit stream of any length and generates bit stream output of anticipated length. The design of the sponge function is based on three factors viz. i) state memory containing specific size of bits, ii) a mathematical function of fixed length to perform permutation and transformation of the state memory, and iii) a padding function. The padding function is incorporated with capability of appending sufficient data bits to the string input so that the size of the padded input becomes the multiple of the bitrate. Hence, adoption of keccak protocol ensures faster hardware implementation as compared to any other cryptographic hash function.

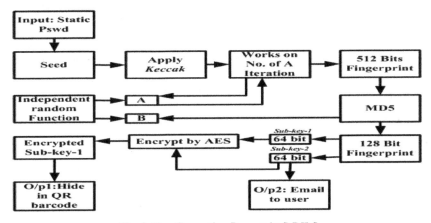

Fig. 3. Key Generation Process in S-DILS

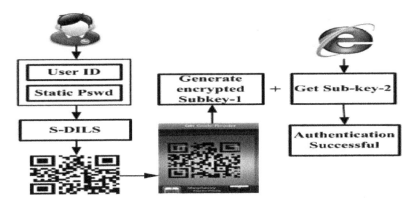

Fig. 4. Key Authentication Process in S-DILS

Fig.4 exhibits the key generation process of proposed S-DILS. In this scheme, the static password of the user is considered as seed that is subjected to keccak protocol. The system also adopts a novel policy of independent random function to generate two random numbers (A and B). The keccak protocol is subjected to A-iteration for the seed i.e. keccak(seed)A. This operation generates only 512 bits of fingerprint as a returned value, which is again subjected to MD5 algorithm. The MD5 algorithm is then applied on B-iteration for the 512 bit of fingerprint to generate only 128 bits of fingerprint. In order to make the security keys light weighted, the algorithm splits the 128 bits into two sub-keys e.g. sub-key-1 and sub-key-2. Both the sub-keys posses the size of 64 bits equally. The first sub-key (sub-key-1) is subjected to AES algorithm for performing encryption, whereas the second sub-key (sub-key-2) is transmitted to the registered e-mail ID of the user. It is to be noted that for optimal security, the proposed S-DILS will use sub-key-2 to perform decryption process towards encrypted sub-key-1. For further security measures, the proposed S-DILS hide the sub-key-1 inside a QR-barcode. The QR barcode is displayed in the interface of the specific user, which will require certain scanner to read it. Hence, the output of the algorithm is basically two fold i.e. i) generation of QR barcode and ii) sub-key-2. The description can be also seen in algorithm-1 as follows:-

Algorithm-1: Key generation in S-DILS

Input: Static password

Output: Generated Keys

Start

1. Initialize Static pswd→Seed

2. Define Independent random function

   $f(x)=[A, B]$

3. Apply keccak protocol on seed

   $keccak(Seed)^A$

4. Generate 512 bits fingerprint

5. Apply MD5

   $MD5 (512 \text{ bits of fingerprint})^B$

6. Generate 128 bits of fingerprint

7. Split 128bit into two sub-keys (S_1 & S_2)

   Split $(128 \text{ bits}) \rightarrow S_1(64 \text{ bits}), S_2(64 \text{ bits})$

8. Apply AES on S_1

   $AES (S1) \rightarrow S_1'$

9. Generate Encrypted keys

   $Enc (S_1')$

10. Transmit S_2 to registered email of user.

11. Hide $Enc (S_1')$ in QR-barcode

End

After the keys are generated, the user will attempt to access their priviledge application using their respective user ID and static password. After feeding the initial credentials, the QR bar code will be displayed on the client-side application, which is required to be read by QR barcode reader. We have used android based QR-barcode reader to read it using our own custom built android application. Once the QR barcode is read, the interface of the mobile device will display the encrypted sub-key-1 that was hidden inside the QR-barcode. The system will instantly prompt the application to enter sub-key-2 from the registered email ID of the user. After the sub-key-2 is feed to the mobile device, the authentication is rendered successful. Hence, the proposed S-DILS ensures optimal security.

Algorithm-2: Key Authentication Mechanism

Input: User ID, static pswd

Output: Successful authentication

Start:

1. Feed user Id and static pswd.

2. Apply Algorithm-1

3. Display the QR-Barcode

4. Extract Enc (S_1')

5. Get the S_2 (64 bits) from registered email ID.

6. Validate using S_2 (64 bits)

End

---

7 Result Analysis

For the purpose of inclusion of networking and security concepts over highly distributed cloud environment, the proposed algorithm of security was designed using Java. S-DILS was implemented on 125 normal desktop pc with 32 bit windows OS to finally confirm the outcomes of the study. In order to do so, a generalized interface for learning management system has been developed with consideration of MySQL as database system to store the static password and the user IDs. Owing to higher degree of novelty of keccak protocol as the security standards, it was quite difficult to benchmark the outcomes of S-DILS. In order to understand the computational complexity of the proposed S-DILS, its outcome is effectively studied and compared with other standard security authentication protocols like SHA2. The chosen parameters are as below:

- **S-DILS Processing Time** (A_T): This parameter records the entire processing time to perform encryption for the user fed data and generation of the keys.
- **Component Processing Time** (C_T): This parameter records the time after A_T to relay the services.

- **Communication Latency (C$_L$):** This parameter records the latency observed during the authentication process owing to networking and server loads.
- **Gross Time (G$_T$):** It is the total time required for successful implementation of the proposed security technique to safely authenticate the user. It can be represented as:

$$G_T = A_T + C_T \tag{1}$$

However, for better evaluation, the proposed system also considers latency in analyzing the time-based complexity (T$_{com}$) of S-DILS as

$$T_{com} = G_T + C_L \tag{2}$$

For optimal analysis of S-DILS complexity, the proposed system considers monitoring the variation of A$_T$, C$_T$, G$_T$, and T$_{com}$ as significant function of number of users and size of user input respectively.

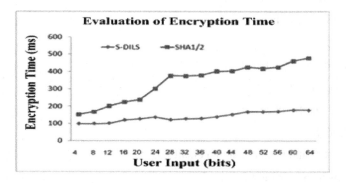

Fig. 5. Analysis of Encryption Time

Fig.5 shows that algorithm capability where the x-axis represents the size of the user input with respect to bits and y-axis represent encryption time by A$_T$ measured in milliseconds. The curve is shown with exponential variation with size of the user inputs. This pattern eventually indicates that maximizing the size of the user input have slight increase in encryption time (from 17.5 ms to 21 ms), which is quite negligible.

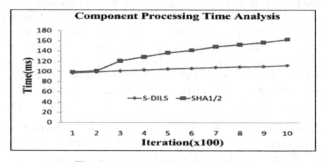

Fig. 6. Analysis of Component Time

Fig.6 shows the variation of C_T with respect to number of user for both proposed system using keccak and existing system using SHA2. It can be seen that curve for S-DILS show a linear ascent with increase in number of users, while slowly it is linearly enhancing almost uniformly. The outcome of Fig. 5 also interprets that both C_T and G_T are quite independent of size as well as user input. This is one of the most important outcomes that show better scalability of S-DILS.

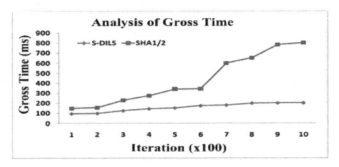

Fig. 7. Analysis of Gross Time

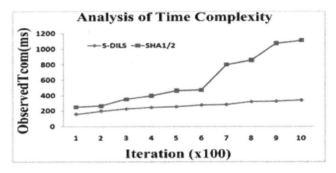

Fig. 8. Analysis of Time Complexity

Fig.7 and Fig.8 exhibits the outcome of time complexity (T_{com}) factor, where it can be seen that gross time (without latency) is superior for proposed S-DILS while T_{com} (with latency) is also found to be superior as compared to conventional SHA1/2 protocols. The prime reason behind this is the keccak protocol ensures the generation of lightweight keys for which reason it doesn't have much adverse effective in processing the algorithm as well as entire application usage with massive data on cloud. Hence, it can be said that S-DILS is highly compatible as well as synchronizable with authentication mechanism on cloud for proposed interactive digital library system.

8   Conclusion

The proposed paper has introduced a novel technique called as S-DILS which not only supports highly interactive digital library system but also ensures optimal security to highest standard. With the inclusion of the most recently launched keccak protocol, the security standards of the proposed S-DILS have become highly resilient against various types of intrusion activities on internet as well as on the applications hosted on cloud. One of the uniqueness in this proposed technique is key-generation process. It can be supported by following statements e.g. i) an intruder cannot access two sub-keys simultaneously, ii) an intruder cannot have access on the registered handheld device of the user (as it is a physical device). Interestingly, if an intruder can possible access subkey-1, that is encrypted with AES and hidden in QR code, it is not possible to decrypt it as it has a dependency on sub-key-2, which is passed on to the registered email ID of the user. However, even if to some extent, it is assumed that email ID of the user is compromised, but the proposed system uses hardware specific mobile device that can scan the QR reader and can generate the encrypted sub-key, which is not at all possible from the intruder's side. Moreover, none of the keys are stored at any point of networking peripherals of cloud environment and becomes volatile after usage. Hence, with the upcoming application of learning management system.

References

1. Kats, Y.: Learning Management System Technologies and Software Solutions for Online Teaching: Tools and Applications: Tools and Applications. Technology & Engineering, 486 (2010)
2. Kats, Y.: Learning Management Systems and Instructional Design: Best Practices in Online Education. Education, 367 (2013)
3. Schreiber, T.: Amazon Web Services Security. Seminar Thesis of Ruhr Universidad Bochum (2011)
4. Long, M., Blumenthal, U.: Manageable one-time password for consumer applications. In: International Conference in Consumer Electronics, Digest of Technical Papers, pp. 1–2 (2007)
5. Hallsteinsen, S., Jorstad, I.: Using the mobile phone as a security token for unified authentication. In: Second International Conference in Systems and Networks Communications, p. 68 (2007)
6. Jeong, J., Chung, M.Y., Choo, H.: Integrated OTP-based User Authentication Scheme Using Smart Cards in Home Networks. In: Proceedings of the 41st Hawaii International Conference on System Sciences (2008)
7. Lifen, L., Zhu, Y.: Authentication Scheme for Substation Information Security Based on Chaotic Theory. In: Asia-Pacific Power and Energy Engineering Conference, APPEEC, pp. 1–3 (2009)
8. Alghathbarand, K., Mahmoud, H.A.: Noisy password scheme: A new one time password system. In: Canadian Conference in Electrical and Computer Engineering, pp. 841–846 (2009)
9. Liao, S., Zhang, Q., Chen, C., Dai, Y.: A Unidirectional One-Time Password Authentication Scheme without Counter Desynchronization. In: International Colloquium on Computing, Communication, Control, and Management (2009)
10. Davaanaym, B., Lee, Y.S., Lee, H.J., Lee, S.G., Lim, H.T.: A Ping Pong based One-Time-Passwords authentication system. In: Fifth International Joint Conference on INC, IMS and IDC (2009)

11. M.-Qing, Z., Bin, D., X.-Yuan, Y.: A New Self-Updating Hash Chain Scheme. In: International Conference on Computational Intelligence and Security (2009)

12. Li, S., Asghar, H.J., Pieprzyk, J.,, Sadeghi, A.-R., Schmitz, R., Wang, H.: On the Security of PAS (Predicate-based Authentication Service). In: Annual Computer Security Applications Conference (2009)

13. Fan, Y.T., Su, G.P.: Design of Two-Way One-Time-Password Authentication Scheme Based on True Random Numbers. In: Second International Workshop on Computer Science and Engineering (2009)

14. Eldefrawy, M.H., Alghathbar, K., Khan, M.K.: OTP-Based Two-Factor Authentication Using Mobile Phones. In: Eighth International Conference on Information Technology: New Generations (2011)

15. Kim, K., Jo, S., Lee, H., Ryu, W.: Implementation for Federated Single Sign-on based on Network Identity. In: 6th International Conference Networked Computing (INC), pp. 1–3 (2010)

16. Hsieh, W.-B., Leu, J.-S.: Design of a time and location based One-Time Password authentication scheme. In: 7th International Wireless Communications and Mobile Computing Conference (IWCMC), pp. 201–206 (2011)

17. Srivastava, V., Keshri, A.K., Roy, A.D., Chaurasiya, V.K., Gupta, R.: Advanced port knocking authentication scheme with QRC using AES. In: International Conference in Emerging Trends in Networks and Computer Communications (ETNCC), pp. 159–163 (2011)

18. Ren, X., Wu, X.-W.: A Novel Dynamic User Authentication Scheme. In: International Symposium on Communications and Information Technologies (2012)

19. Moon, K.Y., Moon, D., Yoo, J.-H., Cho, H.-S.: Biometrics Information Protection using Fuzzy Vault Scheme. In: Eighth International Conference on Signal Image Technology and Internet Based Systems (2012)

20. Shin, S.-H., Kim, D.-H., Yoo, K.-Y.: A Lightweight Multi-User Authentication Scheme based on Cellular Automata in Cloud Environment. In: IEEE 1st International Conference on Cloud Networking, CLOUDNET (2012)

21. Boutin, C.: NIST Selects Winner of Secure Hash Algorithm (SHA-3) Competition (2012), http://www.nist.gov/itl/csd/sha-100212.cfm

22. Bertoni, G., Daemen, J., Peeters, M., Assche, G.V.: The Road from Panama to Keccak via RadioGat, Schloss Dagstuhl - Leibniz-Zentrumfuer Informatik, Germany (2009), http://keccak.noekeon.org/

Artificial Immune System Based Web Page Classification

Aytuğ Onan

Celal Bayar University, Faculty of Engineering,
Department of Computer Engineering,
Manisa, 45140, Turkey
aytug.onan@cbu.edu.tr

Abstract. Automated classification of web pages is an important research direction in web mining, which aims to construct a classification model that can classify new instances based on labeled web documents. Machine learning algorithms are adapted to textual classification problems, including web document classification. Artificial immune systems are a branch of computational intelligence inspired by biological immune systems which is utilized to solve a variety of computational problems, including classification. This paper examines the effectiveness and suitability of artificial immune system based approaches for web page classification. Hence, two artificial immune system based classification algorithms, namely Immunos-1 and Immunos-99 algorithms are compared to two standard machine learning techniques, namely C4.5 decision tree classifier and Naïve Bayes classification. The algorithms are experimentally evaluated on 50 data sets obtained from DMOZ (Open Directory Project). The experimental results indicate that artificial immune based systems achieve higher predictive performance for web page classification.

Keywords: Artificial immune systems, Immunos-1, Immunos-99, Web document classification.

1 Introduction

The advent of World-Wide Web (WWW) has enabled to efficiently disseminate information. Today, Internet users are exposed to a large amount of online content. In order to deal appropriately with online content, it becomes essential to use intelligent software agents responsible for finding, sorting and filtering the available content [1].

Web mining is the process of discovering useful/meaningful patterns from Web documents and services with the application of data mining and machine learning techniques. Web mining is an interdisciplinary research area bringing together the methods of several disciplines, such as data mining, information retrieval, machine learning and natural language processing. Compared to data mining, web mining is a more challenging research direction. The available online data is heterogeneous and less structured, since the online content has been progressively generated in a non-standardized manner [2]. Besides, the online content has been changing rapidly, Internet has been rapidly expanding and much of the online content contains irrelevant and useless information [3]. Web mining can mainly divided into three categories:

© Springer International Publishing Switzerland 2015
R. Silhavy et al. (eds.), *Software Engineering in Intelligent Systems*,
Advances in Intelligent Systems and Computing 349, DOI: 10.1007/978-3-319-18473-9_19

web usage mining, web structure mining and web content mining [1]. Web usage mining is the process of discovering user access patterns from web usage logs, such as web server access logs, proxy server logs, etc.; web structure mining is the process of extracting useful knowledge from hyperlinks in the web; and web content mining is the process of discovering useful information from web contents, such as text, images, audio, video, etc. [4].

Web page classification is the process of identifying to which of a set of categories a Web page belongs based on predefined category labels. The problem of Web page classification can be broadly divided into several fields, such as subject classification, functional classification, sentiment classification, genre classification, search engine spam classification [5]. Subject classification deals with the topic of the Web site, functional classification aims to determine the role of the Web site and sentiment classification deals with opinion orientation of statements within online content. Web page classification is essential to focused crawling, assisted development of web directories, topic-specific Web link analysis, contextual advertising and improvement of web search quality [5]. Hence, it is an essential task of web mining.

Machine learning algorithms have been successfully applied for text classification. Probabilistic classifiers, such as Naïve Bayes classifier, decision tree classifiers, such as ID3, C4.5 and C5, decision rule classifiers, regression methods, neural networks, instance based classifiers, such as K-nearest neighbor algorithm, and support vector machines are among the machine learning methods used for text classification [6].

Artificial immune systems are biologically inspired computing approaches, such as genetic algorithms, neural networks and swarm intelligence. Artificial immune systems are computational systems, inspired by ideas, theories and components of the immune system, which aim to solve complex computational problems, such as pattern recognition and optimization [7]. They have been successfully applied for classification, clustering, anomaly detection, computer security, optimization, medical diagnosis, bioinformatics, image processing, control, robotics, virus detection and web mining [8].

In this paper, we have applied two artificial immune system based classification algorithms, i.e. Immunos-1 and Immunos-99 in order to examine the effectiveness of artificial immune system based approaches for web page classification. The methods are compared to a decision tree classifier (C4.5) and a probabilistic classifier (Naïve Bayes classifier) on 50 data sets which are extracted from DMOZ (Open Directory Project).

The paper is organized as follows: Section 2 reviews related works on web page classification, Section 3 introduces basic concepts of artificial immune systems and algorithms, Section 4 presents evaluation criteria, dataset and experimental results and Section 5 presents concluding remarks.

2   Related Work

This section briefly reviews the existing works on the automated classification of web pages. In [9], adaptive fuzzy learning network was used in conjunction with fair feature selection algorithm for web page classification. The feature selection algorithm was utilized in order to reduce dimensionality of features and network classifier was utilized for fast training and testing. The experimental results indicated that feature selection algorithm can be successfully used for identifying useful keywords and network classifier can achieve extremely fast classification with a high predictive performance. In [10], fuzzy association concept was utilized in order to capture ambiguous use of the same word/vocabulary for referring to different entities. The experimental results were conducted on Yahoo and ODP data sets and compared to vector space model. The average classification accuracy of 67,8% obtained by vector space model was increased to 81,5% for Yahoo data set by fuzzy approach. In [11], a classification system for automated document classification was presented. The presented system uses Naïve Bayes classification algorithm as a base classifier. Hence, multivariate Bernoulli event and multinomial event model of the algorithm were implemented with four different feature selection methods. The experimental results emphasized that the proposed classification model can be used as an effective model for web document classification. Kwon and Lee [12] presented a three-phased approach based on K-nearest neighbor algorithm for web site classification. The approach also utilized a feature selection method and a term weighting scheme. Qi and Sun [13] presented a K-means and genetic algorithm based approach for automated classification of web pages. The genetic algorithm was utilized in order to adjust weights of keywords used for referring categories. Selamat and Omatu [14] presented a classification model for web pages based on neural networks, principal component analysis and profile-based features. The feature set was reduced via principal component analysis. The model was applied for sports news domain with high predictive performance. In [15], a vector space model based on tolerance rough set was presented for Web document classification. The tolerance rough set was used to capture the different index terms and keywords in the documents. The experimental results indicated that higher classification accuracies can be achieved by the proposed approach compared to conventional vector space model. Chen and Hsieh [16] presented a classification model for web pages based on support vector machine, latent semantic analysis and feature selection. The latent semantic analysis was utilized to identify semantic relations among keywords and documents. Besides, text features extracted from web page content were used for classifying web pages into subsequent categories. Support vector machine classifier used semantic relations based features for training, whereas text features for testing. The experimental results indicated that the proposed method can be used as a viable tool for web page classification. Materna [17] examined the performance of four classification algorithms, namely C4.5, k-nearest neighbor classifier, Naïve Bayes classifier, and support vector machines with a number of term representations, term clustering and feature selection methods. The highest predictive performance was achieved by the combination of support vector machines with term frequency representation and mutual information based feature selection. Zhang et al. [18] proposed to use fuzzy k-nearest neighbor classifier for web document classification. Besides, TF-IDF representation method and membership grade were used. The experimental results indicated that fuzzy counterpart outperforms conventional k-nearest neighbor classifier and support vector machines. Chen et al. [19] proposed a fuzzy ranking based feature selection method for web page classification problem. The fuzzy ranking method is utilized in conjunction with discriminating power measure for reducing the input dimensionality. The experimental results indicated that discriminating power measure is useful for reducing redundancy and noise of features. Özel [20] presented a genetic algorithm based approach for web page classification, which takes HTML tags and tags of terms as features for classification. The best weights of each feature were determined with the use of genetic algorithms. The proposed classification schema obtained a classification accuracy of 95% in case of enough negative documents in the training set.

3   Artificial Immune Systems

Artificial immune systems can be defined as a computational intelligence paradigm inspired by theoretical immunology and observed immune functions, principles and models to solve complex engineering and optimization problems [21].

Artificial immune systems are biologically inspired systems, such as evolutionary algorithms, swarm intelligence and neural networks [22]. Human immune system is a complex of specialized cells, molecules and organs which are responsible for protecting the body against attacks of pathogens, such as bacteria and viruses. The defense against the pathogens is handled with the help of innate and adaptive immunity parts of vertebrate immune system. Innate immune system provides immediate host defense against infections, whereas adaptive immune system provides a more complicated response against infections and adapts to the infection [23]. There are two different types of lymphocytes in the immune system: B-cell and T-cell. B-cells are responsible for production and secretion of antibodies, whereas T-cells are responsible for regulation of activities of other cells and attacking to the infected cells [24]. The recognition of antigen, the activation of lymphocytes and effector phase are the three main phases regarding adaptive immune system [23]. Human immune system has some characteristics, such as uniqueness, self/non-self-discrimination, learning, memory, which make it appropriate model to use in development of computational models [8].

Artificial immune system based methods and algorithms are mainly based on three concepts of immune system systems: clonal selection, negative selection and immune network. Clonal selection is the theory in immunology to explain the basic characteristics of adaptive immune system regarding to antigenic stimulus. According to the clonal selection theory, only the cells that are capable of recognizing an antigen will be proliferated. CLONALG algorithm [25], Immunity Clonal Strategy Algorithm [26], Adaptive Clonal Selection algorithm [27], CLONCLAS algorithm [28], Immunos algorithms [29, 30], and Trend Evaluation Algorithm [31] are typical examples of clonal selection theory based algorithms. Negative selection is the process of eliminating T cells that recognize self-cells molecules and antigens. Negative selection algorithm was presented in 1994 to solve computer security problems [32]. It is based on the principles of maturity of T-cells and self/non-self-discrimination in human immune system [33]. Negative selection based algorithms have been successfully applied for anomaly detection, time series prediction, image segmentation and hardware fault tolerance [8, 34]. Immune network consists of immune cells, which are capable of communicating and recognizing antigens and other immune cells and taking part in the process of stimulating and activating the other cells in case of nonappearance of antigens [33]. AiNet (Artificial Immune Network) algorithm [35] is one of the typical examples for immune network based algorithms.

3.1   Immunos-1 Algorithm

The first Immunos algorithm, namely Immunos-81 [29] has been presented by Carter as an attempt to use concepts of artificial immune systems for classification problems. The algorithm was presented by a medical specialist; hence it exhibits some ambiguous points in terms of algorithmic viewpoint. In order to investigate Immunos algorithm, Brownlee [30] presented two different configurations for Immunos-81 algorithm, namely Immunos-1 and Immunos-2 and an extension over the algorithm, which is referred as Immunos-99 algorithm in a technical report.

Immunos-1 algorithm is an artificial immune system based algorithm which uses no data reduction. Therefore, the prepared clone population is kept to classify new data instances. It assumes a single problem domain, which in turn makes the need for T-cells and amino acid library obsolete. For each classification label, a B-cell population is created and for each antigen in the training set, a single B-cell is created. In classification phase of Immunos-1 algorithm, for each B-cell clone population an avidity value is determined. Based on the avidity value, a data instance is assigned to a class of clone with the highest avidity value. This is determined by calculating the affinity between the particular antigen and each single cell of clone, as given by Equation 1 [30]:

$$affinity = \sqrt{\sum_{A}^{i=1} af_i} \qquad (1)$$

where A denotes the total number of attributes in the data vector, af_i represents the affinity value for ith attribute. Affinity measure calculation between numeric attributes is handled by Euclidean distance, whereas the calculation between nominal attributes is handled by a binary distance, where zero is used for matches and one for no matches.

The avidity measure is calculated as given by Equation 2 [30]:

$$avidity = \frac{CS}{CA} \qquad (2)$$

where CS represents the total number of B-cells in the clone population and CA denotes the clone affinity, which is obtained by application of Eq. 1 to all B-cells in the clone population.

3.2   Immunos-99 Algorithm

Immunos-99 is an artificial immune system based algorithm which is inspired from Immunos algorithm and some principles of clonal selection classification algorithm. Immunos-99 algorithm demonstrates some features that are not exhibited in other artificial immune based systems. First, there is a group level competition between classification labels at classification time. Besides, training data is examined for only single pass [30].

The training phase of Immunos-99 algorithm starts with the division of data into antigen groups based on classification labels. Afterwards, a B-cell population is prepared from each antigen group. The creation of populations is handled independently from other created populations. Then, B-cell population is exposed to all antigens for all groups. In this process, affinity measure between each B-cell and the antigen is determined. Based on these affinity values, the population is ordered. Then, fitness scores are calculated. The fitness score is an indicator of a particular B-cell's usefulness, which is determined by dividing the sum rank scores for antigens in the same group by the sum rank scores of all the other groups. Based on calculated fitness measure values, pruning is applied on the population so that only useful cells remain. Once the pruning on the population has been completed, affinity maturation is applied in order to enhance specificity of the population to its particular antigen group. This process is handled by ranked based cloning and somatic hyper-mutation. After performing fitness-rank based cloning and mutation, a randomly selected number of antigens of the same group are inserted. The outlined process has been repeated for each antigen group for a particular number of generations. Afterwards, for each B-cell population, a final pruning is performed and the final B-cell population is presented as the classifier. In the classification phase of the algorithm, the classification schema exhibited in Immunos-1 algorithm is applied. Hence, B-cell populations compete for a particular antigen based on avidity scores and the population with the highest avidity score assigns the classification label [30]. The details of the training and classification phases of Immunos-99 algorithm can be found at [30, 36].

4   Experimental Results

The performance of artificial immune system based algorithms, namely Immunos-1 and Immunos-99 were compared to two well-known classification algorithms, Naïve Bayes classifier and C4.5 decision tree classifier. Naïve Bayes classifier is a probabilistic classifier based on Bayes' theorem and independence assumption of features and C4.5 algorithm is a decision tree classifier which uses information entropy measure to build decision trees.

In order to evaluate the web page classification performance of algorithms, fifty datasets of DMOZ (Open Directory Project) were used [37]. The minimum number of documents among the datasets is 104 and maximum number of documents is 164, the number of topics ranges from 3 to 10 and the number of attribute ranges from 495 to 822. The dataset is available at preprocessed format. Hence, a preprocessing stage is not applied.

In order to evaluate the performance of classification algorithms, two different evaluation metrics, namely classification accuracy and F-measure values were used.

Classification accuracy (ACC) is the proportion of true positives and true negatives obtained by the classification algorithm over the total number of instances as given by Equation 3:

$$ACC = \frac{TN + TP}{TP + FP + FN + TN} \tag{3}$$

where *TN* denotes number of true negatives, *TP* denotes number of true positives, *FP* denotes number of false positives and *FN* denotes number of false negatives.

Precision (PRE) is the proportion of the true positives against the true positives and false positives as given by Equation 4:

$$PRE = \frac{TP}{TP + FP} \tag{4}$$

Recall (REC) is the proportion of the true positives against the true positives and false negatives as given by Equation 5:

$$REC = \frac{TP}{TP + FN} \tag{5}$$

F-measure takes values between 0 and 1. It is the harmonic mean of precision and recall as determined by Equation 6:

$$F - measure = \frac{2 * PRE * REC}{PRE + REC} \tag{6}$$

In the experimental procedure, 10-fold cross validation is used for evaluation. By this schema, the original data set is randomly divided into 10 equal sized subsamples. For each time, a single subsample is used for validation, whereas the other nine subsamples are kept for training. The process is repeated ten times and average results are reported. In the experimental results, C4.5 (J48) and Naïve Bayes implementations of WEKA 3.7.11 and Immunos-1 and Immunos-99 implementations of artificial immune system based extension of the software [38] are utilized.

In Tables 1-2, classification accuracies and F-measure values obtained by the compared four algorithms are presented. As it can be observed from Table 1, the highest classification accuracy is achieved by Immunos-1 algorithm, which is followed by Naïve Bayes classifer. The predictive performance of C4.5 decision tree classifier and Immunos-99 algorithms are close to each other. The comparison of classification algorithms in terms of classification accuracies are illustrated in Figure 1. In Table 2, F-measure results listed exhibit similar patterns with accuracy rates of algorithms. The best (highest) F-measure values are obtained by Immunos-1 algorithm. The average F-measure values for algorithms are 0,9106, 0,8556, 0,741 and 0,739 for Immunos-1, Naïve Bayes, C4.5 and Immunos-99, respectively.

Table 1. Classification Accuracies of Algorithms

Data-set	Naive Bayes	C4.5	Immu-nos-1	Immu-nos-99	Data-set	Naive Bayes	C4.5	Immu-nos-1	Immu-nos-99
1	94.55	95.88	98.26	93.29	26	86.72	72.17	84.11	72.05
2	84.47	68.34	94.27	78.86	27	90.47	75.92	94.03	83.96
3	85.28	75.21	86.19	75.86	28	87.38	78.63	93.72	75.93
4	73.85	70.15	90.69	66.85	29	83.45	71.06	90.85	76.99
5	84.56	71.02	84.61	69.81	30	81.22	76.20	86.68	75.22
6	93.35	84.55	97.32	83.05	31	92.91	84.27	98.18	86.69
7	93.81	76.64	95.26	82.67	32	91.94	81.12	95.84	81.33
8	90.27	77.17	94.80	79.71	33	86.86	72.86	88.97	75.29
9	92.39	80.10	97.79	84.5	34	87.99	78.82	92.41	77.00
10	88.36	77.45	95.64	76.55	35	95.43	85.13	96.43	86.82
11	93.29	82.75	92.86	82.8	36	88.23	85.14	95.95	87.47
12	89.38	81.07	94.72	76.51	37	73.77	70.89	88.87	64.98
13	65.50	56.43	72.79	62.64	38	81.63	73.11	94.43	73.42
14	96.86	82.05	98.73	88.83	39	91.88	81.54	96.13	86.82
15	86.08	81.31	97.21	78.30	40	87.08	73.60	91.41	75.86
16	93.07	84.36	94.79	88.14	41	91.24	80.25	92.43	84.64
17	96.05	80.09	96.11	79.76	42	85.64	78.29	94.29	78.21
18	90.74	87.69	94.00	84.14	43	89.61	83.4	94.73	84.06
19	88.61	75.89	94.89	76.70	44	95.63	86.03	97.09	90.31
20	81.17	75.17	84.34	64.26	45	80.74	76.85	90.69	73.04
21	69.72	61.17	87.00	57.05	46	87.77	84.22	93.90	79.64
22	89.6	86.54	89.75	79.90	47	84.94	71.23	85.75	63.24
23	87.52	81.14	96.01	78.26	48	88.63	78.12	96.93	77.65
24	82.57	71.15	89.63	68.91	49	89.29	78.04	93.42	85.46
25	85.68	75.32	87.64	75.13	50	77.55	68.16	89.26	65.26

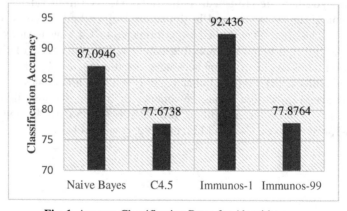

Fig. 1. Average Classification Rates for Algorithms

Table 2. F-measure Values of Algorithms

Data-set	Naive Bayes	C4.5	Immu-nos-1	Immu-nos-99	Data-set	Naive Bayes	C4.5	Immu-nos-1	Immu-nos-99
1	0.96	0.95	0.98	0.97	26	0.87	0.72	0.88	0.75
2	0.72	0.44	0.82	0.49	27	0.90	0.85	0.93	0.86
3	0.98	0.70	0.93	0.86	28	0.93	0.80	0.93	0.72
4	0.63	0.55	0.76	0.61	29	0.87	0.65	0.96	0.67
5	0.73	0.58	0.66	0.57	30	0.65	0.67	0.88	0.71
6	0.90	0.85	0.95	0.75	31	0.97	0.92	0.99	0.95
7	0.98	0.87	0.98	0.87	32	0.93	0.87	0.99	0.86
8	0.78	0.61	0.89	0.59	33	0.93	0.76	0.96	0.85
9	0.86	0.75	0.96	0.78	34	0.89	0.87	0.90	0.75
10	0.93	0.65	0.97	0.74	35	0.97	0.95	0.96	0.78
11	0.87	0.50	0.84	0.62	36	0.89	0.83	0.94	0.87
12	0.91	0.80	0.96	0.81	37	0.78	0.96	0.94	0.68
13	0.75	0.62	0.83	0.75	38	0.94	0.79	0.99	0.90
14	0.93	0.79	0.96	0.77	39	0.98	0.83	0.99	0.90
15	0.85	0.76	0.97	0.71	40	0.95	0.79	0.96	0.87
16	0.89	0.82	0.93	0.83	41	1.00	0.85	0.99	0.93
17	0.97	0.58	0.98	0.76	42	0.88	0.91	0.98	0.81
18	0.95	1.00	0.91	0.83	43	0.89	0.65	0.94	0.80
19	0.73	0.63	0.90	0.60	44	0.95	0.82	0.96	0.89
20	0.77	0.80	0.82	0.53	45	0.50	0.53	0.82	0.53
21	0.50	0.48	0.81	0.44	46	0.87	0.77	0.91	0.76
22	0.74	0.71	0.84	0.57	47	0.61	0.37	0.46	0.21
23	0.98	0.96	0.99	0.87	48	0.95	0.63	0.96	0.77
24	0.95	0.92	0.96	0.86	49	0.89	0.83	0.93	0.87
25	0.94	0.87	0.95	0.71	50	0.59	0.24	0.83	0.37

5   Conclusion

Online content is abundant source of information. Hence, the emerging field of web mining aims to develop efficient and robust models to deal properly with this content. Web page classification is one of the most important issues in web mining, which aims to classify new web pages based on model generated by labeled training instances. Since web page classification is an essential component of other web mining tasks, such as focused crawling, topic-specific Web link analysis, contextual advertising, etc., development of robust classification models become an important research direction. This paper examines the effectiveness of two artificial immune system based algorithms for automated classification of web pages. The experimental results indicate that Immunos-1 algorithm outperforms the existing classifiers for the web page classification domain.

References

1. Fürnkranz, J.: Web Mining. In: Maimon, O., Rokach, L. (eds.) Data Mining and Knowledge Discovery Handbook, pp. 891–920. Springer, Heidelberg (2005)
2. Zhang, Q., Richard, S.: Web Mining: A Survey of Current Research, Techniques, and Software. Int. J. Info. Tech. Dec. Mak. 7, 683–720 (2008)
3. Han, J., Kamber, M., Pei, J.: Data Mining: Concepts and Techniques. Morgan Kaufmann, San Francisco (2011)
4. Bhatia, M.P.S., Kumar, A.: Information Retrieval and Machine Learning: Supporting Technologies for Web Mining Research and Practice. Webology 5(2), Article 55 (2008)
5. Qi, X., Davison, B.D.: Web Page Classification: Features and Algorithms. ACM Computing Surveys 41(2), Article 12 (2009)
6. Sebastiani, F.: Machine Learning in Automated Text Categorization. ACM Computing Surveys 34(1), 1–47 (2002)
7. de Castro, L.N., Timmis, J.: Artificial Immune Systems: A Novel Paradigm to Pattern Recognition. In: Corchado, J.M., Alonso, L., Fyfe, C. (eds.) Artificial Neural Networks in Pattern Recognition, pp. 67–84 (2002)
8. Zheng, J., Chen, Y., Zhang, W.: A Survey of Artificial Immune Applications. Artificial Intelligence Review 34, 19–34 (2010)
9. Lee, H.-M., Chen, C.-M., Tan, C.-C.: An Intelligent Web-Page Classifier with Fair Feature-Subset Selection. In: Joint 9th IFSA World Congress and 20th NAFIPS International Conference, pp. 395–400. IEEE Press, New York (2001)
10. Haruechaiyasak, C., Shyu, M.-C., Chen, S.-C.: Web Document Classification Based on Fuzzy Association. In: 26th Annual International Computer Software and Applications Conference, pp. 487–492. IEEE Press, New York (2002)
11. Wang, Y., Hodges, J., Tang, B.: Classification of Web Documents Using a Naïve Bayes Method. In: 15th IEEE International Conference on Tools with Artificial Intelligence, pp. 560–564. IEEE Press, New York (2003)
12. Kwon, O.-W., Lee, J.-H.: Text Categorization based on K-nearest Neighbor Approach for Web site Classification. Information Processing and Management 39, 25–44 (2003)
13. Qi, D., Sun, B.: A Genetic K-means Approaches for Automated Web Page Classification. In: Proceedings of the 2004 IEEE International Conference on Information Reuse and Integration, pp. 241–246. IEEE Press, New York (2004)
14. Selamat, A., Omatu, S.: Web page feature selection and classification using neural networks. Information Sciences 158, 69–88 (2004)
15. Yi, G., Hu, H., Lu, Z.: Web Document Classification Based on Extended Rough Set. In: PDCAT 2005, pp. 916–919. IEEE Press, New York (2005)
16. Chen, R.-C., Hsich, C.-H.: Web Page Classification Based on a Support Vector Machine Using a Weighted Vote Schema. Expert Systems with Applications 31, 427–435 (2006)
17. Materna, J.: Automated Web Page Classification. In: Proceedings of Recent Advances in Slavonic Natural Language Processing, Masaryk University, pp. 84–93 (2008)
18. Zhang, J., Niu, Y., Nie, H.: Web Document Classification Based on Fuzzy k-NN Algorithm. In: Proceedings of the 2009 International Conference on Computational Intelligence and Security, pp. 193–196. IEEE Press, Washington (2009)
19. Chen, C.-M., Lee, H.-M., Chang, Y.-J.: Two Novel Feature Selection Approaches for Web Page Classification. Expert Systems with Applications 36, 260–272 (2009)
20. Özel, S.A.: A Web Page Classification System Based on a Genetic Algorithm Using Tagged-Terms as Features. Expert Systems with Applications 38, 3407–3415 (2011)

21. de Castro, L.N., Timmis, J.: Artificial Immune System: A New Computational Intelligence Approach. Springer, Heidelberg (2002)
22. Timmis, J., Hone, A., Stibor, T., Clark, E.: Theoretical advances in artificial immune systems. Theoretical Computer Science 403, 11–32 (2008)
23. Sinha, J.K., Bhattacharya, S.: A Text Book of Immunology. Academic Pub., Kolkata (2006)
24. de Castro, L.N., Zuben, F.J.V.: Artificial Immune Systems: Part I- Basic Theory and Applications, Technical report, RT-DCA (1999)
25. de Castro, L., Zuben, F.: Learning and Optimization Using the Clonal Selection Principle. IEEE Transactions on Evolutionary Computation 6(3), 239–251 (2002)
26. Ruochen, L., Haifeng, D., Licheng, J.: Immunity Clonal Strategies. In: Proceedings of the Fifth International Conference on Computational Intelligence and Multimedia Applications, pp. 290–295. IEEE Press, Washington (2003)
27. Garrett, S.: Parameter-Free Adaptive Clonal Selection. In: Proceedings of Congress on Evolutionary Computation, pp. 1052–1058. IEEE Press, Washington (2004)
28. White, J.A., Garrett, S.M.: Improved Pattern Recognition with Artificial Clonal Selection? In: Timmis, J., Bentley, P.J., Hart, E. (eds.) ICARIS 2003. LNCS, vol. 2787, pp. 181–193. Springer, Heidelberg (2003)
29. Carter, J.H.: The immune system as a model for classification and pattern recognition. Journal of the American Informatics Association 7, 28–41 (2000)
30. Brownlee, J.: Immunos-81: The Misunderstood Artificial Immune System. Technical report, Swinburne University (2005)
31. Wilson, W.O., Birkin, P., Aickelin, U.: Price Trackers Inspired by Immune Memory. In: Bersini, H., Carneiro, J. (eds.) ICARIS 2006. LNCS, vol. 4163, pp. 362–375. Springer, Heidelberg (2006)
32. Forrest, S., Perelson, A., Allen, L., Cherukuri, R.: Self-nonself discrimination in a computer. In: Proceedings of the IEEE Symposium on Research in Security and Privacy, pp. 202–212. IEEE Press, New York (1994)
33. Talbi, E.-G.: Metaheuristics: From Design to Implementation. Wiley, New York (2009)
34. Hofmeyr, S.A., Forrest, S.: Architecture for an Artificial Immune System. Evolutionary Computation 8(4), 443–473 (2000)
35. Timmis, J., Neal, M., Hunt, J.: An Artificial Immune System for Data Analysis. Biosystems 55, 143–150 (2000)
36. Kopacek, L., Olej, V.: Municipal Creditworthiness Mlodeling by Artificial Immune Systems. Acta Electrotehnica et Informatica 10(1), 3–11 (2010)
37. DMOZ Open Directory Project Dataset, http://www.unicauca.edu.co/~ccobos/wdc/wdc.htm
38. WEKA Classification Algorithms, http://wekaclassalgos.sourceforge.net/

Ring Algorithms for Scheduling in Grid Systems

A.E. Saak, V.V. Kureichik, and E.V. Kuliev

Southern Federal University, Rostov-on-Don, Russia
{saak,vkur}@sfedu.ru

Abstract. We consider the model of task scheduling in Grid systems that can be represented by the first coordinate quadrant. The process of task scheduling is based on the parity of processor resources and time resources. We previously proposed a classification of sets of multiprocessor tasks that are to be processed by a Grid system that consists of three types: circular-type, hyperbolic and parabolic task queues. We describe the polynomial time ring algorithm, apply it to a circular-type task queue, and compare our results with the optimal packing of rectangles in an enclosing rectangle of minimum area. Our comparative analysis demonstrates the applicability of the proposed ring algorithm in task scheduling in Grid systems.

Keywords: Scheduling, Multiprocessor tasks, Circular-type task queue, Polynomial time ring algorithm for scheduling.

1 Introduction

The performance efficiency of Grid systems to great extent depends on the quality of task scheduling. In this paper we study multiprocessor task scheduling of computational resource when there are many concurrent tasks.

Grid-systems consist of sites which contains parallel systems [1-4]. There are three architectures of Grid-systems: centralized, hierarchical and distributed [1-9]. In centralized structure there is a centralized scheduler, which makes assignments, possessing full information about Grid- system resources and multiprocessor tasks (Grid resource management systems with centralized structure KOALA, EGEE WMS, PlanetLab [10]). Depending on the way of Grid- system resources co-allocation which is used to execute a task, it is distinguished single- site scheduling and multi- site scheduling [5]. Multi- site scheduling can be characterized by the opportunity of task execution on several sites simultaneously (Grid resource management systems, with multi- site scheduling KOALA, GARA, GridARS [11]). In the paper it is considered multiprocessor tasks scheduling centralized Grid- systems with multi- site scheduling.

2 Problem Statement

In [12-14] a computer system is modelled by a horizontal semi-infinite strip with the height that equals the number of processors which are the limiting resource, and the

R. Silhavy et al. (eds.), *Software Engineering in Intelligent Systems*,
Advances in Intelligent Systems and Computing 349, DOI: 10.1007/978-3-319-18473-9_20

unlimited length of time (see Fig. 1(a)), or it is modelled by a vertical semi-infinite strip with a width that equals the number of time units which were the critical resource in this case (see Fig. 1(b)). Observe that both models require only one type of resource to be finite.

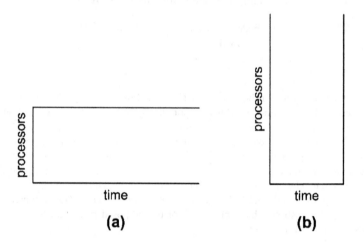

Fig. 1. Model of computer system in which (a) the processor resource is the critical one; (b) the time resource is the critical one

As computing power and the number of processors increases and Grid- systems [15] emerge, scheduling strategy should become more balanced dealing with the distribution of resources on a parity basis. Each multiprocessor task that needs to be scheduled by the Grid- system can be geometrically represented by coordinate resource rectangle with the height and width determined by the number of processors and time required for the task.

Following [16,17] in this paper we consider the model of multiprocessor task scheduling in Grid systems in the form of the first coordinate quadrant (see Fig. 2) with a given set of rectangles with both sides parallel to the horizontal vertical axes must be placed into it according to certain rules [13].

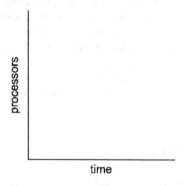

Fig. 2. Model of Grid system based on the concept of parity of resources

In papers [16, 17] the author propose a quadratic classification of a set of multi-processor tasks that are to be processed by a Grid systems. Each one of quadratic types that are termed circular, hyperbolic or parabolic, corresponds to a certain sequence of tasks queue waiting for service i.e. the linear polyhedral of multiprocessor tasks shown in Fig. 3.

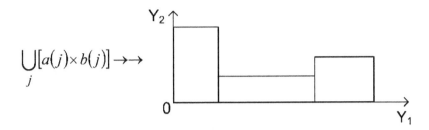

Fig. 3. Linear polyhedral of multiprocessor tasks

The multiprocessor tasks scheduling approach proposed in this paper could be implemented in polynomial time and requires comparison of its quality with the quality of optimal packing. In the paper we describe the polynomial time ring algorithm and apply it to a circular-type task queue.

The problem of rectangular packing in the enclosing rectangle of minimal area were considered in [18] (minimal area rectangular packing problem (MARPP)), [19] (rectangle packing area minimization problem (RPAMP)). We compare our results with optimal packing of rectangles in an enclosing rectangle of minimum area [20].

3   Ring Algorithm for Scheduling

Let $a(j) \times b(j)$ or $[(a(j) \times b(j))]$ denote task j which requires $a(j)$ processors and $b(j)$ time units, where $a(j) \ll M$, $b(j) \ll M$.

In [16,17] a circular-type quadratic ordering refers to an ordering of successive pairs of rectangles $a(j) \times b(j)$, $\Delta a(j) \times \Delta b(j) \geq 0$ such that both the width and height of the smaller rectangle are smaller than those of the larger rectangle for each pair where the rectangles are ordered in decreasing order of $b(j)$ as illustrated in Fig. 4.

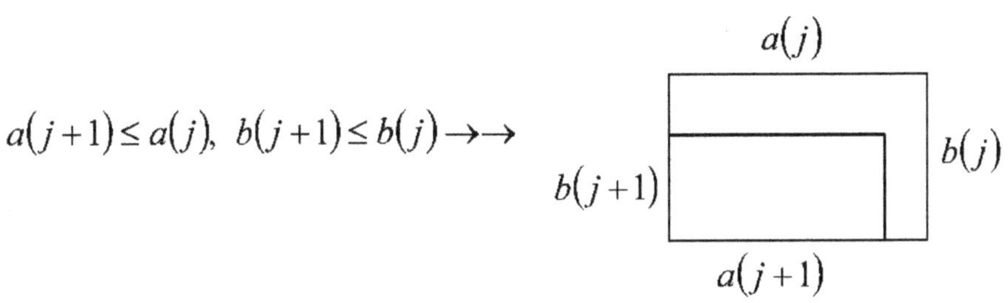

$$a(j+1) \leq a(j), \ b(j+1) \leq b(j) \rightarrow\rightarrow$$

Fig. 4. Comparable pair of resource rectangles

Here we consider packing of rectangles $U_j[(a(j), b(j))]$ sorted in the nesting order.

We start with the resource rectangle of maximum size to form the initial enclosure as shown in Fig. 5.

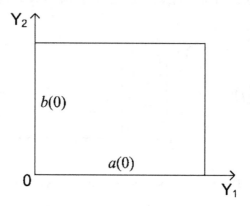

Fig. 5. Initial resource enclosure

On the second stage we allocate successive resource rectangles vertically, along the right-hand side of the enclosure until we reach the top boundary of the enclosure but without extending beyond it. We then form a new resource enclosure shown in Fig. 6.

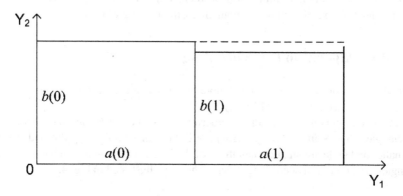

Fig. 6. The first resource enclosure

Next we allocate the rectangles remaining for placing horizontally, one after another on the top of the enclosure and along the top boundary mentioned above. We stop when we can't place a rectangle without exceeding the width of the enclosure as in Fig. 7.

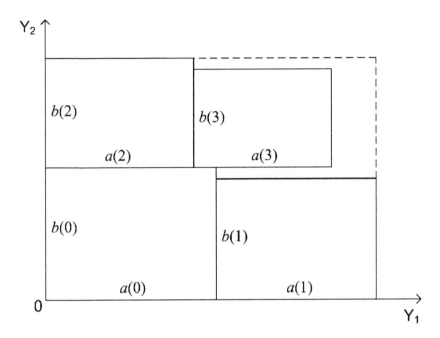

Fig. 7. The second resource enclosure

Fig. 8 presents the ring algorithm for the first resource bin. For the next bins our algorithm works the same way and the only difference is a sequence number of the resource rectangle (i.e. multiprocessor task) from which the algorithm starts packing. Let us use the following notation: $a(j), b(j)$ denotes the parameters of multiprocessor tasks, and k is the quantity of tasks. RA(j), RB(j) are the processor number and the time period from which $a(j), b(j)$ units of the corresponding resource have to be granted for the task j. AL denotes the first resource number along the Y_1 axis which stays the same for all rectangles of the corresponding vertical column, BL denotes the second resource number along the Y2 axis which stays the same for all rectangles of corresponding horizontal row. A, B denote the parameters of the current enclosure on the resources of the first and the second kind respectively.

It is noteworthy that the work of our algorithm takes k addition operations and k comparison operations, thus the algorithm requires polynomial time.

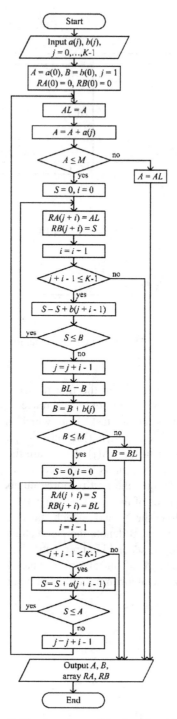

Fig. 8. The flowchart of the ring algorithm

4   Quality of the Ring Algorithm in Comparison with the Optimal One

In [20] the results of optimal packing of a set of natural squares of sizes 1×1 up to $k \times k$ into an enclosing rectangle of minimum area for $k \leq 32$ were presented.

Fig. 9. to Fig. 15. show the work of the ring algorithm for the mentioned set of squares for k=32.

Fig. 9. Initial resource enclosure

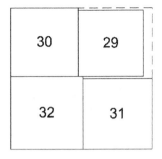

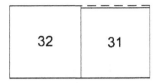

Fig. 11. The first resource enclosure   **Fig. 10.** The second resource enclosure

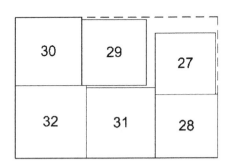

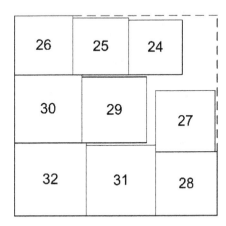

Fig. 12. The third resource enclosure   **Fig. 13.** The forth resource enclosure

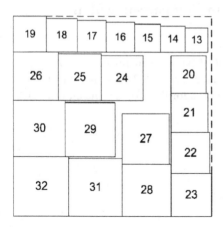

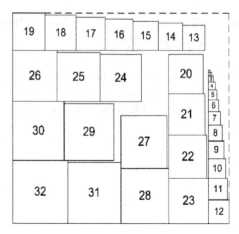

Fig. 14. The fifth resource enclosure   **Fig. 15.** The sixth resource enclosure

We compare the results of the ring algorithm which works in polynomial time with the results of optimal packing into an enclosing rectangle of minimum area presented in [20]: the accuracy doesn't exceed 31%.

Thus the ring algorithm proposed in the paper requires polynomial time and doesn't have great difference in results in comparison with the optimal one. This can serve as a basis for recommending the ring algorithm for scheduling in Grid technology.

5   Conclusions

The developed and explored polynomial ring algorithm which can be used in a Grid-system scheduler is adapted for circular-type tasks scheduling. Comparison of the results of the polynomial ring algorithm and optimal packing shows the advantages of the presented algorithm in consideration of quality of resource enclosure filling compared with the time of the optimal algorithm. For example the time of optimal packing of the 32 consecutive resource squares exceeds 33 days on 2GHz AMD Opteron 246 PC, while the ring algorithm instantly gives the decision with the accuracy which equals to 17,5%.

Acknowledgment. The study was performed by the grant from the Russian Science Foundation (project # 14-11-00242) in the Southern Federal University.

References

1. Magoulès, F., Nguyen, T., Yu, L.: Grid resource management: toward virtual and services compliant grid computing. Numerical analysis and scientific computing. CRC Press, UK (2009)
2. Magoulès, F. (ed.): Fundamentals of grid computing: theory, algorithms and technologies. Numerical analysis and scientific computing. CRC Press, UK (2010)

3. Antonopoulos, N., Exarchakos, G., Li, M., Liotta, A. (eds.): Handbook of research on p2p and grid systems for service-oriented computing: models, methodologies and applications. IGI Global publisher, USA (2010)
4. Li, M., Baker, M.: The grid: core technologies. John Wiley & Sons Ltd., England (2005)
5. Hamscher, V., Schwiegelshohn, U., Streit, A., Yahyapour, R.: Evaluation of job-scheduling strategies for grid computing. In: Buyya, R., Baker, M. (eds.) GRID 2000. LNCS, vol. 1971, pp. 191–202. Springer, Heidelberg (2000)
6. Rahman, M., Ranjan, R., Buyya, R., Benatallah, B.: A taxonomy and survey on autonomic management of applications in grid computing environments. Concurrency Computat.: Pract. Exper. (23), 1990–2019 (2011)
7. Krauter, K., Buyya, R., Maheswaran, M.: A taxonomy and survey of Grid resource management systems for distributed computing. Softw. Pract. Exper. 32(2), 135–164 (2002)
8. Iosup, A., Epema, D.H.J., Tannenbaum, T., Farrellee, M., Livny, M.: Inter-operating Grids through delegated matchmaking. In: 2007 ACM/IEEE Conference on Supercomputing (SC 2007), pp. 1–12. ACM Press, New York (2007)
9. Patel, S.: Survey Report of Job Scheduler on Grids. International Journal of Emerging Research in Management &Technology 2(4), 115–125 (2013)
10. Assunção, M., Buyya, R.: Architectural elements of resource sharing networks. In: Li, K., Hsu, C., Yang, L., Dongarra, J., Zima, H. (eds.) Handbook of Research on Scalable Computing Technologies, vol. 2, pp. 517–550. IGI Global publisher (2010)
11. Netto, M., Buyya, R.: Resource Co-allocation in Grid Computing Environments. In: Antonopoulos, N., Exarchakos, G., Li, M., Liotta, A. (eds.) Handbook of Research on p2p and Grid Systems for Service-oriented Computing: Models, Methodologies and Applications, vol. 1, pp. 476–494. IGI Global publisher (2010)
12. Barsky, A.B.: Parallel information technologies. INTUIT, BINOM Knowledge Lab, Moscow (2007)
13. Lodi, A., Martello, S., Monaci, M.: Two-dimensional packing problems: A survey. European Journal of Operational Research (141), 241–252 (2002)
14. Caramia, M., Giordani, S., Iovanella, A.: Grid scheduling by on-line rectangle packing. Networks 44(2), 106–119 (2004)
15. Schwiegelshohn, U., Badia, R., Bubak, M., Danelutto, M., Dustdar, S., Gagliardi, F., Geiger, A., Hluchy, L., Kranzlmüller, D., Laure, E., Priol, T., Reinefeld, A., Resch, M., Reuter, A., Rienhoff, O., Rüter, T., Sloot, P., Talia, D., Ullmann, K., Yahyapour, R., Voigt, G.: Perspectives on grid computing. Future Generation Computer Systems (26), 1104–1115 (2010)
16. Saak, A.E.: Local-optimal synthesis of schedules for Grid technologies. Journal of Information Technologies 12, 16–20 (2010)
17. Saak, A.E.: Dispatching algorithms in Grid systems based on an array of demands quadratic typification. Journal of Information Technologies 11, 9–13 (2011)
18. Maag, V., Berger, M., Winterfeld, A., Küfer, K.-H.: A novel non-linear approach to minimal area rectangular packing. Annals of Operations Research 179, 243–260 (2010)
19. Bortfeldt, A.: A reduction approach for solving the rectangle packing area minimization problem. European Journal of Operational Research 224, 486–496 (2013)
20. Huang, E., Korf, R.: Optimal rectangle packing: an absolute placement approach. Journal of Artificial Intelligence Research 46, 47–87 (2012)

The Tomas Bata Regional Hospital Grounds – The Design and Implementation of a 3D Visualization

Pavel Pokorný and Petr Macht

Department of Computer and Communication Systems
Tomas Bata University in Zlín, Faculty of Applied Informatics
Nad Stráněmi 4511, 760 05 Zlín, Czech Republic
pokorny@fai.utb.cz,
machtpetr@seznam.cz

Abstract. This paper briefly describes a visualization method of the Tomas Bata Regional Hospital grounds in Zlín. This hospital was founded in 1927 and has grown rapidly to the present day. We also collected all available historical materials of the Tomas Bata Regional Hospital and its surrounding area. We mainly focused on building plans, cadastral maps and historical photos. All the information we collected was chronologically sorted and on this basis, we created a 3D visualization of the urban development of these hospital grounds. To begin with, we created the grounds terrain model based on the Daftlogic database [14]. All buildings and accessories were separately modeled (we used the standard polygonal representation) and textured by the UV mapping technique. After that, we created the more complex 3D scenes from the individual models in these years: 1927, 1930, 1935, 1940, 1950, 1960, 1980, 1990, 2000, 2005 and 2013. The visualization output is performed by rendered images and animations in these years. We used the Blender software suite for this visualization.

Keywords: Computer Graphics, 3D Visualization, Modeling, Texturing, Animation.

1   Introduction

Data visualization is a hot topic. A simple definition of data visualization says: It's the study of how to represent data by using a visual or artistic approach rather than the traditional reporting method [1]. Represented data are most often displayed through texts, images, diagrams or animations to communicate a message.

Visualization today has ever-expanding applications in science, education, engineering (e.g., product visualization), interactive multimedia, medicine, etc. Typical example of a visualization application is the field of computer graphics. The invention of computer graphics may be the most important development in visualization since the invention of central perspective in the Renaissance period. The development of animation also helped the advance of visualization. [17]

With the development and performance of computer technology, the possibilities and limits of computer graphics are still increasing. The consequences of this are 3D

© Springer International Publishing Switzerland 2015   211
R. Silhavy et al. (eds.), *Software Engineering in Intelligent Systems*,
Advances in Intelligent Systems and Computing 349, DOI: 10.1007/978-3-319-18473-9_21

visualizations, which are used more frequently, image outputs get better quality [5] and the number of configurable parameters is rising [6]. As mentioned above, these visualizations are used in many scientific and other areas of human interest [2] [4].

One of the fields is visualizations of history. Based on historical documents, drawings, plans, maps and photographs, it is possible to create 3D models of objects that exist no more - things, products, buildings or extinct animals. If we assign suitable materials and the corresponding textures to these models, we can get a very credible appearance of these historic objects. From these individual objects, we can create very large and complex scenes that can be very beneficial tool for all people interested in history.

This paper describes a 3D visualization method of the Tomas Bata Regional Hospital grounds in Zlín in history and at present.

1.1   The History of the Tomas Bata Regional Hospital in Zlín

The first written record of Zlín dates back to 1322, when it was a center of an independent feudal estate. Zlín became a town in 1397. Until the late 19th century, the town did not differ much from other settlements in the surrounding area, with the population not exceeding 3,000. In 1894, Tomáš Baťa founded a shoe factory in Zlín. The town has grown rapidly since that time. Baťa's factory supplied the Austro-Hungarian army in World War I as the region was part of the Austro-Hungarian Empire. Due to the remarkable economic growth of the company and the increasing prosperity of its workers, Baťa himself was elected Mayor of Zlín in 1923. Baťa designed the town as he saw fit until his death in 1932, at which time the population of Zlín was approximately 35,000.

Fig. 1. The entrance building of the Tomas Bata Regional Hospital in 1927

Due to the population growth, it was also necessary to build a hospital. The history of the Tomas Bata Regional Hospital in Zlín dates back to May 1926, when the project was proposed. The hospital was situated approximately 3 km east of the city center on an almost square plot of land between the Dřevnice River and the forest. This forest was felled some years later and houses were built in its place, so the hospital was directly connected to the entire city.

The first hospital buildings were built in 1927 – i.e. the entrance building (Fig. 1) and two pavilions. By 1938, 16 hospital pavilions had been built, including the original buildings. Each pavilion was designed with the standardized appearance and typical architecture for most buildings in Zlín at this time – the combination of red bricks and a light gray concrete. The structure of the pavilions was also unified. Slight differences were only given by the specific needs of individual departments [8].

New development plans were created for the further expansion of the hospital grounds in 1946. Based on these plans, over the following years, new five pavilions were constructed. For the ensuing almost 30 years, the hospital grounds were without major changes. Only some reconstructions, adaptations and changes of pavilions were carried out over the years.

The next significant development of the hospital grounds was implemented after 1973, when the Surgery Building, Pathology-anatomical Department, District Health Station, Internal Departments and Utility Energy Block were developed in step-by-step phases. The new Ophthalmology pavilion was built in 1984 and the Hospice was built in 1989 [7].

The Tomas Bata Regional Hospital grounds encompasses nearly 60 buildings numerically designated for orientation purposes at present. These buildings have different medical and/or technical functions. Our main objective was to create complex 3D visualizations of the construction phases of the Tomas Bata Regional Hospital grounds in 1927, 1930, 1935, 1940, 1950, 1960, 1980, 1990, 2000, 2005 and 2013. The choice of these years was based on the available documentation, and these were the years that brought the greatest changes in the construction phases of the entire hospital complex.

2   Resources and Software

The first phases we needed to do were to collect all available historical materials of the hospital and select suitable programs for visualization creation.

2.1   Acquiring Resources

The overall progress of this work was initiated by the collation of available historic materials and information about the Tomas Bata Regional Hospital. The main resources were the State District Archive in Zlín – Klečůvka [11], the Moravian Land Archive in Brno and the Tomas Bata Regional Hospital Archive. We mainly focused our attention on building plans, cadastral maps and historical photos.

In the archives in Zlín – Klečůvka and Brno, we mainly found historical photos of this hospital from the foundation to the end of the nineteen-forties. Most resources and information were obtained in the Tomas Bata Regional Hospital Archive, where we found most of the construction plans of the individual buildings. We also held discussions with the hospital staff about the whole hospital grounds and its development over time.

The two books published for the 75th and 80th anniversaries of the Tomas Bata Regional Hospital [7], [8], were the next two important resources. These books contain information about the construction of buildings in the grounds as well as period photographs that were not stored in the archives. A website called Old Zlín [12] was the last important source in the creation of this work. The information on this webpage helped to verify some information which was obtained from the other references.

Based on the materials we obtained, we created a table that contains all of the acquired data on the construction sites, renovations and demolitions of individual buildings. This table was divided into several parts. In the first worksheet, we created a complete list of buildings, from their construction to the present or their possible demolition. For greater clarity, we created additional document pages that contain all construction events in the area in five-year intervals.

From this table, we created a bar graph that shows the number of newly constructed buildings over five-year intervals. This graph is shown in Figure 2.

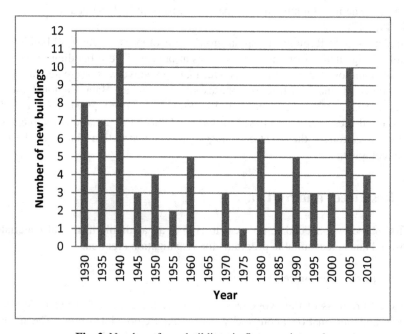

Fig. 2. Number of new buildings in five-year intervals

2.2   Used Programs

We preferred "free to use" software. For 3D modeling, texturing and rendering, we therefore used the Blender software suite [13]. Textures were drawn in GIMP [15]. And Microdem [18] was the last software that we used.

Blender is fully integrated creation suite, offering a broad range of essential tools for the creation of 3D content, including modeling, uv mapping, texturing, rigging, skinning, animation, particle and other simulation, scripting, rendering, compositing, post-production, and game creation [3] [9]. Blender is cross platform, based on the OpenGL technology and it is available under GNU GPL license.

GIMP is an acronym for GNU Image Manipulation Program. It is a freely distributed program under the GNU General Public license. It is mainly a digital image editor and a drawing tool. It allows one to retouch photos by fixing problems affecting the whole image or parts of the image, adjust colors in our photos to bring back the natural look, image compositing or image authoring [10].

Microdem is a freeware microcomputer mapping program designed for displaying and merging digital elevation models, satellite imagery, scanned maps, vector map data or GIS databases [18]. We used this software to convert a landscape elevation map data into a bitmap image (i.e. a heightmap).

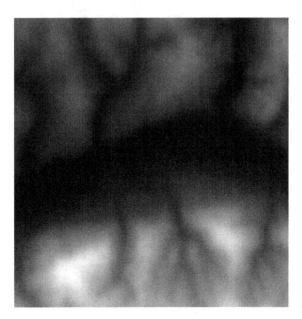

Fig. 3. The heightmap of the Zlín's Region

3   A Landscape Model of Zlín

We used data files which contained text information about the earth elevations to create the landscape model of Zlín (including its hospital grounds) and its vicinity.

We used the Digital Elevation Model data (i.e. DEM) which was provided by the NASA Shuttle Radar Topographic Mission (SRTM) in 2007. The data for over 80% of the globe is stored on [2] and can be freely downloaded for noncommercial use.

So we downloaded the data of Zlín's region, and then we opened it with the Microdem freeware program, which we described above. This software is able to convert the obtained data into a bitmap image. Microdem can clip and convert these images to grayscale (e.g. into a heightmap). We applied that to the Zlín's region. Specifically, we created a heightmap area of 25 km2 (a square with side lengths of 5 km, centered on the center of Zlín – Figure 3).

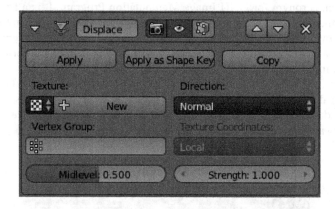

Fig. 4. Settings for the Displace modifier in the Bender environment

We saved this heightmap in PNG format (it is very important to use lossless compression). In Blender, we inserted a square (the Plane object) in the new scene and we divided it several times with the Subdivide tool to get a grid with a density of several thousand vertices. Then, we used the Displace modifier, to which we assigned a texture (for the obtained heightmap). The Displace modifier deforms an object based on the texture and setting parameters (Fig. 4). We got the model of the Zlín's landscape using this method. Although this model is not entirely accurate, but given the scale, the whole scene and the quality of the factory area model buildings, it is sufficient.

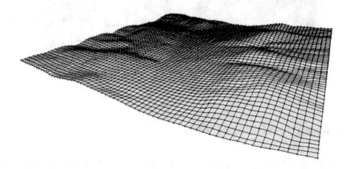

Fig. 5. The final mesh model of the Zlín conglomerate's landscape

This way, we created the landscape model. Its precision is sufficient to the surrounding area of the hospital grounds (within a few meters). But we wanted to obtain higher accuracy for the hospital grounds themselves. We also selected the part of the landscape model where the hospital grounds are, and used the Knife tool to multiply the cuts made to this object in the next step in order to get a lot of new vertices. In the following step, these vertices were placed in suitable positions (i.e. on most roads and paths between hospital pavilions) so as to create the most accurate possible 3D terrain model. The space coordinates (altitude) were measured by a GPS navigation device (Fig. 6).

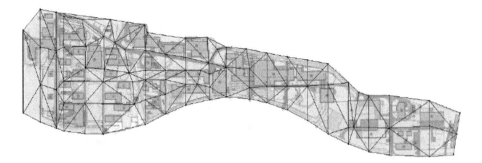

Fig. 6. The hospital ground plan with the inserted ordered vertices of the 3D terrain model

4   Modeling Buildings and Accessories

The modeling of the buildings was always performed according to the same scenario in the Blender program. Before the modeling phase, we put the appropriate construction plans of the selected building on the background screen in Blender. After that, in the window with the floor construction plan, we put the Plane object and then we modified its profile (i.e. the size and shape) according to the floor construction plan of the building. We mainly used two tools to shape this object. The Extrude tool allows us to alter the selected face in a specifically chosen direction and the Subdivide tool which breaks down the selected face into even greater number of smaller parts [9].

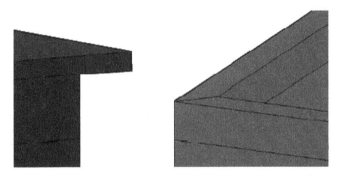

Fig. 7. A flat roof with an overhang (Left) and a flat roof with a raised edge (Right)

When the floor shape was finished, we extruded this profile to a height corresponding to the construction documents. Extrusion was performed for each floor until we reached the total height of the building.

The next step was the necessity to model the building roof. The buildings in the hospital complex have three different types of roofs – a pitched roof, a flat roof with an overhang and a flat roof with a raised edge (Figure 7). The pitched roof was created by dividing the top face into two parts, and the newly created edge was subsequently moved to the required height. The flat roof with an overhang was created by one more extrusion and the newly formed faces were pulled in the normal direction. The flat roof with a raised edge was created by the Inset Faces tool, which creates new smaller faces at the edges of the selected area. After that, these new sub-faces were subsequently extruded in the vertical direction.

To make the windows and doors embedded in the buildings, we used the Subdivide tool again on the relevant faces. After this, we deleted these new sub-faces to model holes. Additionally, we extruded the border edges of these holes to the depth of embedment of windows and doors.

Accessories (e.g. windows, doors, chimneys, railings, etc.) were modeled as separate objects by using the tools described above. All sub-models of each building were subsequently linked to the main building model to unify manipulations with the whole building after completing the model. Overall, we created around 200 models of buildings in this way.

An example of one modelled hospital building is shown in Figure 8. In this picture we can see the 13th pavilion model from 1929 in the Blender environment.

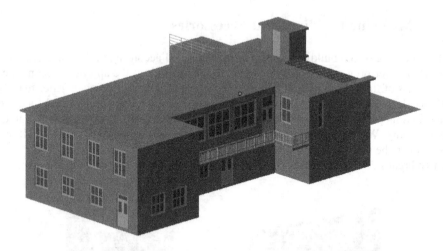

Fig. 8. The 13th pavilion model from 1929 in the Blender environment

We also separately created all models of buildings of the hospital grounds of each period in this way. In addition, we created several simple models of trees and shrubs, which allowed us to make the 3D hospital area model more complex. These models were created by the deformation of the Sphere (treetops) and Cylinder (trunks) objects, which can be simply added into scene in the Blender environment.

5   Texturing

We used the UV mapping technique for texturing objects. This process starts by the decomposition of each object into 2D sub-surfaces (a UV map). At the beginning of the decomposition process, it is necessary to mark the edges, which should be ripped from another one. In Blender, this process is performed by the Mark Seam command. After that, it is possible to finish decomposing by using the Unwrap tool. The UV map created in this way is saved into the .png raster graphic format (it is also possible to save it into another raster graphic format, but we need to use a lossless compression algorithm). We used the resolutions 512 x 512 or 1024 x 1024 pixels for the UV maps.

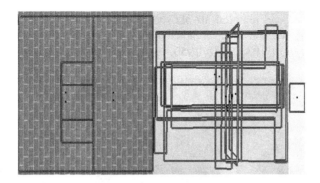

Fig. 9. The UV texture of the Baby box building

All textures were drawn in the GIMP software environment. We also opened all created UV maps in GIMP. In these pictures, the location of each part of the 3D object is visible. With this information, we can fill each individual sub-surface as necessary. Most of these textures were drawn by hand, and in some cases, we used pre-created textures from the CGTextures website [16] – these textures were edited and modified in order to use them on our models. For texture creation and editing purposes, we used standard GIMP drawing, coloring and transforming tools [10]. Once this process was finished, we saved all of the created textures back into same files (it is also possible to use the jpg graphic format with lossy compression) and opened and mapped these on the appropriate 3D models in the Blender environment. An example of the drawn texture is shown in Figure 9.

Once the textures were drawn, they were saved in .jpg format. In this case, we can already use the graphic format with a lossy compression algorithm to save computer memory and the .jpg format is ideal for that. The next step was to re-load these created textures in Blender and to correctly map them into 3D objects. This process is performed by correcting the set parameters in the Texture Mapping panel (Fig. 10).

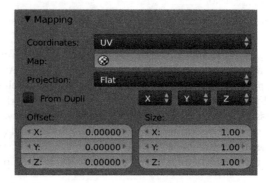

Fig. 10. The user environment for the UV mapping settings in Blender

We also created and textured all 3D sub-models of each period in this way. As mentioned above, we created a completed model of each building of the hospital grounds in 1927, 1930, 1935, 1940, 1950, 1960, 1980, 1990, 2000, 2005 and 2013.

Fig. 11. Setting parameters for the procedural Clouds texture in Blender

6   Rendering and Animation

After the completion of the modeling phase of separate buildings, we imported each of them into a single complex 3D scene with the landscape model. In addition, we set other suitable parameters like the surroundings and lighting. The surrounding area setting parameters are performed by the World window in Blender. It is possible to set simple colors for horizon and zenith and to blend them or to use the internal (procedural) or external texture (any bitmap file). We used the Clouds procedural texture, which looks close to reality in our scene (Fig. 11).

Lighting of the scenes can be realized in several ways in Blender. It is possible to use light objects (called Lamps in Blender) for local lighting or global influences (i.e. Ambient Light, Ambient Occlusion and Environment Lighting and Indirect Lighting). We used the Environment Lighting technique combined with the Emit material settings of selected objects.

The last step before the rendering process was to select a suitable position for the camera. We wanted to create more rendered images from different positions. Due to this, we added more different camera objects and oriented them correctly in order to capture the most graphic images of the whole scene.

The Render command performs the rendering calculation process in the Blender environment. Additionally, we can set many of the accompanying parameters. The basic parameters are the choice of a rendering algorithm, image or animation resolution, type of output file format, antialiasing, motion blur, enable/disable ray-tracing and shadows. We made the decision to use Blender's internal renderer with an image resolution of 1280x720 pixels, 25 frames per second and the MPEG-2 output format to render animations. Figure 12 shows one rendered image of the hospital area from 1980.

Fig. 12. The rendered image of the hospital area from 1980

7   Conclusion

In this paper, we have presented a visualization method of the Tomas Bata Regional Hospital grounds. Based on the historical materials, we created more complex 3D scenes in the Blender software suite in these years: 1927, 1930, 1935, 1940, 1950, 1960, 1980, 1990, 2000, 2005 and 2013.

Our future goal is to expand and improve these models. This process includes creating more detailed pavement models and to add the modeling and texture of the road on the edge of the Dřevnice River (this river is in the surrounding area of the hospital) and more trees model types, etc.

Another improvement would be to create time animation, which visualizes the development of the hospital area over the years of its existence.

A further extension could be the rendering of the current hospital area model in high resolution. After that, this rendered output could be stored on the hospital www pages and converted into an interactive image which could show specific information about each building when the mouse cursor is placed on it.

References

1. Brand, W.: Data Visualization for Dummies. John Wiley & Sons, New Jersey (2014)
2. Qiu, H., Chen, L., Qiu, G., Yang, H.: An Effective Visualization Method for Large-scale Terrain Dataset. WSEAS Transactions on Information Science and Applications 10(5), 149–158 (2013)
3. Skala, V.: Holography, Stereoscopy and Blender 3D. In: 16th WSEAS Int. Conf. on Computers ICCOMP 2012, Kos, Greece, pp. 166–171 (2012)
4. Dozono, H., Kabashima, T., Nakakuni, M., Hara, S.: Visualization of the Packet Flows using Self Organizing Maps. WSEAS Transactions on Information Science and Applications 7(1), 132–141 (2010)
5. Qiu, H., Chen, L., Qiu, G., Yang, H.: Realistic Simultaion of 3D Cloud. WSEAS Transactions on Computers 12(8), 331–340 (2013)
6. Congote, J., Novo, E., Kabongo, L., Ginsburg, D., Gerhard, S., Piennaar, R., Ruiz, O.E.: Real-time Volume Rendering and Tractography Visualization on the Web. Journal of WSCG 20(2), 81–88 (2012)
7. Bakala, J.: Baťova nemocnice ve Zlíně 1927-2002, Zlín, KODIAK print, s.r.o. (2002)
8. Bakala, J.: Baťova nemocnice v obrazech, faktech a dokumentech, Zlín, Finidr, s.r.o. (2002)
9. Hess, R.: Blender Foundations – The essential Guide to Learning Blender 2.6. Focal Press (2010)
10. Peck, A.: Beginning GIMP:From Novice to Professional. Apress (2008)
11. Státní okresní archív Zlín, Moravský zemský archív v Brně, http://www.mza.cz/zlin/
12. Starý Zlín - historie Zlína, pohlednice a video, http://staryzlin.cz/
13. Blender.org - Home, http://www.blender.org
14. Google Maps Find Altitude, http://www.daftlogic.com/sandbox-google-maps-find-altitude.htm
15. GIMP - The GNU Image Manipulation Program, http://www.gimp.org
16. CG Textures – Textures for 3D, graphic design and Photoshop!, http://www.cgtextures.com/
17. Visualization (computer graphics), http://en.wikipedia.org/w/index.php?title=Visualization_(computer_graphics)
18. Microdem Homepage, http://www.usna.edu/Users/oceano/pguth/website/microdem/microdem.htm

Secure End-To-End Authentication for Mobile Banking

Basudeo Singh[1] and K.S. Jasmine[2]

[1] FIS, Bangalore, India
[2] Dept. of MCA, R.V. College of Engineering, Bangalore

Abstract. The advancement in technology makes Mobile Banking transaction more sophisticated. The OTP SMS is generated by the bank server and is handed over to the client's mobile subscriber. To avoid any possible attacks like phishing and other attacks, the OTP must be secured. In order to provide reliable and secure mobile transactions without any compromise to convenience, a reliable m-banking authentication scheme that combines the secret PIN with encryption of the one-time password (OTP) has been developed in this paper. The secured OTP while using eZeeMPay salt algorithm seek to give the more secured m-banking transaction. After the encrypted OTP SMS reaches the client's mobile, the OTP is used again used for decrypting. The plain OTP text should be sent back to the bank will verified at the server to complete the transaction initiated. The combination of OTP with secured OTP provides authentication and security.

Keywords: OTP, eZeeMPay, Salt. SHA, SHA-512, SHA1, SHA1PRNG.

1 Introduction

Security of the banking transactions is the primary concern of the banking industries. The lack of Security may result in serious damages the security issue [1]. SMS (Short Messaging Service) allows users to send and receive text messages on a mobile phone using the numbered keypad on the handset to input characters. Each message can be up to 160 characters long and sent to and from users of different operator networks. All mobile phones available today support SMS. Indeed, SMS has become a global phenomenon, with billions of text messages sent worldwide every week.

SMS Banking requires a registered customer to initiate a transaction by sending a structured SMS message to the Mobile Banking Service. This SMS requires a tag word identifier to instruct the SMS gateway to submit the message to the correct SMS application.

To perform transactions using a mobile device, a customer initiates the request that goes through the mobile network operator (MNO) and terminates at the server application that can be administered by a technology vendor or independently by a financial institution. Between the customer and application server, data is transmitted across GSM network. Because of the security flaws that exist in GSM, the financial institutions are urged to secure the messages with adequate measures prior to sending them. Securing the messages before are sent ensures the end-to-end security, should

© Springer International Publishing Switzerland 2015

R. Silhavy et al. (eds.), *Software Engineering in Intelligent Systems*,

Advances in Intelligent Systems and Computing 349, DOI: 10.1007/978-3-319-18473-9_22

data be intercepted in transit, the intercepted messages are just useless to the interceptor. There are number of security measures have to be taken in order to enhance the security controls of mobile banking systems.

Short Message Service (SMS) [2] based One-Time Passwords were introduced to give secured authentication and authorization of Internet services. The OTP is required to authorize a transaction [3] [4] [5] [6] [7] [8] [9] [10]. SMS-based One-Time Passwords (SMS OTP) were introduced to counter phishing and other attacks against Internet services such as online banking. Today, SMS OTPs are commonly used for authentication and authorization for many different applications. Recently, SMS OTPs have come under heavy attack, especially by smartphone. In this paper, we analyze the security architecture of SMS OTP systems and study attacks that pose a threat to Internet-based authentication and authorization services. We determined that the two foundations SMS OTP is built on, cellular networks and mobile handsets were completely different at the time when SMS OTP was designed and introduced. Throughout this work, we show why SMS OTP systems cannot be considered secure anymore. Based on our findings, we propose mechanisms to secure SMS OTPs against attacks

2   Literature Review

There is several research publications related to security in mobile banking systems and GSM are reviewed and analyzed. The focus is on issues related to provision of end-to-end security, mutual authentication, message integrity, effective server security management policies and mainly on securing financial transactions over mobile networks.

The SHA (Secure Hash Algorithm) is one of a number of cryptographic hash functions. It was introduced by NIST as federal information processing standard (FIPS) [14] and it was first introduced in 1993. A cryptographic hash is like a signature for a text or a data file. SHA-256 algorithm generates an almost-unique, fixed size 256-bit (32-byte) hash. It is a compression function operates on a 512-bit message block and a 256-bit intermediate hash value. It is essentially a 256-bit block cipher algorithm which encrypts the intermediate hash value using the message block as key [15] [16].

The SHA-512 compression function operates on a 1024-bit message block and a 512-bit intermediate hash value. It is essentially a 512-bit block cipher algorithm which encrypts the intermediate hash value using the message block as key. Table 1 shows the different function strength.

In general salt is a random block of data. Computer languages provide different random number generation classes or functions are used to generate random numbers and bytes, but these classes and functions are not able to generate cryptographically secure random numbers. They are pseudo random number generators (PRNG) algorithms which are used by classes and functions in any language because the random value is completely dependent on data used to initiate the algorithm. So cryptographically secure pseudo random number generator (CSPRNG) algorithm is to be required which must produce statically random number and they must hold up against attack. In some highly secure application such as banking application where these salt algorithm need to be used to produce true random number. A general Steps to generate Salt Hash password.

1. Get user's password
2. Generate Salt using trusted random method
3. Append salt (system defined) to original password
4. Generate Salt Hash password using appropriate hash function
5. Store salt hash in the database

Table 1. Popular cryptographic hashes [11]

Function	2010	2011	2012
MD5	Broken	Broken	Broken
SHA0	Broken	Broken	Broken
SHA1	Week	Week	Week
SHA2	Week	Week	Week
SHA3	unbroken	unbroken	unbroken

2.1   SHA-512 Performance

It is more cost effective to compute a SHA-512 than it is to compute a SHA-256 over a given size of data. The attacks against SHA-1 reversed this situation and there are many new proposals being evaluated in response to the NIST SHA-3 competition. In the consequence of the SHA-1 attacks the advice NIST produced was to move to SHA-256 [14].

The SHA-512 compression function operates on a 1024-bit message block and a 512-bit intermediate hash value. It is essentially a 512-bit block cipher algorithm which encrypts the intermediate hash value using the message block as key[15]. SHA-512 is faster than SHA-256 on 64-bit machines is that has 37.5% less rounds per byte (80 rounds operating on 128 byte blocks) compared to SHA-256 (64 rounds operating on 64 byte blocks), where the operations use 64-bit integer arithmetic. The adoption across the breadth of our product range of 64 bit ALU's make it possible to achieve better security using SHA-512 in less time than it takes to compute a SHA-256 hash.

3   Proposed Secured Password System Flow

The Proposed system allows the user to use a secured password while providing greater reassurance and security. The proposed method guarantees the highly secured online banking transactions while using eZeeMPay Salt Algorithm.

eZeeMPay Salt Algorithm been generated using a Cryptographically Secure Pseudo-Random Number Generator (CSPRNG). CSPRNGs are designed to be cryptographically secure, meaning they provide a high level of randomness and are completely unpredictable. The salt needs to be unique per-user per-password. The password should be hashed using a new random salt. Never reuse a salt, Date Time help to never reuse the older salt that is per user and per password. The salt also needs to be long, so that there are many possible salts. Thumb rule of eZeeMPay Salt Algorithm to make salt at least as long as the hash function's output. The salt should be stored in the user's account table alongside the hash.

A hacker cannot attack a hash when he doesn't know the algorithm, but as per note from Kerckhoffs's principle [13], that the hacker will usually have access to the source code (especially if it's free or open source software), that is the reason our proposed solution has used the two way security, hence properly salting the hash solves the problem.

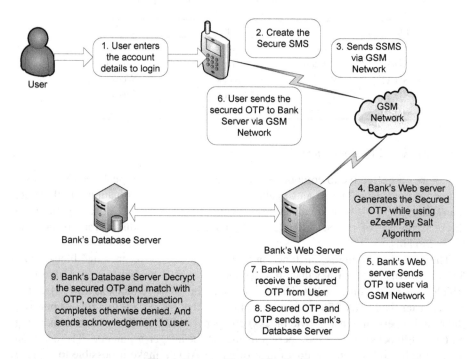

Fig. 1. Proposed System Design flow

3.1 Proposed System Flow

1. User visit to the Bank's Application to make the transaction using bank's user name and password.

2. If user name and password is correct then bank's application allows the user to access his/her account.

3. User fill the transaction detail and proceed next step for the making the transaction.

4. Bank sends the OTP to verify and complete the transaction.

5. User enters the OTP password which goes to the bank's Webserver.

6. Bank's Webserver generate the secured OTP password while using eZeeMPay salt Algorithm, which includes SHA 512 + eZeeMPay salt, eZeeMPay salt algorithm has been generated with the help of User's user name, DOB, ATM PIN and Date Time. And sends the OTP and Secured OTP to bank's Database server.

7. As soon as secured OTP and OTP reaches Database server, secured OTP been decrypted by eZeeMPay salt Algorithm and compare with OTP.

8. At the Database server, if decrypted OTP and OTP match then transaction will be completed, otherwise it gets denied and acknowledgement sent to User's Mobile via bank's Webserver.

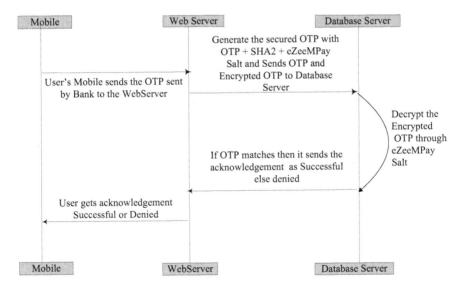

Fig. 2. Proposed Secured OTP generation flow

3.2   Pseudo Code of eZeeMPay in Java

The pseudo code of the eZeeMPay Salt algorithm where encryptEZeeMPayPassword method encrypt the password while passing the two parameter one is userPassword and other is ezeeMpaySalt.

```
public static String encryptEZeeMPayPassword(final
String userPassword, final String eZeeMPaySalt) throws
NoSuchProviderException {

MessageDigest md;

String eZeeMPayPassword = "";

try {

md = MessageDigest.getInstance("SHA-512");

md.update(eZeeMPaySalt.getBytes());

final byte[] bytes =
md.digest(userPassword.getBytes());

.


.


.


}

eZeeMPayPassword = sb.toString();

} catch (final NoSuchAlgorithmException e) {

System.out.println("ERROR: " + e.getMessage());

}


return eZeeMPayPassword;

}
```

The pseudo code of the eZeeMPay Salt algorithm where decryptEZeeMPayPassword
method decrypts the password while passing the two parameter one is userPassword
and other is ezeeMpaySalt.

```
public String decryptEZeeMPayPassword(final String
eZeeMPayPassword) {

// Pass the encrypted eZeeMPay password

// Use the eZeeMPay encrypted password algorithm to
decrypt the password.

return eZeeMPayPassword;

}
```

The pseudo code of the eZeeMPay Salt algorithm where getSecureRandom method generate the secure random number while using SHA1PRNG, "SHA1PRNG" is a sun.security.provider.SecureRandom implementation which generate the Seed, set the Seed and initialises the object if not already seeded otherwise mixes it in the additional data to the existing PRNG seed.

```
public static String getSecureRandom() throws
NoSuchAlgorithmException, NoSuchProviderException {

// SecureRandom generator

final SecureRandom secureRandom =
SecureRandom.getInstance("SHA1PRNG", "SUN");

.

return saltByte.toString();

}
```

The pseudo code of the eZeeMPay Salt algorithm where getSecureRandom method create the user salt while using user id, dob, ATM pin and Date time.

```
public static String createUserSalt() throws
NoSuchAlgorithmException, NoSuchProviderException {

return userSalt.toString();

}
```

3.3   Results and Verification

Proposed eZeeMPay Salt algorithm does not have any secret hashing, still attacker won't be able to know the exact salt been applied. More on that at the Database end Algorithm again decrypt and verify. This make the algorithm more secure.

Below result has shown the applied user's salt, secure random number and complete eZeeMPay salt password. Fig.3 has shown the result's screenshot.

User Salt: basu1234523/01/19771234Sat Jan 03 03:32:00 IST 2015

Secure Random Number: [B@5fcf29

eZeeMPay Algorithm Salt password:
bec00b386761f48735a6d0d81fc54a2bb2baca64582ff3f0bad6057acdbce2aa2c834c
d7e2354c68dde34e35b33b28496c6c62d91895295e23b3dd16021f8eba

User's Password:basu12345
User Salt with Secured Random: basu1234523/01/19771234Sat Jan 03 03:32:00 IST 2015[B@5fcf29
eZeeMPayPassword : bec00b386761f48735a6d0d81fc54a2bb2baca64582ff3f0bad6057acdbce2aa2c834cd7e2354c68dde34e35b33b28496c6c62d91895295e23b3dd16021f8eba

Fig. 3. Proposed Secured eZeeMPay Algorithm Salt password

Generated the SHA512 hash on online tool md5#ashing.net [12] and miniwebtool [17], have observed the same hash been generated, shown in Fig.4 and Fig.5, but it was not same hash when we have use the eZeeMPay salted algorithm, this has been shown in Fig.3.

Also decode the hash generated by the md5#ashing.net [12] and miniwebtool, as well decode the hash, shown in Fig.6, but when tried to decode the eZeeMPay salted hash password it shows failed, shown in Fig.7.

SHA512:

c5f79315cdf5265501a53f9bc39b5268b7a71ab1f5c7f0146cfda1a42f8d2cf3b061d8
b732d5e1e70440d1bbadb9a7965ca3451c914e90430a9e7666c2f19b2f

Fig. 4. Generated SHA512 hash on md5#ashing.net [12]

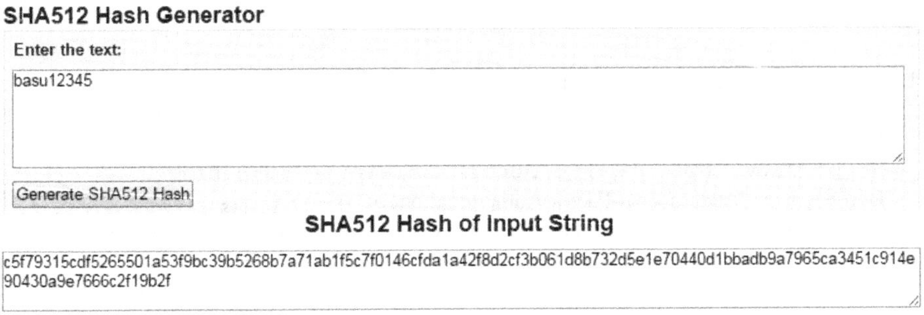

SHA512 Hash Generator

Enter the text:

basu12345

Generate SHA512 Hash

SHA512 Hash of Input String

c5f79315cdf5265501a53f9bc39b5268b7a71ab1f5c7f0146cfda1a42f8d2cf3b061d8b732d5e1e70440d1bbadb9a7965ca3451c914e
90430a9e7666c2f19b2f

Fig. 5. Generated the SHA512 through [17]

Decoded sha512 (?) **Hash**

Decoded Value:

basu12345

Select Decoded Value

Your hash is decoded! You're very lucky

sha512 hash: c5f79315cdf5265501a53f9bc39b5268b7a71ab1f5c7f0146cfda1a42f8d2cf3b061d8b732d

Fig. 6. Decoded the SHA512 Hash which is generated through md5#ashing.net [12] and miniwebtool [17]

Searching sha512 (?) **Hash**

Reverse decoding Failed for:

sha512 hash:
586b9103905a82666192eff60b0541c9cbf9de1fba41eab8600b397202c58533c3fa00be0b3e

Fig. 7. Decoded the SHA512 Hash which is generated through eZeeMPay Salt Algorithm

4   Conclusions

In this current research work, we seek to improve the security of SMS-based one-time passwords, while implementing the eZeeMPay Salt algorithm on Bank's Webserver and Bank's Database server. In today's world, one would expect that OTPs are transported using end-to-end security. Our work shows that this is not true anymore, unless we have same security implemented on back end server.

References

1. Managing the Risk of Mobile Banking Technologies, Bankable Frontier Associates
2. 3rd Generation Partnership Project. 3GPP TS 23.040 - Technical realization of the Short Message Service (SMS) (September 2004), http://www.3gpp.org/ftp/Specs/html-info/23040.htm
3. Duo Security. Modern Two-Factor Authentication, http://duosecurity.com
4. PhoneFactor, Inc. Comparing PhoneFactor to Other SMS Authentication Solutions, http://www.phonefactor.com/sms-authentication
5. Yang, R.: SMS Text Message Based Authentication. Citrix Developer Network: http://community.citrix.com/display/xa/SMS+Text+Message+Based+Authentication
6. VISUALtron Software Corporation. 2-Factor Authentication - What is MobileKey? http://www.visualtron.com/products-mobilekey.htm
7. SMS PASSCODE A/S. Two-factor Authentication, http://www.smspasscode.com/twofactorauthentication
8. Google Inc. SMS Verification for App Creation, https://developers.google.com/appengine/kb/sms
9. Google Inc. Verifying your account via SMS or Voice Call, http://support.google.com/mail/bin/answer.py?hl=en&answer=114129
10. Blizzard Inc. Battle.net SMS Protect FAQ (September 2012), https://us.battle.net/support/en/article/battlenet-sms-protect
11. http://valerieaurora.org/hash.html
12. http://md5hashing.net/
13. https://en.wikipedia.org/wiki/Kerckhoffs%27s_principle
14. http://csrc.nist.gov/publications/fips/fips180-2/fips180-2.pdf (February 10, 2014)
15. NIST, "NIST Brief Comments on Recent Cryptanalytic Attacks on Secure Hashing Functions and Continued Security Provided by SHA-1" (August 25, 2004), http://csrc.nist.gov/groups/ST/toolkit/documents/shs/hash_standards_comments.pdf
16. NIST, "Descriptions of SHA-256, SHA-384, and SHA-512", http://csrc.nist.gov/groups/STM/cavp/documents/shs/sha256-384-512.pdf
17. http://www.miniwebtool.com/hash-and-checksum/

Improving Mutation Testing Process of Python Programs

Anna Derezinska and Konrad Hałas

Institute of Computer Science, Warsaw University of Technology
Nowowiejska 15/19, 00-665 Warsaw, Poland
A.Derezinska@ii.pw.edu.pl,
halas.konrad@gmail.com

Abstract. Mutation testing helps in evaluation of test suite quality and test development. It can be directed to programs of different languages. High cost of a mutation testing process limits its applicability. This paper focuses on mutation testing of Python programs, discussing several issues of mutant creation and execution. It was showed how they can be effectively handled in the Python environment. We discuss introduction of first and higher order mutation in an abstract syntax tree with use of generators, dealing with code coverage with AST, executing mutants via mutant injection into tests. The solutions were used in reengineering of MutPy - a mutation testing tool for Python programs. The improvements were positively verified in mutation experiments.

Keywords: mutation testing, Python, dynamically typed language.

1   Introduction

Python is an interpreted high-level programing language [1] gaining more and more concern in different application areas, including web development, internet protocols, software engineering tools, or scientific computing [2]. Python programs, like in other general purpose languages, can be companioned with a test suite, especially unit tests. Test quality evaluation and their development can be assisted by mutation testing [3].

In mutation testing, a small change, so-called *mutation*, is introduced into a program code. A type of changes is specified by *mutation operators*. If a mutated program, a *mutant*, is modified by one mutation operator at one place, we speak about first order mutation. Introducing of many mutations into the same mutant is called higher order mutation. Mutants are run against tests. A mutant status is determined by the output of its test runs, a mutant can be *killed* by tests, *alive*, or *equivalent* – not to be killable by any tests. *Mutation score* of a test set is calculated as the number of killed mutants divided by the number of all considered nonequivalent mutants.

Usage of mutation operators in Python is restricted in comparison to other languages, like Java or C#. Python belongs to dynamically typed languages which raises a problem of *incompetent mutants,* mutants with possible type inconsistencies [4]. We proposed a pragmatic approach, in which some incompetent mutants are generated and run against tests. A set of mutation operators applicable in this context to Python

© Springer International Publishing Switzerland 2015
R. Silhavy et al. (eds.), *Software Engineering in Intelligent Systems,*
Advances in Intelligent Systems and Computing 349, DOI: 10.1007/978-3-319-18473-9_23

is discussed in [5]. It was experimentally shown that the number of incompetent mutants was limited (about 5% on average) and the mutation results were accurate.

Details of a mutation testing process depend on different factors, such as a program level at which mutations are introduced (e.g. source code, syntax tree, intermediate code), selection of mutation operators, limitation of a mutant number due to different reduction criteria, selection of tests to be run, etc. [6,7]. The main problem of a mutation testing process is its high labor consumption. There are many attempts to reduce the cost, focusing on different aspects of the process [3], [8-12]. Some methods of mutant number reduction are associated with the less accurate results of the mutation testing process. One of approaches, in which less mutants are created and run, is higher order mutation [13]. Experiences of higher order mutation of Python programs are reported in [14].

Dynamic typing makes it possible to encounter incompetent mutants. Therefore, mutation in Python can be done either in the run-time or we can deal with incompetent mutants during test execution. In this work, based on our experiences [5], [14], the second approach is continued. Mutations are introduced into the program code and during code interpretation incompetent mutants are discarded.

In this paper, we focus on the improvements to the mutation testing process that do not decline the mutation score. We present how using different Python structures, selected steps of the process could be realized more effectively. The approaches were applied in refactoring of the primary version of MutPy – a mutation testing tool for Python [15]. Experiments confirmed benefits for mutation testing of Python programs.

This paper is organized as follows: we describe selected features of Python programs in the next Section. Section 3 presents various problems and their solutions for the mutation testing of Python programs. Tool support for the process and experimental results are discussed in Section 4. Finally, Section 5 describes related work and Section 6 concludes the paper.

2   Fundamentals of Python Programs

The Python programing language [1] supports the notions of procedural and object-oriented programming. Python is an interpreted language. A program is translated into its intermediate form that is interpreted during a program run. During the translation process, at first an Abstract Syntax Tree (AST) of a Python program is created. The tree is further converted into the bytecode form. Below, we recall briefly selected characteristics of the Python language.

2.1   Iteration and Generators

In Python, loop *for* is used in iterating over collection elements. An object used as an argument in a *for* loop is required to have a special *iterator* method. An iterator returns a next element of the collection or a special exception at the collection end.

Using this schema, a *generator function* can be created. It is designated by usage the keyword *yield* instead of the *return* operation. Each calling of the function, its iterator is returned, so-called *generator iterator,* or *generator* in short. An iteration step corresponds to execution of the function body till encountering of a *yield* statement. The function status is saved. In the next iteration step, the function state is resumed, and the code is executed until the subsequent *yield* statement or the function end occurs. In this way, an iterated object (generator) is created and can be used in a *for* loop.

2.2   Module Import

A module is a basic unit of a program execution. Each module has its namespace. Elements of a namespace, defined in a module or imported from other modules, are accessed by a module *dictionary.* Import of Python modules is supported by the interpreter system. The import algorithm consists of the following main steps:

1. A module is imported via *import* statement. If the module was already imported, its reference exists in the appropriate dictionary and can be further used. Finish.
2. Otherwise, the Python interpreter searches for the desired module by iterating over a system list of *meta_paths.* For each element of the list a so-called *finder* is called. The finder returns a *loader* object if the module was found. If all finders were asked for the module and none found it, the procedure is finished. The appropriate exception (*ImportError*) is raised indicating the missing module.
3. After the module was found, the *loader* is used to load the module. The reference of the loaded module is added to the dictionary. The procedure is completed and the module can be used.
4. It can happen that a loader cannot load the module. This situation is also pointed out by the exception *ImportError.*

A Python user can expand this functionality by designing appropriate *finder* and *loaders* objects.

3   Mutation Testing Process Improvements

In this section we discuss the improvements in the selected steps of the mutation testing process: introduction of mutations into the code, execution mutants with tests, and application of code coverage data in the mutant creation and execution.

3.1   Introducing First Order Mutations into Python AST

Application of a mutation into a code is based on a mutation operator. An operator specifies a kind of a source code that can be mutated and a kind of a code change that should be introduced.

One idea of application of mutation operators is manipulation of a code on the Abstract Syntax Tree (AST) level. Using the standard *ast* module we can manipulate the tree and apply mutation operators. A simple process of a mutant creation can consist of the following steps:

1. An original code is converted into its AST form (once per a program).
2. For each mutation operator, a copy of the AST is made:
   (a) a place of a mutation is found in the tree,
   (b) information about the mutation place is stored,
   (c) the tree is modified according to a given mutation operator,
   (d) the modified tree is adjusted - the line numbers are revised,
   (e) the tree is passed to the further translation steps and the intermediate code of the mutant is created.

Considering one place of AST many mutants can be created due to different mutation operators. The basic input of a module accomplishing a mutation operator is an AST node. The output of the mutation is a modified node, or information about a node deletion, or information about resining of the mutation (e.g. raising of an appropriate exception). A node that was already used by all mutation operators is marked as a recently visited code. Therefore, the same node will not be modified again during creation of subsequent mutants.

This simple introduction of mutations into an AST has disadvantages. The first one is copying of a syntax tree. It is performed in order to keep the original tree unmodified to be used for creation of subsequent mutants. However this coping is memory and time-consuming. An alternative to creation of a copy can be a reverse transformation of the tree. The modified tree fragments are restored to the original form.

The next problem is traversing of the whole tree while searching of the mutation place. This operation can take a considerable time amount in dependence of the number of tree nodes. Straightforward searching in a tree results in visiting of nodes modified in mutants that have already been generated during the mutation process.

Therefore, an improved and more efficient algorithm of mutation injection can solve these problems. A syntax tree is not to be copied for each step. The search engine of mutation places starts with the recently visited place. These ideas can be efficiently realized in Python with assistance of the *generator* concept (Sect. 2.1).

A new mutation engine can be served by a *MutationOperator* base class which is specialized by various classes responsible for particular mutation operators. A mutation method of the base class is a generator function. Therefore, we can iterate other the results of this operation. Input of the mutation method is the original AST to be mutated. Using the generator mechanism, succeeding mutants in the AST form are returned and a context is stored. The context is used by the mutation algorithm to memorize a mutation place in a tree. After a subsequent call of the generator function, the AST can be restored and the searching for a next mutation place can be continued.

At first, a possibility of being mutated is checked for a current node. If a mutation is allowed at this place the node is mutated. Then, the *yield* method returns a new version of the mutated node and suspends mutant generation process according to the *generator* principle. If a node cannot be mutated, its children nodes are visited. For each child a mutation generator is recursively activated in the *for* loop. A child node is substituted by a mutated node returned by the recursive call of the mutation generator. After the next iteration of the generator the AST is reconstructed.

More complex cases, are also taken into account. A node can be inspected for the next time, even if it was already modified. Moreover, an examined child of the current node can consist not only of a single node but also of a whole collection of nodes.

Other functionality performed on an AST is mutation of nodes based on the context information. A context of a node is specified mainly by its predecessor. Node successors are simply accessible in a tree. In order to access a node predecessor an AST supported by the standard Python library should be extended. Working on a node context, we can easily omit selected nodes that should not be mutated, e.g. documentation nodes that are equal to string nodes but are placed in a different context.

3.2 Higher Order Mutation

Mutations discussed in the previous subsection referred to the first order mutation (FOM). In higher order mutation, two or more changes are provided to the same mutant. A higher order mutant (HOM) can be created using several algorithms, such as FirstToLast, Between-Operators, Random or Each-Choice [13], [16]. For example, using a FirstToLast strategy, a 2nd order mutant is a combination of a not-used FOM from the top of the list with a not-used FOM from the list end.

Higher order mutation can also be applied to Python programs [14]. While building a HOM, changes of a program are also forced into the AST form. At the beginning, a list of all first order mutations is generated. The general idea of creating a HOM is based on merging of AST node changes defined by the first order mutation. If different HOM algorithms are handled, the strategy design pattern can be used.

In general, conflict situations where more than one mutation have to modify the same node are possible. We assumed that an AST node is modified only once in a given HOM. In this way, a situation is avoided when a next modification overwrites the previous one. Therefore, the merging algorithms were modified, taking into account the next first order mutation from the list.

3.3 Mutant Injection into Tests

Execution of mutants with tests can be effectively realized by a mutant "injection" into tests. Based on this operation, a test will be re-run with the mutated code, not an original one. It can be performed in the operational memory, without storing a mutant to a disk.

During evaluation of mutants, two program modules are available: a mutant module and a test module. If a mutant has to be tested, a reference in the test namespace should refer to the mutant functions instead of those from the original program. A valid reference to a mutated function is resolved by a concept of mutant injection.

A simple approach to mutant injection can be based on the comparison between namespaces of a mutant module and a test module. If an object belongs to both workspaces, its mutated version is injected into the namespace of the test module. This operation can be implemented by a simple assignment. However, this solution is not sufficient when a mutated module is imported inside a function or a method. Import mechanism is often used in such a context for different purposes. During a re-run of tests, a test method will import an original code but not its mutated version.

A new approach on the mutant injection was based on the extension of the import module mechanism (Sect. 2.2). Additional *finder* and *loader* were created that can be activated after an attempt to import a mutant module. Therefore, a mutation engine overrides importing a mutated module in a function or method. A dedicated class can combine the functionality of a finder and a loader. The class is treated as an *importer* in the Python system. Before a test is launched, our new importer should be added at the beginning of a system list (*meta_path*). If so, during the test execution this importer will be called at first in the import routine. Therefore, a mutated version of the module will be fetched instead of the original one.

3.4   Code Coverage Evaluation on AST

Code coverage is an important criterion of test quality. Reachability of a mutation place by a test is a one of necessary conditions to kill a mutant [17].

Performing mutation on the AST level is reasonable to resolve the code coverage also on the same program level. The solution is inspired by an instrumentation of AST [18]. The idea of the   algorithm can be summarized in the following steps:

1. Numbering of all nodes of an AST.
2. Injection a special nodes to the AST. Calling of these nodes results in memorizing a number of a considered node.
3. Transformation of the modified AST into the intermediate form of a program.
4. Running the program (i.e., interpreting it) and collecting information about reachable nodes. A code of a reached AST node is counted to be covered.

Injected nodes are related to instructions that add a node number to a special set. These so-called *coverage instructions* have various types and are injected into different places in accordance to a considered original AST node:

— *a simple expression (line)* – a coverage instruction is injected just before the node; it adds a set of the numbers of all nodes included in the expression,
— *a definition of a function, method or a class* – a coverage instruction is injected at the beginning of the body of the node; it adds the number of the node definition,
— *a loop or conditional instruction* - a coverage instruction is injected just before the node; it adds a set of the numbers of all nodes included in the loop or conditional instruction,
— *other entities (e.g. nodes inside expressions)* – no instruction is injected, as the node number is already included in an instruction injected for a simple expression.

A mutant generation engine can use information about coverage of AST nodes. Mutations that would modify non-covered nodes can be omitted. Hence, the number of mutants to be generated is reduced.

Analyzing code coverage we can also obtain information by which test cases a particular node is covered. Using this information we can run only those tests that cover the node modified by a considered mutant. Hence the number of test runs is lowered.

3.5   Reduction of Test Runs

If all tests are run with each mutant (or only tests covering the mutation) we can gather detailed information about test effectiveness [19]. Though, if we are only interested in the mutation score, the testing process can be performed quicker. The testing of a mutant is interrupted if any test kills the mutant or a mutant is classified as an incompetent one. The remaining tests can be omitted for this mutant.

4   Tool and Experiments

The solutions discussed in Sect. 3 were implemented in the newest version of the MutPy tool [15]. MutPy was the first mutation testing tool supporting object-oriented mutations of Python programs and written in the Python language. Its earlier version (v. 0.2) was published in 2011. Based on the gathered experience, the tool was considerably refactored in order to assist the more effective mutation testing process. The main functionality supported by the current version of the MutPy tool (v. 0.4) is the following:

— generation of mutants with selected mutation operators, 20 operators including structural, object-oriented, and Python specific operators [5],
— first- or high-order mutation operators of a selected algorithm of HOMs: FirstToLast, Between-Operators, Random and Each-Choice [14].
— mutation of all code, or mutation of the code covered by tests of a given test suite,
— selection of a code that can be omitted while a mutation is introduced, e.g. a code of test cases,
— storing of mutants on disk (optionally),
— view of original and mutated Python source code and intermediate code,
— selection of tests that cover the mutated parts of code,
— running mutants with unit tests,
— categorization of mutants after test runs, detection of mutants alive, incompetent, killed by tests and by timeout,
— evaluation of mutation score and generation of reports on mutation results.

Results of two versions of the mutation testing process are compared in Table 1. The table summarizes the mutation testing of four projects: *astmonkey, bitstring, dictsect* and *MutPy* (v. 0.2) [14]. The data refers to first order mutation without considering code coverage data, as in the previous process only this kind of mutation was supported. The second column distinguishes a version of the mutation testing process with MutPy 0.2 (old), and the new one.

The first two rows present the total time of the mutation testing process. This time includes time of mutant generation, execution of test runs and evaluation of results. We can observe that in the new version the time is 3 to 7 times shorter. The highest benefits were for the biggest project - *bitstring*, with the highest number of tests and the longest execution time. However, there are several factors that contribute to these times. The first one is the number of mutants caused by the revised set of mutation operators and avoidance of creation many incompetent mutants [5]. The number of mutants influences both the time of all mutant creation and all mutant execution. The second factor is the number of tests. It was lowered due to the strategy of deciding a mutant status after killing it by any test (Sect. 3.5).

Another factor is a cost of a mutant creation. The impact of this factor can be calculated independently. Average times of a mutant generation are compared in the bottom rows of the table. This improvement is independent on the mutant number and follows from the new algorithm of mutation introduction. The average time of a mutant generation was from few to even more than hundred times shorter that previously. In the former solution, generation time depends strongly on the size of a program, therefore, for the biggest project (*bitstring*) the improvement is the most substantial. The current algorithm of an AST manipulation caused that this average time is of the similar magnitude for different projects (0.06-0.02s).

The new approach to mutant injection into tests has not visible effect in the performance measures, but has made it possible to mutate some projects that were previously not allowed.

Table 1. Comparison of two versions of mutation testing process performed with MutPy

Measure	Version	astmonkey	bitstring	dictset	MutPy_old	Sum
Time of mutation	old	635.6	19405.4	297.8	337.8	20676.6
process [s]	new	98.7	2700.0	60.4	108.1	2967.2
Mutant number	old	1010	5444	572	703	7729
	new	521	3774	311	270	4876
Test run number	old	66660	2482464	98956	60458	2708538
	new	22230	733260	28500	16843	800833
Average time of	old	0.53	2.42	0.38	0.23	-
mutant generation [s]	new	0.05	0.02	0.05	0.06	-

Other improvements in the process performance are obtained by usage of code coverage information and higher order mutation. Usage of code coverage data reduced the mutant number of 6 % on average. The higher impact has the code coverage criterion in selecting of tests. The number of test runs dropped about 74-94 % for these projects. Summing up, only due to the code coverage inspection both in mutant generation and execution the mutation time was lowered, taking 44-88 % of the time without this data.

The detailed results of 2nd and 3rd order mutation with different HOM algorithms concerning these projects are given in [14]. In general, the number of mutants were halved for the 2nd order and about 33 % for the 3rd order. The resulting mutation time was equal 40-83 %, and 25-90 % of the 1st order mutation time, for 2nd and 3rd order mutation accordingly.

5 Related Work

Manipulation of the syntax tree in the mutant generation step was applied for different languages. Mutation of C# programs was performed at AST of the source code [20], but also at a syntax tree of the Intermediate Language form of .NET [21]. In this paper, the general ideas are combined with the Python-specific utilities. The most of other tools dealing with mutation testing of Python programs has a limited functionality and is restricted to few standard mutation operators, with no support to object-oriented ones. Their discussion can be found in [14]. Simple application of AST for introduction of Python mutations was also reported in the Elcap tool [22].

The problems of mutating dynamically typed languages were addressed in [4]. Mutations introduced in run-time were analyzed and illustrated by JavaScript examples. In the contrast, this paper was focused on the mutation in the source code, as a solution giving practically justified results.

An interesting approach to verification of Python programs was presented in [23], where mutations are introduced into the intermediate code. However, the purpose of mutation is different from discussed here. The authors compare error location abilities of two tools: UnnaturalCode for Python and the Python translator. UnnaturalCode locates errors in an instrumented program during its run-time. Many mutations introduced to the code could be detected during translation of the source code, and in this sense they are not suitable for evaluation of tests, as in the typical mutation testing.

Application of coverage data in the step of mutant generation was beneficially performed in the CREAM tool for mutation testing of C# programs [20]. In PROTEUM, the mutation environment for C programs, the code instrumentation was used and the reachability of mutation statements by tests were observed [6]. However, there is no information about an prior reduction of a test set based on the coverage data, like performed in the process discussed here. The integration of code coverage with mutation testing approach was also promoted in Pitest – a mutation tool for Java programs.

6   Conclusion

In this paper, we proposed improvements to the mutation testing process of Python programs. They concerned mutant generation based on modifications of an abstract syntax tree and execution of mutants with tests. First and higher order mutations were used. We explained how different Python language structures, like iteration with generators and an importer class that extends module import, were applied in the mutation tool reengineering. The advances implemented in the MutPy tool resulted in the considerable lowering the mutation time.

References

1. Python Programming Language, http://python.org
2. Pérez, F., Granger, B.E., Hunter, J.D.: Python: An Ecosystem for Scientific Computing. Computing in Science & Engineering 13(2), 13–21 (2011)
3. Jia, Y., Harman, M.: An Analysis and Survey of the Development of Mutation Testing. IEEE Trans. Softw. Eng. 37(5), 649–678 (2011)
4. Bottaci, L.: Type Sensitive Application of Mutation Operators for Dynamically Typed Programs. In: Proceedings of 3rd International Conference on Software Testing, Verification and Validation Workshops (ICSTW), pp. 126–131. IEEE Comp. Soc. (2010)

5. Derezińska, A., Hałas, K.: Analysis of Mutation Operators for the Python Language. In: Zamojski, W., Mazurkiewicz, J., Sugier, J., Walkowiak, T., Kacprzyk, J. (eds.) DepCos-RELCOMEX 2014. AISC, vol. 286, pp. 155–164. Springer, Heidelberg (2014)
6. Vincenzi, A.M.R., Simao, A.S., Delamro, M.E., Maldonado, J.C.: Muta-Pro: Towards the Definition of a Mutation testing Process. J. of the Brazilian Computer Society 12(2), 49–61 (2006)
7. Mateo, P.R., Usaola, M.P., Offutt, J.: Mutation at the Multi-Class and System Levels. Science of Computer Programming 78, 364–387 (2013)
8. Usaola, M.P., Mateo, P.R.: Mutation Testing Cost Reduction Techniques: a Survey. IEEE Software 27(3), 80–86 (2010)
9. Offut, J., Rothermel, G., Zapf, C.: An Experimental Evaluation of Selective Mutation. In: 15th International Conference on Software Engineering, pp. 100–107 (1993)
10. Zhang, L., Gligoric, M., Marinov, D., Khurshid, S.: Operator-Based and Random Mutant Selection: Better Together. In: 28th IEEE/ACM Conference on Automated Software Engineering, Palo Alto, pp. 92–102 (2013)
11. Derezińska, A., Rudnik, M.: Quality Evaluation of Object-Oriented and Standard Mutation Operators Applied to C# Programs. In: Furia, C.A., Nanz, S. (eds.) TOOLS Europe 2012. LNCS, vol. 7304, pp. 42–57. Springer, Heidelberg (2012)
12. Bluemke, I., Kulesza, K.: Reduction in Mutation Testing of Java Classes. In: Holzinger, A., Libourel, T., Maciaszek, L.A., Mellor, S. (eds.) Proceedings of the 9th International Conference on Software Engineering and Applications, pp. 297–304. SCITEPRESS (2014)
13. Polo, M., Piattini, M., Garcıa-Rodrıguez, I.: Decreasing the Cost of Mutation Testing with Second-Order Mutants. Software Testing Verification & Reliability 19(2), 111–131 (2009)
14. Derezinska, A., Hałas, K.: Experimental Evaluation of Mutation Testing Approaches to Python Programs. In: Proceedings of IEEE International Conference on Software Testing, Verification, and Validation Workshops, pp. 156–164. IEEE Comp. Soc. (2014)
15. MutPy, https://bitbucket.org/khalas/mutpy
16. Mateo, P.R., Usaola, M.P., Aleman, J.L.F.: Validating 2nd-order Mutation at System Level. IEEE Trans. Softw. Eng. 39(4), 570–587 (2013)
17. DeMillo, R.A., Offutt, A.J.: Constraint based automatic test data generation. IEEE Transactions on Software Engineering 17(9), 900–910 (1991)
18. Dalke, A.: Instrumenting the AST, http://www.dalkescientific.com/writings/diary/archive/2010/02/22/instrumenting_the_ast.html
19. Derezinska, A.: Quality Assessment of Mutation Operators Dedicated for C# Programs. In: Proceedings of 6th International Conference on Quality Software, QSIC 2006, pp. 227–234. IEEE Computer Society Press, California (2006)
20. Derezińska, A., Szustek, A.: Object-Oriented Testing Capabilities and Performance Evaluation of the C# Mutation System. In: Szmuc, T., Szpyrka, M., Zendulka, J. (eds.) CEE-SET 2009. LNCS, vol. 7054, pp. 229–242. Springer, Heidelberg (2012)
21. VisualMutator, http://visualmutator.github.io/web/
22. Elcap, http://githup.com/sk-/elcap
23. Campbell, J.C., Hindle, A., Amaral, J.N.: Python: Where the Mutants Hide, or Corpus-based Coding Mistake Location in Dynamic Languages, http://webdocs.cs.ualberta.ca/~joshua2/python.pdf

Task Allocation Approaches in Distributed Agile Software Development: A Quasi-systematic Review

Marum Simão Filho[1,2], Plácido Rogério Pinheiro[1], and Adriano Bessa Albuquerque[1]

[1] University of Fortaleza, Fortaleza, Brazil
`{marum,placido,adrianoba}@unifor.br`
[2] 7 de Setembro College, Fortaleza, Brazil
`marum@fa7.edu.br`

Abstract. Increasingly, software organizations are investing in distributed software development. However, this new scenario introduces a number of new challenges and risks. Organizations have sought alternatives to the traditional software development models by applying agile software development practices to distributed development. The key point in a distributed scenario is related to task allocation. This paper conducts a quasi-systematic review of studies of task allocation in distributed software development projects that incorporate agile practices, trying to establish issues for additional research. The study allows us to conclude that there are few works on task classification and prioritization what suggests a fertile area for work.

Keywords: Distributed Software Development, Agile Methods, Task Allocation.

1   Introduction

The software development globalization is a reality more and more present in the world market for the supply of software. The business expansion into new markets, the need to expand the capacity of the workforce and the search for cost reduction are among the main factors of the increasing number of Distributed Software Development (DSD) projects [20]. However, the distribution in software development brings many challenges, such as cultural and time zone differences. In addition, communication, coordination, and control throughout the project have increased its complexity due to distribution [19].

This dynamic business environment requires organizations to develop and evolve software at much higher speeds than in the past. Agile methods seek to answer this demand by reducing the typical formalities of traditional processes of software development commonly used. As a result, software development companies have sought alternatives to traditional software development models applying agile software development methods to distributed development, in order to reap the benefits of both. However, agile approaches and distributed development differ significantly in their fundamental principles. For example, while agile methods are mainly based on informal processes to facilitate coordination, distributed software development is usually based on formal mechanisms [23]. Many studies have been conducted to incorporate agile practices to distributed software development aiming at minimizing the incongruities, such as is the case [8, 9, 2, and 28], to name a few.

Task allocation is critical to decision-making in project planning. Its complexity increases in distributed development software environment, where we must consider additional factors inherent to distribution [18]. The assignment of tasks can be seen as a critical activity for success in project management of distributed development. However, this assignment of tasks is still an important challenge in global software development due to insufficient understanding of the criteria that influence task allocation decisions [15].

Many other reviews were developed but they covered just a part of our desired goal: some of them aimed at exploring task allocation in DSD projects, other ones aimed at investigating about DSD with agile methods. Furthermore, some studies described approaches applied to the allocation of tasks; however, prioritization and task classification seem to be a little explored field.

In this paper, we describe a quasi-systematic review of studies on task allocation in distributed software development projects that incorporate agile practices, trying to establish issues of prioritization and classification of tasks to maximize team performance and to improve customer satisfaction.

The rest of the paper is organized as follows: Section 2 provides a brief description of the used method. Section 3 discusses the results of our research. Finally, in Section 4, the conclusions and suggestions for further work are provided.

2   Research Approach Used

Our goal is to survey on work involving allocation of tasks in distributed software development projects with agile characteristics. We divided the work into two stages: first, we searched for works, which involved distributed software development and agile methods. Lastly, we examined works that dealt with task allocation in distributed software development.

The whole methodology is detailed as follows.

RESEARCH QUESTION

Main Question
Problem: What are the existing models to support the task allocation process in distributed software development based on any agile method?
Secondary Question
What are the most common agile practices in distributed software development?

SAMPLE

Papers published in conferences and journals reporting models, approaches, frameworks and processes for allocating tasks to agile methods projects executed in distributed software development environments.

RESULTS

We hoped to obtain a set of existing models for task allocation in distributed software development based on agile methods.

EVALUATION AND TESTING APPROACH

No assessment was made regarding the validity of the models found. That is; the interest was restricted to the identification of models and criteria influencing the allocation.

RESEARCH SCOPE

To delineate the scope of the search, the primary sources of information to be searched were established. We performed the study based on searching in digital libraries, through their respective search engines.

The criteria adopted for the selection of libraries were as follows:

- It must have a search engine that allows the use of logical expressions or equivalent mechanism;
- It must include in its base exact or related area of publications that has a direct relation to the topic being researched;
- Search engines must allow searching the full text of publications;
- The search is restricted to the analysis of publications obtained exclusively from digital libraries. Access to these publications should not incur the cost to the research.

In our particular case, we used the search scope the following sources: articles and IEEE publications, articles and publications of the ACM, Science Direct, and Scopus.

SOURCES

The survey covered all the publications from publishers selected according to the criteria defined in the search scope. There are restrictions regarding the start date, but not the end date of publication in the sources of information. We looked for publications from 2000 to the present day.

LANGUAGE

For this research, we selected the English and Portuguese languages. We chose English language because it is adopted by most international conferences and journals related to research topic and for being the language used by most publishers related to the topic listed in the Brazilian Journal Portal. We also chose the Portuguese language because it is adopted by major national conferences and journals of Software Engineering.

SEARCH EXPRESSIONS

For articles in English, you should use the search expression below:

(distributed software development OR global software development OR collaborative software development OR global software engineering OR globally distributed work OR collaborative software engineering OR distributed development

OR distributed team OR global software teams OR globally distributed development OR geographically distributed software development OR offshore software development OR offshoring OR offshore OR offshore outsourcing OR dispersed teams) AND (task allocation OR task distribution OR task assignment) AND (Scrum OR agile).

SELECTION AND DATA EXTRACTION PROCEDURES

Selection Procedure

The study selection occurred in three steps:

i) Selection and preliminary cataloging of publications. The initial composition of the papers made from the application of the search expression in the search engines provided by the selected sources. We cataloged each publication in a spreadsheet and stored it in a repository for further analysis;

ii) Selection of relevant publications - 1st filter. The preliminary selection using the search expression did not ensure that the collected results would be useful in the context of research since the application of search expressions in engines is restricted to the syntactic aspect. To exclude irrelevant documents in the preliminary set, the following criteria were applied:

• Does the publication mention models, approaches, frameworks, algorithms or processes for allocating tasks to projects using agile methods that run on distributed software development environments?

• Does the publication mention factors that influence the allocation of tasks in agile projects in distributed software development environments?

Then, we moved the publications that met these criteria to a second repository, which was used in the third stage of selection.

iii) Selection of publications from quality criteria - 2nd filter. Much of the documents comprising the repository obtained from the application of the first filter cannot contain data to be collected. To minimize the collection effort, we applied a second filter to give a repository containing only documents that have data to be collected.

The criteria which was applied was as follows:

• Models, approaches, frameworks, algorithms or processes for allocating tasks in agile projects development environments run on distributed software. Publication should present some plan to support the allocation of teams in this context. The procedures identified were registered in the database relating it to the publication, the work context and the reference model used.

To reduce the risk of excluding a publication prematurely in one of the stages of the study, where there was doubt, the publication was not deleted.

2.1   Selection of the Reviewed Papers

One evaluator under orientation of two researchers performed the review during the course of 8 months. During the selection and preliminary cataloging of publications step, we got 148 publications. In the next step, after applying the first filter, we got the selection of relevant publications. This round of selection resulted in 80 research papers. Having applied the second filter, in the selection of publications from quality criteria step, we came up with 30 full research papers.

3   Results

3.1   Distributed Software Development (DSD) and Agile Methods

This section discusses task allocation in distributed software development projects that incorporate agile practices.

Much of the work involving DSD and agile methods deal with the incorporation of one or more agile practices to distributed software development process. Among these, many approach the practice of pair programming. Pair programming is a practice of Extreme Programming (XP) where two programmers work side by side on the same computer to produce a software artifact [29]. This technique has been shown to produce better quality code in less time than it would take for an individual programmer. Typically, high emphasis on communication is observed in this activity, and closeness between the couple seems inevitable. How developers who tend to be strangers around the world, connected through the Internet, can benefit of pair programming [8]?

Kircher et al. [14] proposed the concept of "Distributed Extreme Programming", which inherits the merits of XP and applies them to DSD. Natsu et al. [21] presented the COPPER system, a synchronous source code editor that allows two software engineers to write a program using distributed pair programming. COPPER implements characteristics of groupware systems, mechanisms of communication, collaboration, concurrency control, and a "radar view" of documents, among others. Flor [8] performed a study on XP and DSD which concludes that, for a couple of remote programmers to get the same properties as a close pair, and so benefit from XP, you must use remote collaboration tools to provide an information infrastructure and workspace similar to the one when programmers work closely.

Rather than just sticking to the adoption of a particular practice by DSD, other studies are more comprehensive and discuss the agility in DSD more generally. Fowler [9] reported their experiences and lessons learned in doing offshore agile development. He concludes that it is possible to do this combination of agility with overseas work, although the benefits are still open to debate.

Ramesh et al. [21] promoted a debate trying to answer the following question: can distributed software development be agile? According to him, the careful incorporation of agility in distributed software development environments is essential to address various challenges of distributed teams. The studied practices demonstrate how a balance between agile and distributed approaches can help meet these challenges. Another work that raises the question about the benefits of agility for Global Software Development (GSD) can be found in the work of Paasivaara and Lassenius [22], where the potential benefits and challenges of adopting agile methods in GSD are discussed.

Several studies arise addressing another topic related to this one: the management of distributed projects. Sutherland et al. [28] analyzed it and recommended best practices for globally distributed agile teams, based on the Scrum method. Berczuk [7] addressed how a product's team used creativity, communication tools, and basic good engineering practices to building a successful product. In this work, he adopted Scrum since the requirements were changing fast, and the team needed to deliver very quickly. Hossain et al. [11] presented a systematic review on the use of Scrum in the global software development, where they reported a growing interest in the subject and concluded that literature demands a more empirical study to understand the use of Scrum practices in globally distributed projects. Moreover, they stated that Scrum practices needed to be extended or modified to support globally distributed software development teams. Jimenez et al. [13], which focused on the challenges and improvements in distributed software development, developed another systematic review on this topic in the study. Ten DSD's success factors were related, among which standout the use of maturity models, incremental integration, and deliveries and frequent revisions.

Jalali and Wohlin [12] developed a systematic literature review and presented the results on the use of agile practices and Lean software development on Global Software Engineering (GSE). They pointed to the need to conduct further study on the challenges and benefits of combining agile methods with GSE in the form of evaluation research. They also concluded that there is insufficient evidence to conclude that agile is efficiently applicable to large distributed projects. In many studies that were reviewed, agile practices have been customized, and a modified agile method was applied. Sriram and Mathew [26] also drew up a systematic review of DSD with agility, in which the following themes were identified: the performance of global software development, issues related to governance and issues related to the software engineering process.

Table 1 summarizes the distribution of agile practices/methods by searched works. We detected a concentration of work about XP practices, in particular, pair programming. But when the issue is the method of management, Scrum is almost consensus.

Table 1. Agile Practices/Methods X Studies

Agile Practices/Methods	Study(ies)		
Pair Programming	[8], [14], [21], [29]		
Many XP Practices		[9], [13], [21], [22]	
Scrum Practices			[7], [11], [28]
Lean Practices			[12]

3.2   Task Allocation in Distributed Software Development

Lamersdorf et al. [16] developed an analysis of the existing approaches of distribution of duties. The analysis was comprehensive and involved procedures for the distributed development, distributed generation, and distributed systems areas. Finally, they pointed out the Bokhari algorithm as a promising model for applying to the DSD.

Lamersdorf et al. [17] conducted a survey on the state of practice in DSD in which they investigated the criteria that influence task allocation decisions. Lamersdorf and

Münch [15] presented TAMRI (Task Allocation based on Multiple cRIteria), a model based on multiple criteria and influencing factors to support the systematic decision of task allocation in distributed development projects. TAMRI builds upon the Bokhari algorithm and Bayesian networks.

Ruano-Mayoral et al. [24] presented a methodological framework to allocate work packages among participants in global software development projects. The framework consists of two stages: definition and validation. Lastly, the paper also presented the results of framework application.

Marques et al. [19] performed a systematic mapping, which enabled us to identify models that propose to solve the problems of allocation of tasks in DSD projects. As future work, they intended to propose a combinatorial optimization-based model involving classical task scheduling problems. Marques et al. [18] also performed a tertiary review applying the systematic review method on systematic reviews that address the DSD issues. The study revealed that most of the work focuses on summarizing the current knowledge about a particular research question, and the amount of empirical studies is relatively small.

Galviņa and Šmite [10] provided an extensive literature review for understanding the industrial practice of software development processes and concluded that the evidence of how these projects are organized is scarce. Despite, they presented the models deducted from the development processes found in the selected literature and summarized the main challenges that may affect the project management processes.

Babar and Zahedi [8] presented a literature review considering the studies published in the International Conference in Global Software Engineering (ICGSE) between 2007 and 2011. There was a growing tendency to propose solutions rather than just mention problems in GSD. It was also found that the vast majority of the evaluated studies were in software development governance and its sub-categories, and much of the work had focused on the human aspects of the GSD rather than technical aspects. There was also a growing trend to provide solutions to the problems identified in the area. However, many of the proposed solutions have not been rigorously evaluated. It is also noted that while the GSD search production is increasing, there are significant gaps that must be filled, for example, to study the specific challenges and offer solutions for the different software development phases, such as design and testing.

Almeida et al. [3] presented a multi-criteria decision model for planning and fine-tuning such project plans: Multi-criteria Decision Analysis (MCDA). The model was developed using cognitive mapping and MACBETH (Measuring Attractiveness by a Categorical Based Evaluation Technique) [6]. In [4], Almeida et al. applied (MCDA) [3] on the choice of DSD Scrum project plans that have a better chance of success. The application of the model was demonstrated and based on initial results, seemed to significantly help managers to improve the success rates of their projects. The proposed model was based on the input and judgment of decision makers, giving a subjective character to the model. The work sustained by the MCDA methodology allowed for a new project replacing old decision processes that worked in the past without any multi-criteria method and that were mainly focused on cost reduction and task allocation. According to the authors, the MCDA methodology proved to be suitable for modeling the problem setting being handled.

The two works above were the base of a masters dissertation [2] where the author concluded that, in terms of practical contribution, the proposed model allowed for a better rational for advising organizations undergoing conversations for starting globally distributed projects. Furthermore, it had the potential to improve the success rates of the client company currently applying the model and prospective companies.

As we can see, there are many published studies addressing the task allocation in distributed environments, but few task allocation methods in such environments have been actually proposed and tested. Furthermore, most models found in this research do not address agility together with the distribution.

Table 2 outlines task allocation methods found in the research.

Table 2. Task Allocation Methods

Task Allocation Methods	Summary
Bokari Algorithm [16]	Based on procedures for the distributed development, distributed generation, and distributed systems areas.
TAMRI [15]	Based on multiple criteria and influencing factors to support the systematic decision of task allocation in distributed development projects.
Ruano-Mayoral et al. [24]	A methodological framework to allocate work packages among participants in global software development projects.
MCDA [3]	A multi-criteria decision model for planning and fine-tuning such project plans.

4   Conclusion and Future Works

We developed a quasi-systematic review considering task allocation approaches in distributed software development projects which adopt some agile practice. We can see that agility in DSD is a theme in evidence, and we are far from exhausting it, given the diversity of issues involved. Besides, although the DSD is a subject studied for some time, few concrete works on the allocation of tasks and the results of its application exist in the literature. If we consider aspects of prioritization and classification, the numbers are even scarcer. Thus, a promising area of research is open.

As future work, we propose the development of a multi-criteria model to support task prioritization and classification in agile distributed software development projects in order to help decision makers to get the best teams' performance. The application of the above proposed model in real scenarios seems like another very useful future work. Finally, the application of an action-research [30] to one of the scenarios in order to calibrate the model also would add enough value to the field of study, when monitoring and adjustments in the application of the model could be made, collecting the results, and comparing them with the results of other scenarios.

Acknowledgments. The first author is thankful for "Coordination for the Improvement of Higher Level-or Education- Personnel" (CAPES) and 7 de Setembro College for the support received from this project. The second author is grateful to National Counsel of Technological and Scientific Development (CNPq) via Grants #305844/2011-3. The authors would like to thank Edson Queiroz Foundation/University of Fortaleza for all the support.

References

1. Ågerfalk, J., Fitzgerald, B.: Flexible and distributed software processes: Old petunias in new bowls? Communications of the ACM 49(10), 27–34 (2006)
2. Almeida, L.H.P.: McDSDS: A Multi-criteria Model for Planning Distributed Software Development Projects with Scrum. Dissertation (Master of Postgraduate Program in Applied Informatics) - University of Fortaleza (2011)
3. Almeida, L.H., Albuquerque, A.B., Pinheiro, P.R.: A Multi-criteria Model for Planning and Fine-Tuning Distributed Scrum Projects. In: Proceedings of the 6th IEEE International Conference on Global Software Engineering (2011)
4. Almeida, L.H., Pinheiro, P.R., Albuquerque, A.B.: Applying Multi-Criteria Decision Analysis to Global Software Development with Scrum Project Planning. In: Yao, J., Ramanna, S., Wang, G., Suraj, Z. (eds.) RSKT 2011. LNCS, vol. 6954, pp. 311–320. Springer, Heidelberg (2011)
5. Babar, M.A., Zahedi, M.: Global Software Development: A Review of the State-Of-The-Art (2007 – 2011), IT University Technical Report Series. IT University of Copenhagen (2012)
6. Bana e Costa, C.A., Sanchez-Lopez, R., Vansnick, J.C., De Corte, J.M.: Introducción a MACBETH. In: Leyva López, J.C. (ed.) Análisis Multicriterio para la Toma de Decisiones: Métodos y Aplicaciones, Plaza y Valdés, México, pp. 233–241 (2011)
7. Berczuk, S.: Back to Basics: The Role of Agile Principles in Success with an Distributed Scrum Team. In: Agile 2007. IEEE Computer Society (2007)
8. Flor, N.V.: Globally distributed software development and pair programming. Communications of the ACM 49(10), 57–58 (2006)
9. Fowler, M.: Using an Agile Software Process with Offshore Development (2006)
10. Galviņa, Z., Šmite, D.: Software Development Processes in Globally Distributed Environment. In: Scientific Papers, University of Latvia, vol. 770, Computer Science and Information Technologies (2011)
11. Hossain, E., Babar, M.A., Paik, H.-Y.: Using Scrum in global software development: A systematic literature review. In: Fourth IEEE International Conference on Global Software Engineering, ICGSE 2009, pp. 175–184 (2009)
12. Jalali, S., Wohlin, C.: Agile Practices in Global Software Engineering: A Systematic Map. In: International Conference on Global Software Engineering (ICGSE), Princeton (2010)
13. Jimenez, M., Piattini, M., Vizcaino, A.: Challenges and improvements in distributed software development: A systematic review. Advances in Software Engineering, Article ID 710971, 1–14 (2009)
14. Kircher, M., Jain, P., Corsaro, A., Levine, D.: Distributed eXtreme Programming. In: Proceedings of the International Conference on EXtreme Programming and Flexible Processes in Software Engineering, Sardinia, Italy, pp. 20–23 (May 2001)
15. Lamersdorf, A., Münch, J.: A multi-criteria distribution model for global software development projects. The Brazilian Computer Society (2010)
16. Lamersdorf, A., Münch, J., Rombach, D.: Towards a Multi-Criteria Development Distribution Model: An Analysis of Existing Task Distribution Approaches. In: IEEE International Conference on Global Software Engineering, ICGSE 2008 (2008)
17. Lamersdorf, A., Münch, J., Rombach, D.: A Survey on the State of the Practice in Distributed Software Development: Criteria for Task Allocation. In: Fourth IEEE International Conference on Global Software Engineering, ICGSE 2009 (2009)
18. Marques, A.B., Rodrigues, R., Conte, T.: Systematic Literature Reviews in Distributed Software Development: A Tertiary Study. In: ICGSE 2012, pp. 134–143 (2012)

19. Marques, A.B., Rodrigues, R., Prikladnicki, R., Conte, T.: Alocação de Tarefas em Projetos de Desenvolvimento Distribuído de Software: Análise das Soluções Existentes. In: II Congresso Brasileiro de Software, V WDDS – Workshop de Desenvolvimento Distribuído de Software, São Paulo (2011)
20. Miller, A.: Distributed Agile Development at Microsoft patterns & practices, Microsoft patterns & practices (2008)
21. Natsu, H., Favela, J., Moran, A.L., Decouchant, D., Martinez-Enriquez, A.M.: Distributed pair programming on the Web. In: Proceedings of the Fourth Mexican International Conference, pp. 81–88 (2003)
22. Paasivaara, M., Lassenius, C.: Could Global Software Development Benefit from Agile Methods? In: International Conference on Global Software Engineering, ICGSE 2006, Florianópolis (2006)
23. Ramesh, B., Cao, L., Mohan, K., Xu, P.: Can distributed software development be Agile? Communication of the ACM 49(10), 41–46 (2006)
24. Ruano-Mayoral, M., Casado-Lumbreras, C., Garbarino-Alberti, H., Misra, S.: Methodological framework for the allocation of work packages in global software development. Journal of Software: Evolution and Process. (2013)
25. Sarker, S., Sarker, S.: Exploring Agility in Distributed Information Systems Development Teams: An Interpretive Study in an Offshoring Context. Journal of Information Systems Research Archive 20(3), 440–461 (2009)
26. Sriram, R., Mathew, S.K.: Global software development using agile methodologies: A review of literature. In: 2012 IEEE 6th International Conference on Management of Innovation and Technology, ICMIT 2012, art. no. 6225837, pp. 389–393 (2012)
27. Sureshchandra, K., Shrinivasavadhani, J.: Adopting Agile in Distributed Development. In: Proceedings of the 2008 IEEE International Conference on Global Software Engineering, pp. 217–221 (2008)
28. Sutherland, J., Viktorov, A., Blount, J., Puntikov, N.: Distributed Scrum: Agile project management with outsourced development teams. In: Proceedings of the Hawaii International Conference on System Sciences (HICSS'40), pp. 1–10 (2007)
29. Teles, V.M.: Extreme Programming: Aprenda como encantar seus usuários desenvolvendo software com agilidade e alta qualidade, Novatec (2004)
30. Tripp, D.: Action research: a methodological introduction. Educ. Pesqui. 31(3), 443–466 (2005), http://dx.doi.org/10.1590/S1517-97022005000300009, ISSN 1517-9702

Network Data Analysis Using Spark

K.V. Swetha, Shiju Sathyadevan, and P. Bilna

Amrita Center for Cyber Security Systems & Networks
Amrita University, Kollam, India
{swe0523,bilna.p}@gmail.com,
shiju.s@am.amrita.edu

Abstract. With the huge increase in the volume of network traffic, there is a need for network monitoring systems that capture network packets and provide packet features in near real time to protect from attacks. As a first step towards developing such a system using distributed computation, new system has been developed in Spark, a cluster computing system, which extracts packet features with less memory consumption and at a faster rate. Traffic analysis and extraction of packet features are carried out using streaming capability inherent in Spark. Analysing the network data features provide a means for detecting attacks. This paper describes a system for the analysis of network data using Spark streaming technology which focuses on real time stream processing, built on top of Spark.

Keywords: Spark, Spark Streaming, Network Monitoring.

1 Introduction

Packet sniffing and analysis of the captured packets form an integral part in maintaining and monitoring the network activity. The rate of traffic flow has increased to such an extent that processing thousands of gigabytes of information per second has become difficult to deal with. A new solution is developed to solve the problem of handling large amount of network data to a very big extent. It involves computation across clusters of machines with shared in-memory dataset and less use of network and disk I/O. Here, we propose a system that processes the captured packets coming across a network using real time streaming computation framework called Spark Streaming. Spark is an in-memory data analytics cluster computing framework. Spark streaming is an extension to Spark developed for real time stream processing. Spark streaming is used in this system with the goal of providing near real time response along with fault-tolerance capability. It provides faster recovery from faults compared to the existing systems through parallel recovery across nodes. This system extracts the packet features on the fly which can be further used for detecting attacks in real time.

© Springer International Publishing Switzerland 2015
R. Silhavy et al. (eds.), *Software Engineering in Intelligent Systems,*
Advances in Intelligent Systems and Computing 349, DOI: 10.1007/978-3-319-18473-9_25

2   Spark and Spark Streaming

2.1   Spark

Spark is an open-source in-memory analytics cluster computing engine that supports operations in parallel steps. It is a fast map reducing engine and efficient due to its in-memory storage and fault-tolerance. It has rich APIs in scala, java and python. Also, it provides an interactive shell to the user. Two abstractions provided by Spark for programming are Resilient Distributed Datasets (RDDs) and parallel operations on the datasets. RDD is a read-only collection of objects partitioned across a set of machines that can be rebuilt if a partition is lost. RDDs achieve fault tolerance through a notion of lineage: if a partition of an RDD is lost, the RDD has enough information about how it was derived from other RDDs to be able to rebuild just that partition. Each RDD is represented by a scala object. Several parallel operations can be performed on the RDDs like count, reduce, foreach etc. RDDs support two types of operations: transformations, which create a new dataset from an existing one, and actions, which return a value after running a computation on the dataset. Map is a transformation and reduce is an action.

2.2   Spark Streaming

Spark streaming is an extension to Spark for continuous stream processing of data. It is a real time processing tool that runs on top of the Spark engine. It unifies batch programming and stream processing model. Data can be ingested from various sources Kafka, Twitter, Flame or TCP sockets and can be processed using complex algorithms. The processed results can be pushed to file system, databases and live dashboards. Spark streaming receives live input data streams and divides the data into batches. The batches of input data are processed by the Spark engine and generate the batches of processed data.

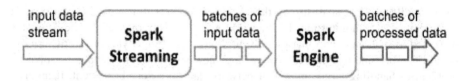

Fig. 1. Spark Streaming

Spark streaming utilizes Spark primary data abstraction item, resilient distributed datasets. Spark streaming provides a high level abstraction called discretized stream (DStream) which is a continuous stream of data. Input data is stored replicated in multiple nodes. Spark treats input as a series of deterministic batch computations performed on some interval. The dataset of a particular interval is processed after the time interval and final output or new intermediate state is produced. This state is stored as RDD. A sequence of RDDs is treated as DStream. DStream can be created from input sources or by applying operations on other DStreams.

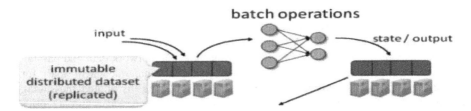

Fig. 2. Discretized Stream Processing

There are two types of operators for building Spark streaming programs. They are transformation operators and output operators. Transformation operators produce a new DStream from one or more parent streams. Output operators allow the program write data to external systems.

2.3   Scala – The Programming Language Used

Spark is implemented in scala language. Scala is an acronym for "Scalable Language". Scala is an object-functional programming language. Scala has full support on functional programming features like type inference and pattern matching. Thus programs written in scala appear to be concise and smaller. Scala is a pure-bred object-oriented language. Conceptually, every value is an object and every operation is a method-call. The language supports advanced component architectures through classes and traits. Even though its syntax is fairly conventional, Scala is also a full-blown functional language. Scala code is compiled to be java byte code so that the executable runs on a java virtual machine. This facilitates scala code to call java libraries. Java and Scala can even mutually refer to each other; the Scala compiler contains a subset of a Java compiler to make sense of such recursive dependencies.

3   Packet Capturing in Spark: Methodology

3.1   Reading Packets Using Packet Capturing Library

There is no packet capturing library available in Spark or scala. There are two packet capturing libraries in java that can be used: JPcap and JNetPcap. JPcap requires Sun java which is not available now. Also, Jpcap had defects in capturing ICMP packets and its development seemed to be stopped long ago. So JNetPcap library is used in this system to capture packets from the network interface. JNetPcap is a java software development kit that provides a java wrapper to popular libpcap library for capturing network packets. There are two parts to JNetPcap SDK. The first part is the libpcap wrapper, which provides nearly all of the functionality of libpcap native library, in a java environment. The second part is the packet decoding framework which has packages included for processing and decoding the captured packets. In JNetPcap, a packet is a raw data buffer that has been delivered by the network interface in the system. The entry point to the entire decoding packet API is PcapPacket and JPacket classes in the package. If a packet has been captured using the library, objects of type PcapPacket are obtained.

3.2   Integrating Queuing System with Spark

Spark Streaming can receive streaming data from any arbitrary data sources like Kafka, Twitter, Flume, sockets. A source is required for receiving network packets in Spark. The source should be capable of receiving the entire network data and passing it to Spark without fail. Hence, Kafka queuing system was integrated with Spark for the transfer of network traffic. Kafka is a publish-subscribe distributed messaging system which is fast, scalable and durable. Kafka maintains messages in categories called topics. In Kafka, the processes that publish messages to a topic are called producers and the processes that subscribe to that topic are called consumers. The producer sends network packets captured using JNetPcap library to the Kafka cluster which in turn is received by the Spark Streaming Kafka consumer. Data streams are partitioned and spread over a cluster of machines to allow data streams larger than the capability of any single machine. Kafka can handle terabytes of network packets without performance impact. Kafka uses zookeeper as a co-ordination service between producer and receiver.

Fig. 3. Integrating Kafka with Spark streaming

3.3   Transfer of Packets from Producer to Receiver

When a packet is captured, it is passed to the Kafka API. Objects of type PcapPacket are published by the Kafka producer. The producer uses an encoder to serialize the objects. Kafka provides a means to declare a specific serializer for a producer. Since the objects are of type PcapPacket , the default serializer class or the String serializer class that comes with Kafka library cannot be used. The property "serializer.class" defines what serializer to use when preparing the message for transmission to the Kafka broker. In this system, PcapPacket encoder is written as serializer and it is fed to the producer properties. The PcapPacket encoder takes the packets and encodes it to byte stream. The method 'public int transferStateAndDataTo(byte[] buffer)' of PcapPacket class will copy the contents of the packet to the byte array . The byte[] is defined with size as packet.getTotalSize() which gives the total size of packets. Kafka Spark consumer will pull messages from Kafka cluster using Spark streaming to generate DStreams from Kafka and apply output operation for every messages of the RDD. Inorder to receive objects of specific type, i.e. PcapPacket, a custom decoder is implemented. The decoder deserializes the byte array into PcapPacket format. The PcapPackets from the receiver are then treated as DStreams which is a sequence of RDDs. The DStreams are then processed by the Spark engine for extracting packet features.

4   Architecture of the Proposed System

Packets from network interface are captured using JNetPcap library. JNetPcap is an open source java library that contains a java wrapper for nearly all libpcap library native calls. It decodes captured packets in real-time and provides a large library of network protocols. Spark streaming requires data to be ingested from various sources like Kafka, Twitter etc. Here Kafka is used. Kafka is run as a cluster comprised of one or more servers each of which is called a broker. Kafka producer pushes the packets captured and the Kafka Spark streaming receiver receives the packets from producer. Spark streaming receiver receives the packets as DStreams which are transformed and batched into a sequence of RDDs. RDDs are then processed using Spark operations by the Spark engine. Inorder to process the received streams as packets as such, encoding and decoding is implemented. Packets are encoded into streams and are then converted back to packets at the receiver side. Analysis of packets is carried out at receiver after decoding the received packets. The Spark engine starts packet analysis by extracting the packet features such as protocol, received time, source ip, destination ip, source port and destination port and utilizing the packet decoding framework provided by JNetPcap library.

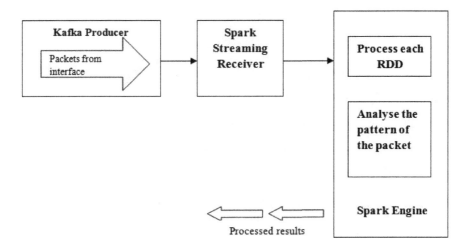

Fig. 4. Architecture of the system

5   Experiment

On a Kafka installed and running machine, connection is established to Kafka to a particular topic using one of the producer interfaces (perhaps Kafka.javaapi. producer.Producer). Testing the setup involves capturing the live packets over a network and passing it to Spark streaming receiver inorder to extract the packet features. Following are the configuration details required for the live implementation of the system.

- Start the zookeeper (Co-ordination service) server
  bin/zookeeper-server-start.sh config/zookeeper.properties
- Start the Kafka server
  bin/Kafka-server-start.sh config/server.properties
- Start the producer with all configuration settings
  Run the Kafkaproducer
- Start the Kafka Spark streaming receiver
  Run the Kafka Spark streaming receiver

The receiver on receiving each packet, after decoding it to PcapPacket format, passes to a function for extracting fields such as protocol, received time, source ip, destination ip, source port and destination port. A log file with all the details is obtained as output. The packet loss is significantly reduced since the computation is distributed.

```
TCP      429    Mon Dec 15 12:09:32 IST 2014    74.125.236.85    10.30.56.184     443      46721
TCP      330    Mon Dec 15 12:09:32 IST 2014    10.30.56.184     74.125.236.85    46721    443
UDP      547    Mon Dec 15 12:10:01 IST 2014    10.30.56.178     10.30.59.255     138      138
TCP      429    Mon Dec 15 12:10:02 IST 2014    74.125.236.85    10.30.56.184     443      46721
TCP      330    Mon Dec 15 12:10:02 IST 2014    10.30.56.184     74.125.236.85    46721    443
TCP      411    Mon Dec 15 12:10:02 IST 2014    74.125.236.85    10.30.56.184     443      46721
TCP      330    Mon Dec 15 12:10:02 IST 2014    10.30.56.184     74.125.236.85    46721    443
TCP      411    Mon Dec 15 12:10:02 IST 2014    10.30.56.184     74.125.236.85    46721    443
TCP      1788   Mon Dec 15 12:10:02 IST 2014    10.30.56.184     74.125.236.85    46721    443
ICMP     402    Mon Dec 15 12:10:05 IST 2014    10.30.56.184     10.30.56.109
ICMP     402    Mon Dec 15 12:10:05 IST 2014    10.30.56.109     10.30.56.184
ICMP     402    Mon Dec 15 12:10:06 IST 2014    10.30.56.184     10.30.56.109
ICMP     402    Mon Dec 15 12:10:06 IST 2014    10.30.56.109     10.30.56.184
ICMP     402    Mon Dec 15 12:10:07 IST 2014    10.30.56.184     10.30.56.109
ICMP     402    Mon Dec 15 12:10:07 IST 2014    10.30.56.109     10.30.56.184
ICMP     402    Mon Dec 15 12:10:08 IST 2014    10.30.56.184     10.30.56.109
DNS      394    Mon Dec 15 11:55:05 IST 2014    8.8.8.8          10.30.56.184     53       60963
TELNET   324    Mon Dec 15 12:10:19 IST 2014    10.30.56.109     10.30.56.184     23       57290
```

Fig. 5. Log file with statistics of network data

6   Future Work

The features obtained from the packets with this experiment can be utilized to detect network level attacks in near real time due to the efficiency of Spark engine. The source ip, destination ip and port numbers can be monitored with real time streaming facility in Spark to identify and report different network attacks like DOS attacks. This can be extended to build a signature based and anomaly based intrusion detection system that can overcome the limitations of existing systems in dealing with bulk amount of network data. The data mining algorithms built on top of Spark can be used to build an anomaly detection module.

7   Conclusion

The statistics of real time network data is obtained with the developed system using Spark streaming. The features of network data such as protocol, source ip, destination ip, time at which packet is received and port numbers are extracted with the developed system using spark streaming to provide response in near real time. Providing real time reporting by network data monitoring add a tremendous value in the present world since the number of network level attacks emerge rapidly. Spark Streaming is designed with the goal of providing near real time processing with approximately one second latency to such programs. Intrusion Detection System could be the most needed application with such requirements. Development of IDS with the features extracted can detect and report live network level attacks on the fly.

References

1. Zaharia, M., Chowdhury, M., Franklin, M.J., Shenker, S., Stoica, I.: Spark: Cluster Computing with Working Sets. In: Hotcloud Conference. USENIX, the Advanced Computing Systems Association (2010)
2. Zaharia, M., Das, T., Li, H., Hunter, T., Shenker, S., Stoica, I.: Discretized Streams: Fault-Tolerant Streaming Computation at Scale. In: Hotcloud Conference. USENIX, the Advanced Computing Systems Association (2012)
3. Asrodia, P., Patel, H.: Analysis of Various Packet Sniffing Tools for Network Monitoring and Analysis. International Journal of Electrical, Electronics and Computer Engineering 1(1), 55–58 (2012)

Optimization of Hadoop Using Software-Internet Wide Area Remote Direct Memory Access Protocol and Unstructured Data Accelerator

V. Vejesh, G. Reshma Nayar, and Shiju Sathyadevan

Amrita Center for Cyber Security Systems & Networks, Amrita University, Kollam, India
{vejeshv,reshmagovilla}@gmail.com, shiju.s@am.amrita.edu

Abstract. Over the last few years, data size grew tremendously in size and thus data analytics is always geared towards low latency processing. Processing of Big Data using traditional methodologies is not cost effective and fast enough to meet the requirements. Existing socket based communication (TCP/IP) used in Hadoop causes performance bottleneck on the significant amount of data transfers through a multi-gigabit network fabric. To fulfill the emerging demands , the underlying design should be modified to make use of data centre's powerful hardware. The proposed project include integration of Hadoop with remote direct memory access (RDMA).For data-intensive applications, network performance becomes key component as the amount of data being stored and replicated to HDFS increases. RDMA is implemented in a commodity hardware through software ,namely, Soft-iWARP (Software-Internet Wide Area Protocol). Hadoop employs a Java-based network transport stack on top of the JVM . JVM introduces a significant amount of overhead to data processing capability of the native interfaces which constrains use of RDMA. The usage of plug-in library for data shuffling and merging part of Hadoop can take advantage of RDMA . An optimization for Hadoop in data shuffling part can be thus implemented.

Keywords: soft-iWARP, Hadoop, UDA, RDMA.

1 Introduction

Every year the amount of information in the electronic world increase tremendously. Traditional data storage and retrieval mechanisms that have come of age over it has to face issues in storing and indexing large collections of unstructured data and their inability to effectively retrieve and distribute the data. The Big Data is entering into a new era where data analysis system should take care of the processing of increasingly large volume of data as well as the response time in the range of minutes or seconds. In order to deal with their increasing load of unstructured information, Hadoop came into existence.

Hadoop is an open-source framework that lightens the fast process of massive amount of data via parallel processing on distributed clusters of servers. Main parts of Hadoop includes HDFS (Hadoop Distributed File System) and Map-reduce. HDFS

© Springer International Publishing Switzerland 2015   261
R. Silhavy et al. (eds.), *Software Engineering in Intelligent Systems,*
Advances in Intelligent Systems and Computing 349, DOI: 10.1007/978-3-319-18473-9_26

manages the storage of files across the cluster and enables global access to them , while map-reduce have the ability of parallel processing of a cluster in order to quickly process large data-sets in a fault-tolerant and scalable manner. Usually a Hadoop job consists of number of phases which produce intermediate results that should be instantly consumed by the next phase and it includes lots of read-write operations that delays the process as a whole. Emergence of new group of cloud application has its key factor as response time also. To meet the above requirement ,at first the system should make use of the multi-gigabit network fabric, and remove slow storage devices from critical path.

Remote direct memory access (RDMA) communication model can play a role in the optimization .By deploying a fully software based iWARP(Internet Wide Area RDMA Protocol) called Soft-iWARP can provide standards-compliant iWARP RDMA functionality. The direct data placement feature of RDMA enables its efficient networking. For enabling Hadoop to use RDMA, a plug-in called UDA (Unstructured Data Accelerator) can be used. A Java-based network transport stack placed on top of the Java Virtual Machine (JVM) in Hadoop for its data shuffling and merging purposes. JVM constrains the use of high-performance networking mechanisms such as RDMA. UDA removes JVM from the critical path of intermediate data shuffling. UDA acts as a novel data shuffling protocol, available for Hadoop to enable it to take advantage of RDMA in the network technologies and improves the scaling of Hadoop clusters executing data-analytics intensive applications. UDA requires using proprietary RDMA capable NIC(Network Interface Card)or RNIC to accelerate Hadoop data shuffling. By combining a pure software implementation of RDMA called soft-iWARP (soft internet Wide Area RDMA Protocol)and UDA we can overcome the limitations posed by UDA requiring the use of RNIC.

2   Methodology

Recent trends in Big Data calls for innovative solutions that can help datacenters to handle the chaos created by the large volume of data that needs to be processed for a business day. The cause of bottleneck in the analytics needs to be identified and classified to propose an efficient solution. Hadoop jobs require split, map, sort, shuffle, merge and reduce operations. The sort, shuffle and merge operations are often associated with disk operations. These disk operations can be identified as a reason for bottleneck.

Hadoop employs Java based network transport stack on top of the Java Virtual Machine (JVM) for its data shuffling and merging purposes. JVM introduces significant amount of overhead to data processing capability of the java native interfaces. Furthermore, JVM constrains the use of high performance networking mechanisms such as RDMA which has established itself as an effective data movement technology in many networking environments because of its low-latency, high bandwidth, low CPU utilization and energy efficiency.

The overhead due to the two factors mentioned above actually results in underutilization of the high network bandwidth available in multi-gigabit fabric which is being used in Hadoop clusters. Introducing RDMA into Hadoop shuffling will help in reducing the overhead and effectively freeing up CPU resource for utilization in analytics.

2.1   RDMA

Remote Direct Memory Access is the ability of one computer to directly place information in another computer's memory with minimal demands on memory bus bandwidth and CPU processing overhead, while preserving memory protection semantics. RDMA mechanism allows a system to place the data directly into its final location in the memory of the remote system without any additional data copy. This salient feature of RDMA is called Direct Data Placement or Zero Copy and enables efficient networking.

Direct Data Placement is made possible using RNIC (RDMA enabled Network Interface Card) which provides an RDMA-ed network stack to the Operating System. RDMA communication model differs significantly from socket based communication model. Typically the following operations are involved in an RDMA communication :

* Applications in both end register to their RNICs which memory location should be exchanged.

* RNIC fetches the data from that buffer without CPU involvement using DMA and transfers them across the network.

* RNIC at the receiver side places the incoming data directly to the destination buffer.

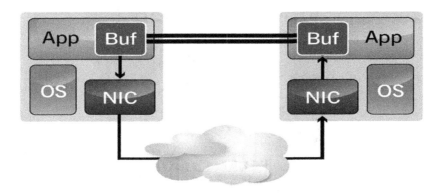

Fig. 1. RDMA Communication model

2.1.1   Soft-iWARP

Soft-iWARP expanded as Software-Internet Wide Area Protocol, implements an iWARP protocol suite completely as a software kernel module. Soft-iWARP exports the RDMA verbs and connection manager interfaces useable for user level applications. A software kernel module is inserted on top of TCP kernel sockets and this will be kernel specific one. This layer provides standards-compliant iWARP RDMA functionality at a decent performance level.

Soft-iWARP provides a software based RDMA stack for Linux. Soft-iWARP project comes in two parts a kernel module and a user space library. The kernel module is used by applications for interfacing with RDMA API. RDMA semantics are available to application level via RDMA API.

RDMA operations require registering of memory with RDMA driver to pin the memory in place thus preventing swapping to disk by Operating System. The driver assigns an identifier to memory region called STag. RDMA data exchange happens without the user space process at the server being scheduled and without an extra copy.

Software based iWARP stack provides several benefits:

 * As an RNIC independent device driver, it immediately enables RDMA services on conventional Ethernet protocol based adapters which lack RDMA hardware acceleration.

 * Soft-iWARP can act as an intermediate step when migrating applications and systems to RDMA APIs.

 * Soft-iWARP seamlessly supports outstanding work requests and RDMA operations in either direct or asynchronous transmission.

 * A software-based iWARP stack can enable any available hardware assists for performance-critical operations such as MPA CRC checksum calculation and zero copy. The resulting performance levels can match or approximate to that of a fully offloaded iWARP stack.

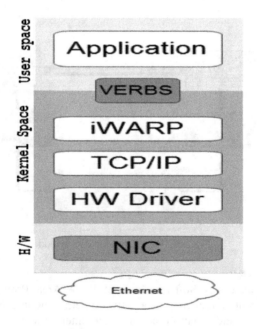

Fig. 2. Soft-iWARP stack layout

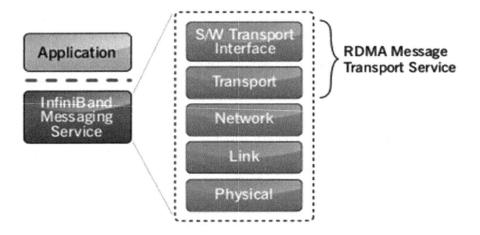

Fig. 3. Soft-iWARP service layer

2.2   Hadoop

Apache Hadoop is an open source software project that enables the distributed processing of large data sets across clusters of commodity servers. It is designed to scale up from a single server to thousands of machines while providing a very high degree of fault tolerance.

Hadoop enables a computing solution that is :

* Scalable: Addition of new nodes to the cluster can be done as per need without requiring any need to change the data formats, how data is loaded, how jobs are written, or the applications on top.

* Cost effective: Massively parallel computing can be made possible by the use of commodity servers. This results in a sizeable decrease in the cost per TB (Terabyte) of storage and makes the system affordable to model your data.

* Flexible: Hadoop is schema-less, meaning it can deal with almost any type of data, be it structured or not from any number of sources

* Fault tolerant: when a node goes offline , the system redirects work to another location of the data and continues processing.

Hadoop has two major subsystems : the Hadoop Distributed File System (HDFS) and a distributed data processing framework called MapReduce.

2.2.1   HDFS

HDFS manages the storage of files and enables global access to them. It is designed to be fault-tolerant through a replication mechanism. One namenode and many datanode are included in HDFS. The namenode holds and creates the mapping between file blocks and datanodes. The datanode sends heartbeat messages to the namenode to confirm that the data remains safe and available.

2.2.2   MapReduce

MapReduce model is implemented through two components : jobtracker and tasktracker. The jobtracker accepts job requests, splits the data input, define tasks required for the job, assigning tasks for parallel execution and for handling errors. The tasktracker executes tasks as ordered from jobtracker, which can be either map or reduce. Steps followed while processing a Hadoop job : split, map ,sort, shuffle , merge, reduce and then final output. When a job is submitted, at first job is split into different tasks then each task is given into each mapper. Each MOF (Map Output File) is given into sort phase where the divided tasks are sorted and this will be the input of the next phase, that is shuffle phase. In shuffle phase, the data are arranged randomly and this output is given as input to reducer .Number of mappers and reducers user for a particular job can be determined by user. Then in reduce stage the data are managed to store in a perfect way so that a user searching for a particular data can easily retrieve it. Here the final output from reducer stage   is stored in different parts.

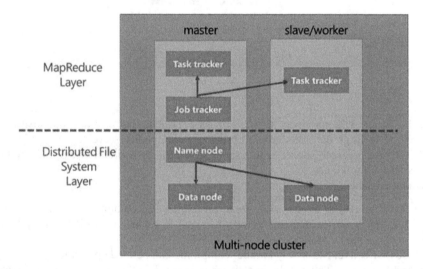

Fig. 4. Hadoop architecture

2.3   UDA

UDA (Unstructured Data Accelerator) is a software plug-in that is developed by Mellanox Technologies under Apache 2.0 license. UDA accelerates Hadoop and improves the scaling of Hadoop clusters executing data analytics intensive applications. The plug-in provides Hadoop with a novel data shuffling protocol that enables Hadoop to take advantage of RDMA network technology. UDA plug-in when combined with Linux inbox driver and Mellanox RNIC will accelerate file transfer tasks associated with Map/Reduce tasks.

UDA plug-in is able to help Hadoop data shuffling by avoiding Java based transport protocols effectively bypassing the JVM itself. This bypass can avoid overhead caused by transport protocols without changing the user programming interfaces like map and reduce functions. This library integrated RDMA as underlying mechanism for data transfer and now Hadoop can work on Ethernet and Infiniband. MOFSupplier and

NetMerger are two components of plug-in library and these two are standalone native C processes. These will be launched by local tasktracker and they communicate through loopback sockets. These two above mentioned components replace HttpServlets and MOFCopiers. An HttpServer is embedded inside each tasktracker, and this server spawns multiple HttpServlets to answer incoming fetch requests for segment data. On the other side of the data shuffling, within each reducetask, multiple MOFCopiers are running concurrently to gather all segments. MOFSupplier is designed as pipelined pre-fetching scheme and provides index cache (for quick identification of MOF segments) and data cache (pre-fetch MOF segments for request processing). NetMerger undertakes shuffling and merging of intermediate data and for this network-levitated merge algorithm is used. This part consolidates data networking fetching requests from all reducetasks on a single node. Thus an RDMA connection can be established within a Hadoop framework. Main module of shuffle bypasser is UdaShuffleProviderPlugin. UdaShuffleProviderPlugin allocates buffers for reading MOFs from the disk and for writing them using RDMA to satisfy reduce task shuffle requests. Therefore, UdaShuffleProviderPlugin's buffer size determines the max buffer size to be used also by reduce tasks. When tasktracker is spawned and the UdaShuffleProviderPlugin is initialized, it is essential that the mapred.rdma.buf.size parameter is properly configured to satisfy reducers. RDMA buffers from each reducer are allocated from mapred.child.java.opts, mapred.job.shuffle. input.buffer.percent.

3 Benefits of Integrating UDA Enabled Hadoop with Soft-iWARP

UDA plug-in is capable of accelerating Hadoop if the underlying network supports RDMA. But RDMA support can be made available by the use of a dedicated hardware (RDMA capable network card) or by opting for the software implementation of RDMA. Our work concentrates on soft-iWARP project and its integration which will enable RDMA using a pure software implementation.

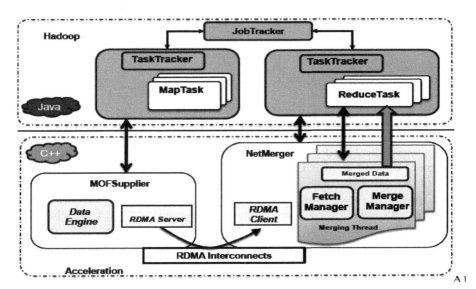

Fig. 5. Architecture of Hadoop with UDA

By opting for the software based approach we can cut costs by avoiding the requirement of expensive RDMA capable network adapters.

4   Challenges of Integrating UDA Enabled Hadoop with Soft-iWARP

UDA plug-in is recommended to be used with Linux inbox driver supplied by Mellanox. Linux inbox driver is intended to be used with Mellanox network adapters. Soft-iWARP project requires installation of RDMA libraries. Mellanox inbox driver comes with substitutes of these libraries which will remove the existing ones (including the libraries required by soft-iWARP). Thus installation of Linux inbox driver from Mellanox will cripple soft-iWARP project based RDMA.

5   Implementation

Hardware setup :

* Intel Core i7 3770 CPU @ 3.4 GHz to 3.9 GHz
* 8 GB DDR3 @ 1600MHz
* Intel 82579LM Gigabit Ethernet adapter

Software setup (testing) :

* Ubuntu 12.04 AMD64 desktop with kernel 3.2.0-74-generic
*   soft-iWARP
* UDA plug-in v3.4.1
* Oracle Java Development Kit (JDK) 1.7.0_72.
* Cloudera Hadoop cdh 4.7.0 with MRv1 pseudo distributed mode
* Alien –tool to convert   rpm package to debian   package

Software setup (building uda package) :

* Fedora 21 server 64-Bit
* Oracle jdk 1.7.0_72
* libaio package
* librdmacm-devel package

Implementation starts with installation and testing of soft-iWARP RDMA connectivity. Soft-iWARP installation is divided to two parts : compiling and installation of soft-iWARP kernel driver module followed by soft-iWARP user space library. RDMA connectivity was tested using rping utility and RDMA client server programs. Integration of UDA and soft-iWARP requires testing for interoperability without any source code modifications and studying their behavior to understand the issues posing the challenge. Our study revealed that UDA could not register RDMA memory in the localhost (since we used pseudo distributed mode for Hadoop with just one datanode). Examination of the UDA logs (integrated to the tasktracker log) revealed the reason to be 'FUNCTION NOT IMPLEMENTED' (errno=38).

```
org.apache.hadoop.mapred.ShuffleConsumerPlugin: Successfully init'ed
device (DataNet/RDMAComm.cc:368)
2014-12-15 16:55:33,720 INFO
org.apache.hadoop.mapred.ShuffleConsumerPlugin: Going to register
memory with contig-pages (DataNet/RDMAComm.cc:121)
2014-12-15 16:55:33,720 INFO
org.apache.hadoop.mapred.ShuffleConsumerPlugin: Going to register RDMA
memory. size=37748736 (DataNet/RDMAComm.cc:69)
2014-12-15 16:55:33,720 ERROR
org.apache.hadoop.mapred.ShuffleConsumerPlugin: ibv_reg_mr failed for
memory of total_size=37748736 , MSG=Function not implemented
(errno=38), memory-pointer=140650188156000 (DataNet/RDMAComm.cc:96)
2014-12-15 16:55:33,720 ERROR
org.apache.hadoop.mapred.ShuffleConsumerPlugin: raising com/mellanox/
hadoop/mapred/UdaRuntimeException to java side, with info=ibv_reg_mr
failure (UdaBridge.cc:91)
```

Fig. 6. Log generated for RDMA memory registration failure

```
org.apache.hadoop.mapred.ShuffleConsumerPlugin: Going to register RDMA
memory. size=3407872 (DataNet/RDMAComm.cc:69)
2014-12-29 12:23:16,021 INFO
org.apache.hadoop.mapred.ShuffleConsumerPlugin: Registering
memory....**** (DataNet/RDMAComm.cc:92)
2014-12-29 12:23:16,021 INFO
org.apache.hadoop.mapred.ShuffleConsumerPlugin: After RDMA memory
registration. size=3407872 (DataNet/RDMAComm.cc:106)
2014-12-29 12:23:16,021 INFO
org.apache.hadoop.mapred.ShuffleConsumerPlugin:  After RDMA buffers
registration: buffer1 = 131072 bytes , buffer2 = 131072 bytes ,
buffers count = 13 , total = 3407872 bytes) (DataNet/RDMAClient.cc:448)
2014-12-29 12:23:16,022 INFO
org.apache.hadoop.mapred.ShuffleConsumerPlugin: C++ THREAD STARTED (by
start) and attached to JVM tid=0x7e23d700 (CommUtils/UdaUtil.cc:46)
2014-12-29 12:23:16,022 INFO
org.apache.hadoop.mapred.ShuffleConsumerPlugin: Merge online (Merger/
MergeManager.cc:186)
2014-12-29 12:23:16,023 INFO
org.apache.hadoop.mapred.ShuffleConsumerPlugin: fetchOutputs - Using
UdaShuffleConsumerPlugin
```

Fig. 7. Log generated after integration

Further examination revealed that RDMA device attribute querying and initialization was successful. UDA source code needed studying and modifications to enable successful memory registration. RDMA memory registration failure was identified to be inappropriate access permission (read, write, remote read, remote write etc.) on the region to be registered. Local access permissions when used for memory registration will result in memories being successfully registered. Changed the UDA source code to correct the access permission. RDMA communication parameter values needed tuning especially in the area of queue pair attributes to facilitate queue pair creation. Queue pair attribute values were changed to enable successful queue pair creation and registration.

Modified source builds are based on RPMs which required conversion to Debian packages for testing in Ubuntu. RPM package obtained after building of the modified UDA source code was converted to a proper Debian package for use in Ubuntu.

6   Conclusion

Integration of UDA with soft-iWARP for Hadoop jobs was able to successfully register the memory regions and create the queue pair for shuffling using RDMA. Hadoop was able to use RDMA operations without requiring expensive RDMA capable network card. This implementation can thus have the advantages of RDMA such as freeing up some CPU resources which can improve MapReduce processing (using Direct Data Placement / zero copy). Hadoop operations can thus be accelerated by this integration.

7   Future Works

RDMA post receive   operation was unsuccessful. RDMA operations that follow post receive needs testing for functionality. Performance comparison (performance gain measurement) of UDA accelerated Hadoop using soft-iWARP based RDMA connectivity and vanilla Hadoop.

References

1. Konstantinos, K.: An In-Memory RDMA-Based Architecture for the Hadoop Distributed Filesystem. Swiss Federal Institute of Technology in Zurich
2. Islam, N.S., Rahman, M.W., Jose, J., Rajachandrasekar, R., Wang, H., Subramoni, H., Murthy, C., Panda, D.K.: High Performance RDMA-based Design of HDFS over InfiniBand. Department of Computer Science and Engineering, The Ohio State University and IBM T.J Watson Research Center Yorktown Heights, NY
3. Wang, Y., Xu, C., Li, X., Yu, W.: JVM-Bypass for Efficient Hadoop Shuffling. Department of Computer Science, Auburn University, AL 36849, USA
4. Fenn, M., Calderin, L., Nucciarone, J., Argod, V.: Evaluation of iWARP versus InfiniBand Performance. White paper by Pennstate Computer Science and Service System, CSSS 2012, Washington, DC, USA, pp. 574–577 (2012)
5. Wang, Y., Que, X., Yu, W., Goldenberg, D., Sehgal, D.: Hadoop Acceleration Through Network Levitated Merge. In: SC 2011, November 12-18, Seattle, Washington, USA (2011)
6. Mellanox Technologies: Unstructured Data Accelerator Rev 3.4.0
7. Shainer, G.: RDMA based Big Data Analytic. Technion (March 2014)
8. Mellanox Technologies: Deploying Hadoop with Mellanox End-to-End 10/40Gb Ethernet Solutions (2012)
9. Mellanox Technologies: Driving IBM BigInsights Performance Over GPFS using Infiniband+RDMA (April 2014)
10. The OpenFabrics Alliance: A Guide to Installing OFED on Linux (October 2011)

Towards Cooperative Cloud Service Brokerage for SLA-driven Selection of Cloud Services

Elarbi Badidi

College of Information Technology, United Arab Emirates University,
P.O. Box 15551, Al-Ain, United Arab Emirates
ebadidi@uaeu.ac.ae

Abstract. Enterprises are showing growing interest in outsourced cloud service offerings that can help them increase business agility and reduce operational costs. Cloud services are built to permit easy, scalable access to resources and services, and are entirely managed by cloud services' providers. However, enterprises are faced with the inevitability of selecting suitable cloud services that can meet their functional and quality requirements. We propose, in this paper, a federation of Cloud Brokers (CBs) for cloud services' selection. Cloud brokers of the federation cooperate to assist service consumers select suitable cloud services, which can offer their required service and fulfill their quality-of-service (QoS) requirements. Service selection is part of our integrated framework for SLA management, service provisioning, and SLA compliance monitoring.

Keywords: Cloud services, Quality-of-service, Cloud federation, Cloud broker, Service level agreement.

1 Introduction

Cloud computing involves the provisioning over the Internet of dynamically scalable and virtualized services that users can access from anywhere in the world on a subscription basis at competitive costs. The three principal models of cloud services are Infrastructure-as-a-Service (IaaS), Platform-as-a-Service (PaaS), and Software-as-a-Service (SaaS) [11]. To access cloud services, users typically use various devices from desktops, laptops, tablets to smartphones. Besides, with the proliferation of mobile internet-enabled handheld devices over the last few years, mobile users are increasingly demanding services while they are on the move. However, mobile devices still lack the necessary resources, in terms of storage and computing power, when compared with a conventional information processing device on which large-sized applications could be deployed.

One way for overcoming the limitations of mobile devices is using mobile cloud computing (MCC) [15][7]. MCC allows using the resources of a cloud computing platform such as Amazon EC2, Microsoft Azure, and Google AppEngine. As large-sized applications could not be deployed on mobile devices, a well-matching solution is to deploy reusable services on the service provider servers, and to let mobile devices access these services over the Internet and maybe adapt the interface according to the category of the mobile user.

© Springer International Publishing Switzerland 2015
R. Silhavy et al. (eds.), *Software Engineering in Intelligent Systems*,
Advances in Intelligent Systems and Computing 349, DOI: 10.1007/978-3-319-18473-9_27

The success of cloud services will deeply depend on the satisfaction of service consumers concerning the service performance and delivered quality-of-service (QoS). Both cloud service providers and service consumers have an interest in coping with QoS. For service providers, QoS is extremely significant when they offer several levels of service and implement priority-based admission mechanisms. The Service Level Agreement (SLA) represents the agreement between the cloud service provider and the service consumer. It describes agreed service that the service provider has to deliver, the service cost, and qualities [1].

Given the abundance of cloud service offerings from increasing number of cloud providers, enterprises are facing the challenges of discovering potential cloud services, that can meet their functional and non-functional requirements, and ranking them based on their quality requirements and cloud services quality offerings. Another challenge is the lack of heterogeneity in terms of quality indicators and SLA indicators used by cloud service providers in their SLAs. Few efforts about this issue have started to emerge such as the Service Measurement Index (SMI) developed by the Cloud Services Measurement Initiative Consortium (CSMIC) [3].

In this paper, we describe our proposed framework for SLA-driven selection of cloud services. The main component of the framework is a federation of Cloud brokers (CBs). These brokers are in charge of selecting suitable cloud services and mediating between service consumers and cloud services to reach agreements that explicitly describe expected service delivery and QoS levels.

The rest of this paper is organized as follows. The next section presents background information on SLA performance indicators for the main cloud delivery models. Section 3 describes related work and motivation for this work. Section 4 presents an overview of the proposed framework. Section 5 describes the SLA-driven algorithm for cloud services' selection. Finally, Section 6 concludes the paper.

2   SLA Performance Indicators

When it comes to SLA quality indicators, every cloud provider uses its own key performance indicators (KPIs). There is no standard at this level yet even though there are some efforts with this regards by the Cloud Services Measurement Initiative Consortium (CSMIC) [3].

1)   IaaS KPIs

The IaaS delivery model is typically segmented into three offerings: compute, storage and network. Each of these offerings is delivered in a variety of usage models. Meegan et al. [8] considered that the following metrics are the typical indicators that an IaaS consumer may expect finding in the SLA.

- *Compute indicators: availability, outage length, server reboot time*
- *Network indicators: availability, packet loss, bandwidth, latency, mean/max jitter*
- *Storage indicators: availability, input/output per second, max restore time, processing time, latency with internal compute resource*

Frey et al. [13] also provided a list of KPIs that may be considered for network services, storage services, and compute services. IaaS refers not only to the service itself but also to the virtual machines (VMs) that are used. Therefore, additional VM KPIs need to be considered. These KPIs are CPU Utilization, VM Memory, Memory Utilization, Minimum Number of VMs, Migration Time, Migration Interruption Time, and Logging.

2)   PaaS KPIs

PaaS KPIs vary significantly across cloud providers. Standards initiatives as in [3] have just started to emerge for the PaaS service delivery model. Meanwhile, service consumers need to identify the PaaS services that are extremely vital to their business and make sure that their negotiated SLAs with PaaS providers contain clear and measurable KPIs.

3)   SaaS KPIs

Meegan et al. [8] consider that service consumers should expect having general SaaS indicators in their SLA: monthly cumulative downtime, response time, persistence of consumer information, and automatic scalability. Burkon L. [9] summarizes the SLA dimensions and their associated quality indicators for the SaaS delivery model. These dimensions are:

- *Availability: Uptime*
- *Performance: Response Time, Throughput, and Timeliness*
- *Reliability: Uptime History*
- *Scalability: Granularity, Elasticity*
- *Security: Compliance Certifications, Authentication, Authorization, Attack Prevention, Backup Coverage, Backup Periodicity, Recovery Velocity, and Infrastructure Architecture*
- *Support: Resolution Rate, Resolution Time, RFC Implementation Rate, and Community*
- *Interoperability: API Coverage, API Form*
- *Modifiability: Data Layer Modifiability, Logic Layer Modifiability, and Presentation Layer Modifiability*
- *Usability: Satisfaction, Accessibility, and Efficiency*
- *Testability: QoS Testability, Application Testability*

3   Related Work and Motivation

Only few research efforts investigated the issue of cloud services' selection. Garg et al. [12] proposed a framework, called SMICloud that allows comparing and ranking IaaS cloud services based on a set of defined attributes. The ranking mechanism uses the Analytical Hierarchical Process (AHP) that permits assigning weights to the attributes to reflect the interdependencies among them. Wang et al. [14] proposed a cloud-based model for the selection of Web services. The model computes QoS uncertainty and finds the most appropriate Web services by using mixed integer programming. Hussain et al. [6] used a multi-criteria decision-making methodology for the selection of cloud services. Ranking of services is achieved by matching the consumer requirements against each service offering and for each criterion. To decide on the most suitable service, they used the weighted difference or the exponential weighted difference methods. Likewise, Li et al. [2] developed a promising system, called CloudCmp, to compare various offers from cloud providers with regard to performance and cost. CloudCmp uses ten metrics in the comparison of the service offerings of cloud providers. Results of the system enable service consumers forecast the cost and performance of their applications before their actual deployment on the cloud. Rehman et al. [16] proposed a system, which is using information collected by existing cloud users in order to select appropriate cloud offerings.

The Cloud Services Measurement Initiative Consortium (CSMIC) [3] developed a comprehensive framework to allow service consumers to describe the performance and quality provisions they require. CSMIC also developed a hierarchical measurement framework called the Service Measurement Index (SMI). SMI aims at becoming a standard method for measuring any cloud service (IaaS, PaaS, SaaS, BPaaS, and Big Data). SMI involves the application of significant measures designed to allow comparison of cloud services available from multiple providers. The top level categories of SMI are *Accountability*, *Agility*, *Assurance*, *Financials*, *Performance*, *Security and Privacy*, and *Usability*. Each category is then refined into several attributes to reach a total of 51 attributes.

Our work aims at facilitating mediation between service consumers and service providers and providing support for automated selection of suitable cloud services based on their QoS capabilities. Finding the right service offering is not an easy task for service consumers given the variety of service offerings. Moreover, dealing with a service provider requires knowledge of its operating environment, the availability of service management tools, and the service conditions and terms. Collecting this information from numerous service providers is likely to be a difficult task that is expensive and time-consuming. The Cloud broker with its knowledge and value-added services will assist service consumers in carrying out the following tasks: (a) expressing their service requirements (functional and non-functional) (b) selecting appropriate service offerings, (c) negotiating the terms and conditions for service delivery, (d) monitoring and assessing the implementation of SLAs, and (e) offering a single interface to interact with multiple service providers.

4   Framework Overview

Fig. 1 depicts our proposed framework for cloud services' selection. It is founded on a cloud federation. The research community has already recognized the need for cloud federation given the heterogeneity in cloud offerings [4][10]. A cloud federation extends the reach of both service providers and service consumers. A service is no longer accessible only to requestors of its cloud. Instead, requestors may request services across cloud borders (similar to the cellular network). The main components of the framework are: Service consumers (SCs), Cloud Brokers (CBs), and Service providers (SPs). Service providers deploying their cloud services on different clouds of the federation may provide similar or different cloud services.

4.1   Cloud Brokers' Federation

The federation consists of cloud brokers, from several clouds, which cooperate to achieve intermediation between service consumers and cloud services at a large scale. Federated routes connect disparate brokers across various clouds, while service consumers connect to brokers on their cloud or found in a brokers' directory. Cloud brokers may share information about cloud services and the kind of service that these services can provide. If a cloud broker can not satisfy the request of a service consumer for a particular service type, it forwards the request to other cloud brokers of the federation that may have the knowledge about cloud services offering the demanded service type.

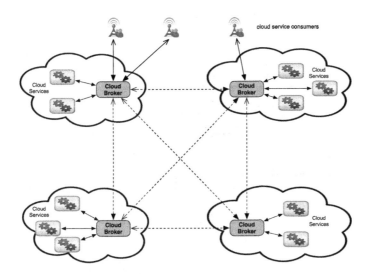

Fig. 1. Cloud brokers federation

Cloud brokers may be organized by geography or service type. Here are some use cases of the federation:

- Geography: service consumer's requests may be routed to a cloud broker close to the consumer.
- Service type: high-value service consumers may be routed to cloud brokers that have the knowledge about cloud services providing requested service type.
- Load balancing: interactions among brokers may be triggered dynamically to cope with changes in actual or anticipated load of brokers.
- High availability: requests may be routed to a new cloud broker if a solicited one becomes unavailable.

4.2   Cloud Brokers

Cloud brokers deployed on several clouds collaborate to provide services across cloud borders. Specifying the links between them creates the federation. A link contains information about the target broker, such as the reference to its Coordinator component. As we mentioned earlier, a Cloud broker is a mediator service that decouples service consumers from service providers. Given that service consumers do not usually have the competence to negotiate, manage and monitor QoS, they delegate management tasks, such as selection of appropriate service providers and SLA negotiation, to the Cloud broker. The architecture of the Cloud broker, depicted by Fig. 2, and the SLA negotiation process have been described in our previous work [5]. The SLA specifies the service information and the agreed upon level of QoS that the service-provider shall deliver. We have extended the architecture of the Cloud broker with a Link Manager component. The *Link Manager* component is responsible for managing links with other brokers of the federation. It provides the following link management operations: *add_brokerlink(), remove_brokerlink(), modify_brokerlink(), describe_brokerlink(), and list_brokerlinks()*.

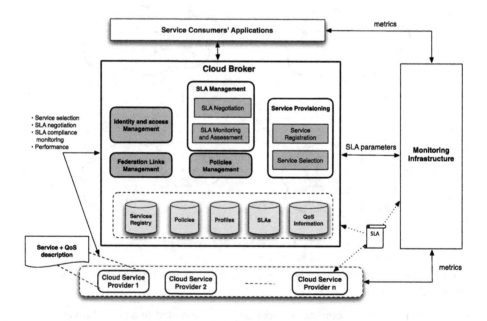

Fig. 2. Cloud Broker management operations

5   Algorithm for SLA-Based Cloud Service Selection

In this section we describe our proposed cloud service selection algorithm. The algorithm permits ranking the potential cloud services with respect to the service consumer's required QoS. First, we explain how the algorithm works in the case of a single cloud. Then, we show how the selection is done at the level of the federation.

Let $C = \{C_1, C_2, \ldots, C_m\}$ represent the set of clouds of the federation F. One or several cloud brokers are deployed on each cloud of the federation. Let $B = \{B_1, B_2, \ldots, B_l\}$ be the list of Cloud brokers of the federation F. For each type of service (IaaS, PaaS, SaaS), the brokers of the federation use the same quality metrics according to the Service Measurement Index (SMI) [3].

5.1   Selection in a Single Cloud

When submitting a request to a cloud broker B_i of cloud C_i, the service consumer specifies her preferences with regard to the QoS indicators that she can tolerate. Let $Q^i = \{Q_1^i, Q_2^i, \ldots, Q_n^i\}$ be the list of QoS indicators considered by the cloud broker B_i for the type of service (IaaS, PaaS, or SaaS) requested. B_i initiates, then, an auction in order to get bids from relevant cloud services that can handle the service-consumer request. Let $CS_i = \{CS_{1i}, CS_{2i}, \ldots, CS_{ki}\}$ be the set of potential cloud services of the cloud C_i, with $1 \leq i \leq m$. Each bid is a vector, which describes the level of QoS for each of the quality indicators that the cloud service may assure. This model corresponds to a *multiple-items auction*. After a predefined period, the broker B_i closes the auction and ranks the bids according to the weight specified by the service consumer for each quality indicator. It, then, selects the winner of the auction as the most suitable cloud service to handle the service consumer request. The auction models, proposed in the auction theory, are *English auction*, *Sealed-Bid auction*, *Dutch auction*, and *Vickrey auction*. As the bidders should not be aware of the bids of each other, the sealed-bid auction is the most relevant in our case. In the following, we describe the process of determining the winner of the auction.

As we have seen in previous sections, quality indicators come in different units. To rank the potential cloud services that can provide the service requested by the service consumer, we use the normalization process. The goals of this process are to: (a) use a uniform measurement of cloud service qualities independent of units, (b) provide a uniform index to represent service qualities for each cloud service provider, and (c) allow setting thresholds for the various quality indicators based on the requirements of the service consumer in terms of quality of service. Using these thresholds, QoS indicators are normalized into values between 0 and 1. To rank the offers, B_i evaluates the utility function associated with each QoS attribute and the global utility function for each individual offer. A global utility U is a function of the individual utility functions U_j. $1 \leq j \leq n$

$$U = f(U_j), 1 \leq j \leq n \tag{1}$$

U_j represents the individual utility function associated with the QoS indicator Q_j. If we assume that the QoS attributes are independent, the global utility function U associated with the offer of cloud service CS_{si} can be expressed by the additive linear utility function as:

$$U^s = w_1 U_1^s + w_2 U_2^s + \cdots + w_n U_n^s \tag{2}$$

w_j is the importance weight that the service consumer assigns to that attribute. Each weight is a number in the range $(0,1)$ and $\sum_1^n w_j = 1$. The most suitable cloud service is the one that maximizes the above additive utility.

Several utility functions may be considered to evaluate the offers of cloud services depending on the cloud service delivery model (IaaS, PaaS, and SaaS). To illustrate our approach, we consider the case of a SaaS service selection and only two quality attributes, *availability* and *response time*. We define in equation (1) $U_V(v)$, the utility function for the *availability* quality attribute v. This function reaches its maximum value of 1 when v=1 and decreases to 0 when v is decreasing to 0.

$$U_V(v) = \frac{v^{\beta_V}\,(1+\alpha_V)}{1+\alpha_V v^{\beta_V}} \tag{3}$$

v is the offer of the cloud service for the availability and α_V is the desired value of the service consumer for v. β_V is a measure of the service consumer sensitivity to the *availability* quality attribute. When $\beta_V = 0$, the service consumer is indifferent to the *availability*. When $\beta_V = 1$, the service consumer is moderately sensitive to the *availability*. When $\beta_V > 1$, the service consumer is increasingly sensitive to the *availability*. As β_V increases, the service consumer is expressing increasing concern about it. For $\beta_V < 1$, as v decreases to approach 0, the service consumer is expressing increasing indifference to the *availability*. Fig. 3 shows $U_V(v)$ for $\alpha_V= 0.75$, and $\beta_V=1, 3,$ and 5 respectively.

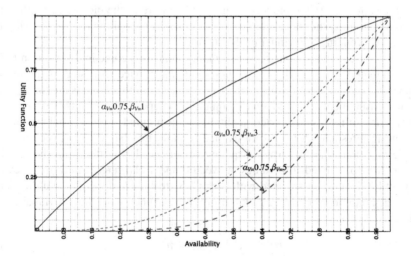

Fig. 3. Utility vs. Availability

We consider also $U_R(r)$ the utility function for the *response time* quality attribute r. It reaches its maximum, which is 1, when r is 0 and decreases to 0 when r reaches 1 as we are considering the normalized values.

$$U_R(r) = \frac{1 - r^{\beta_R}}{1 + \alpha_R r^{\beta_R}} \qquad (4)$$

r is the offer of the cloud service for the *response time* quality attribute and α_R is the desired value of the service consumer for that quality attribute. β_R is a measure of the service consumer sensitivity to the response time quality attribute similar to β_R. Fig. 4 shows $U_R(r)$ for $\beta_R=2$ and $\alpha_R= 0.4, 0.8$, and 0.95 respectively.

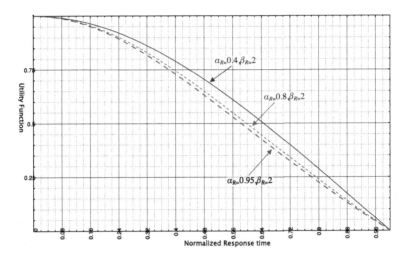

Fig. 4. Utility vs. response time

As a proof of concept, we consider in Table 1 a scenario with 4 cloud service offerings and the following values for the SLA values, the sensitivity parameters, and the weights associated with the quality indicators.

Table 1. Example of cloud service consumer parameters

	Response time	Availability
SLA value	$\alpha_R= 0.95$	$\alpha_V= 0.98$
Sensitivity	$\beta_R= 2$	$\beta_V= 4$
Weight	$w_R= 0.4$	$w_V= 0.6$

Table 2. Cloud service offers

Cloud service	r	v	U_R	U_V	U
CS_1	0.94	0.99	0.06	0.98	0.61
CS_2	0.95	0.97	0.052	0.940	0.58
CS_3	0.96	0.98	0.042	0.96	0.59
CS_4	0.96	0.94	0.042	0.88	0.54

The results in Table 2 show that the offer of CS_1 is the best cloud service offer.

5.2 Selection in Multiple Clouds

The previous subsection describes how the ranking of cloud services is achieved within a single cloud. To find out the most suitable cloud service within the federation, the initial cloud broker, which receives the service consumer request, forwards her QoS requirements to each cloud broker of the federation. Each of these brokers executes, then, the selection algorithm to find, within their cloud, the best cloud service capable of handling the service consumer request. Selected cloud services from the various clouds of the federation are then ranked to find out the best cloud service, which maximizes the global utility function expressed by Eq. (2).

6 Conclusion

Given the growing number of available cloud services offerings, enterprises are facing the challenge of finding appropriate cloud service providers that can supply them with required services (IaaS, PaaS, and SaaS) and meet their quality requirements.

In this paper, we have presented a framework for cloud services' selection that relies on a federation of cloud brokers. These brokers are in charge of selecting appropriate cloud services that can satisfy service consumers' requests, and mediating between service consumers and selected cloud services by carrying out, on behalf of service consumers, the negotiation of the terms and conditions of service delivery. The selection algorithm ranks the offers of potential cloud services by evaluating a global utility function for each offer. The best offer is the one that maximizes the utility function. To illustrate our approach, we have considered the weighted additive utility function of the individual utility functions of two quality-of-service attributes, response time and availability.

References

1. Dan, A., Davis, D., Kearney, R., Keller, A.: Web services on demand: WSLA-driven automated management. IBM Systems Journal 43(1), 136–158 (2004)
2. Li, A., Yang, X., Kandula, S., Zhang, M.: Comparing public-cloud providers. IEEE Internet Computing 15, 50–53 (2011)
3. CSMIC, The Cloud Services Measurement Initiative Consortium. "The Service Measurement Index (SMI)", http://csmic.org/understanding-smi/
4. Villegas, D., Bobroff, N., Rodero, I., Delgado, J., Liu, Y., Devarakonda, A., Fong, L., Sadjadi, S.M., Parashar, M.: Cloud federation in a layered service model. Journal of Computer and System Sciences 78(5), 1330–1344 (2012)
5. Badidi, E.: A Cloud Service Broker for SLA-based SaaS Provisioning. In: International Conference on Information Society (i-Society 2013), Toronto, Canada, pp. 61–66 (2013)
6. Hussain, F.K., Hussain, O.K.: Towards Multi-Criteria Cloud Service Selection. In: Proc. of the 2011 Fifth International Conference on Innovative Mobile and Internet Services in Ubiquitous Computing (IMIS), pp. 44–48 (2011)
7. Dinh, H.T., Lee, C., Niyato, D., Wang, P.: A survey of mobile cloud computing: architecture, applications, and approaches. Wirel. Commun. Mob. Comput. 13(18), 1587–1611 (2011)

8. Meegan, J., et al.: Practical Guide to Cloud Service Level Agreements. Version 1.0. Cloud Standards Customer Council, April 10 (2012)
9. Burkon, L.: Quality of Service Attributes for Software as a Service. Journal of Systems Integration, 38–47 (2013)
10. Toosi, N., Calheiros, R.N., Buyya, R.: Interconnected Cloud Computing Environments: Challenges, Taxonomy, and Survey. ACM Computing Surveys (CSUR) 47(1), 7 (2014)
11. Mell, P., Grance, T.: The NIST Definition of Cloud Computing, National Institute of Standards and Technology, Special Publication 800-145 (2011)
12. Garg, S.K., et al.: Smicloud: A framework for comparing and ranking cloud services. In: Proc. of the 2011 Fourth IEEE International Conference on Utility and Cloud Computing (UCC), pp. 210–218 (2011)
13. Frey, S., Reich, C., Lüthje, C.: Key Performance Indicators for Cloud Computing SLAs. In: The Fifth International Conference on Emerging Network Intelligence, EMERGING 2013, pp. 60–64 (2013)
14. Wang, S., et al.: Cloud model for service selection. In: Proc. of the 2011 IEEE Conference on Computer Communications Workshops (INFOCOM WKSHPS), pp. 666–671 (2011)
15. Xu, Y., Mao, S.: A survey of mobile cloud computing for rich media applications. IEEE Wirel. Comm. Mag. 20(3), 46–53 (2013)
16. ur Rehman, Z., et al.: A Framework for User Feedback Based Cloud Service Monitoring. Presented at the 2012 Sixth International Conference on Complex, Intelligent and Software Intensive Systems, CISIS (2012)

A Framework for Incremental Covering Arrays Construction

Andrea Calvagna and Emiliano Tramontana

Dipartimento di Matematica e Informatica
Università di Catania, Italy
{calvagna,tramonta}@cs.unict.it

Abstract. In this paper we first show that the combinatorial task of enumerating t-wise tuples out of a given set of elements is the recursive iteration of the simpler task of computing pairs out of the same set of elements. We then show how to apply this result to the design of a general framework for incremental CIT test suite construction, that is, producing at each iteration an additional set of tuples required to increase the coverage interaction degree just by one. As a last contribution, we show that in the presented construction framework, the covering array minimization problem and its inherent complexity can be encapsulated inside a delegate task of merging two given smaller covering array, while minimizing rows redundancy.

Keywords: Software engineering, covering arrays, combinatorial interaction testing, incremental strength coverage.

1   Introduction

A covering array $CA_\lambda(N; t, n, \boldsymbol{r})$ is an $N \times n$ array of symbols with the property that in every $N \times t$ sub-array, each t-tuple of symbols occurs at least λ times, where t is the *strength* of the interaction, n is the overall number of array components (the *degree*), and $\boldsymbol{r} = (r_1, r_2, ...r_n)$ is a vector of positive integers defining the number of possible symbols for each component. When applied to combinatorial interaction testing (CIT) only the case when $\lambda = 1$ is of interest, that is, where every t-tuple of symbols is covered at least once. The general task of computing a CA is NP-complete [1, 2]. Although exact solutions to compute it by *algebraic* construction do exist [3], they are *not generally applicable* due to strict requirements on the considered task geometry (that is, the number of parameters, their ranges, and the interaction strength are not free variables). As a consequence, researchers have addressed the issue of designing generally applicable algorithms and tools to compute CIT test suites, most of which are based on *greedy* heuristic searches [4–10]. While it is easy to see that a CA of strength t is also a CA for all lower strengths too, generally these techniques don't allow to distinguish which subset of rows of the CA is actually contributing to each specific strength of coverage in the $[1..t]$ range.

© Springer International Publishing Switzerland 2015
R. Silhavy et al. (eds.), *Software Engineering in Intelligent Systems*,
Advances in Intelligent Systems and Computing 349, DOI: 10.1007/978-3-319-18473-9_28

In this paper a new framework algorithm for the construction of t-wise covering arrays is presented, which is instead designed to be inherently incremental with respect to increasing the strength of interaction, and whose CA rows can thus be individually marked with the additional level of strength they are required for. The proposed approach is the direct application of an introduced set theoretic relation between combinatorial enumerations of consecutive strengths. This relation shows how the combinatorial problem of enumerating all t-wise tuples of a set of objects is the recursive iteration of the simpler task of enumerating *pairs* of a set of objects. Based on this relation, a family of recursive t-wise CA construction algorithms can be devised, capable of generating not only the CA at the desired strength, but also the whole set of CAs for all lower interaction strengths, as known subsets of rows of the whole main CA. Simply put, the generated CA comes already divided in t ordered partitions, corresponding to increasing strength of interaction coverage. This allows different strengths of CIT testing to be applied progressively, at different times or on-demand, by applying only the additional set of (rows) test cases required to achieve next strength of coverage. Large sized combinatorial models, or models requiring high strengths of interaction, which are expensive to compute, would take great advantage from having the burden of compute the test suite just once, to actually obtain the whole set of incremental sub-suites from strength 2 up to the desired t, instead of having to compute separately many suites, which together are also expected to amount to a overall bigger number of tests.

The paper is structured as follows: section 2 introduces related work, section 3 presents the recursive relation between two combinatorial set; section 4 based on results from previous section, derives a new strategy to implement incremental t-wise covering arrays; section 5 contrasts the proposed approach to existing literature; section 6 discusses candidate scenarios to apply the proposed technique, and eventually section 7 draws our conclusive statements.

2   Related Work

To the best of our knowledge, there is no other generally applicable CA generation technique currently available which computes at once a CA of a given strength and is also able to tell exactly which subset of rows in the CA is required to address each specific intermediate degree of interaction. The only similar work with respect to this, to the best of our knowledge, is from M.B. Cohen et al. [11], who proposed a technique consisting in supplying a previously computed CA as a *seed* for the generation of a new test suite of higher strength. However, their approach to incremental CAs delivery does not have any particular analytic foundation, like this work does, and is just based on the general idea that computing a covering array not starting from scratch but from a pre-computed set of *seeded* rows is possible with any combinatorial algorithms. This is in fact a well known feature of many CIT tools, called *seeding* of test cases, and which is completely irrespective of the type of construction algorithm used. Not being it a peculiarity of the construction algorithm, any one could adopt this strategy to enable incremental strength coverage support. Conversely, the construction framework presented in this paper supports incremental strength delivery as an inherent feature. In fact, one run of the framework algorithm for a given strength already produces all incremental subsets of rows for each strength at once, without requiring the seeding of any precomputed and saved, lower strength CA. Moreover, we are here proposing not a specific covering array minimization technique but a *framework* algorithm for incremental test suite delivery, where the CA minimization task is encapsulated into the task of effectively *merging* two existing covering arrays.

3   Coverage Enumeration

Given a reactive system under test, with n input parameters, all ranging in discrete and distinct set of values, let $P = \{P_1, P_2, ..P_n\}$ be a set of indexed objects modeling the n system input parameters. Let P_i be in turn modeled itself as a set of r_i distinct symbols, that is, the i-th parameter values, where r_i is the range of that parameter. Please note that as all the symbols are distinct by assumption, all P_i are disjunct. As a consequence, their elements might be conveniently represented by a unique integer enumeration, e.g.: $P_1 = \{0, 1\}$, $P_2 = \{2, 3, 4\}$, $P_3 = \{5, 6\}$. An *assignment* is a set of symbols mapping a *value* to each element in a given subset of P. I.e., the set $\{1, 5\}$ is an assignment of the parameters set $\{P_1, P_3\}$.

Let be a *pair* of two objects A and B the set union of exactly those two objects: $\{A, B\}$.

Definition 1. [*pairing relation*] We define the $\cdot$ (*dot*) *pairing* as the set of all possible pairs of the elements of two given set.

Note that this is quite different from the cartesian product, since pair objects are not required to be ordered, and they are themselves set. Moreover, in contrast to cartesian product, $A^2 \equiv A$. As an example:

$$P_1 \cdot P_2 = \{\{0, 2\}, \{0, 3\}, \{0, 4\}, \{1, 2\}, \{1, 3\}, \{1, 4\}\}$$
$$P_1 \cdot P_2 \cdot P_3 = \{\{0, 2, 5\}, \{0, 3, 5\}, \{0, 4, 5\}, \{1, 2, 5\}, \{1, 3, 5\},$$
$$\{1, 4, 5\}, \{0, 2, 6\}, \{0, 3, 6\}, \{0, 4, 6\}, \{1, 2, 6\}, \{1, 3, 6\}, \{1, 4, 6\}\},$$

The product $P_1 \cdot P_2 \cdot \ldots P_n$ then describes the set of *exhaustive* combinations of values of the n factors, that is all their n-wise assignments. As a shorthand notation for that product we will write $P_{1...n}$. Generalizing, let's write $P_{1...n}^t$, or simply P_n^t, to denote the set of all t-wise values combinations of the first n parameters, with $t < n$. As an example, we shall write P_3^2, to mean the set of pairwise combinations of $\{P_1, P_2, P_3\}$. It is then easy to check, by the respective definitions, that $P_2^2 = P_1 \cdot P_2$, and similarly $P_3^2 = P_1 \cdot P_2 \cdot P_3$, because the introduced *pairing* relation defines exactly the same set of elements required by the left side terms.

It is important to highlight that the term P_n^t denotes a unique set of t-wise assignments, of the given n parameters. On the other hand, the term $CA(N; t, n, r)$

denotes *any* $N \times n$ array of values actually covering that set of required assignments. Hence, we will refer to the therm P_n^t as the t-wise *coverage requirements* to be met by a valid covering array implementation. The following *recursive* equation holds true:

$$P_n^t = P_{n-1}^t \cup P_{n-1}^{t-1} \cdot P_n \tag{1}$$

Proof. It suffices to note that the left side is the set of assignments required for t-wise coverage. On the right side, since the t-wise assignments of the first $n-1$ parameters are already included, then only the t-wise assignments including P_n are missing to the count. These latter are just obtained by pairing each $(t-1)$-wise assignment of the previous $n-1$ parameters with every value of P_n.

The implication of this result is that to compute P_n^t we only need to be able to compute the two right side terms, which have lower strength and/or less parameters. By recursively applying equation 1 to both right side terms, as visually depicted in the graph of Figure 1, the leaves eventually reached will be terms of the trivial kind: P_k^1 (flat union of all values of the first k parameters) and P_k^k (exhaustive assignments of the first k parameters). The left side term can thus be computed simply by knowledge of the trivial set union and pairing relation, and the set P of course.

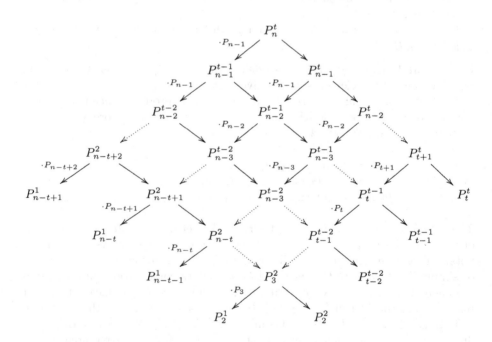

Fig. 1. Recursive definition of P_n^t ($\equiv P_{1..n}^t$)

As an example, the recursion graph for a term of the type $P_{1..5}^3$ will be:

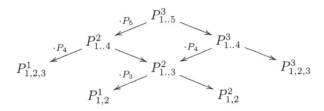

It is easy to check that visiting the graph in post-order will produce the following expression, which is exactly the set of all required 3-wise factor combinations:
$P_1 P_2 P_3 \cup P_1 P_2 P_4 \cup P_1 P_3 P_4 \cup P_2 P_3 P_4 \cup P_1 P_2 P_5 \cup P_1 P_3 P_5 \cup P_2 P_3 P_5 \cup P_1 P_4 P_5 \cup P_2 P_4 P_5 \cup P_3 P_4 P_5$. (the dot sign has been omitted for convenience).

4   Coverage Implementation

Let C_n^t be a shorthand for any array of size N, whose elements are tuples, covering all assignments in the set P_n^t at least once. Although C_n^t is actually a $N \times n$ array it shall be treated here as a one dimensional array of N atomic objects. Also note that C_n^t is a $CA_1(N; t, n, r)$. It is possible to build a C_n^t based on the recursion of equation 3. In fact, being by definition C_{n-1}^t and C_{n-1}^{t-1} two real arrays covering respectively P_n^t and P_{n-1}^t, the following will also be true:

$$C_{1..n}^t = C_{1..n-1}^t + C_{1..n-1}^{t-1} \times C_n \qquad (2)$$

where the equivalence is in terms of the coverage achieved by the arrays resulting from the right and left side expressions; the sum sign represent the array append operator; CA_n is an r_n sized array covering P_n, that is, listing all values of the n-th parameter; and the $\times$ operator is a simple implementation of the semantics of the aforementioned *pairing* relation, applied to the array elements, as shown by the following:

Example 1

$$C_2^1 \times C_3 = \begin{bmatrix} \{0, 2\} \\ \{1, 3\} \\ \{x, 4\} \end{bmatrix} \times \begin{bmatrix} \{5\} \\ \{6\} \end{bmatrix} = \begin{bmatrix} \{0, 2, 5\} \\ \{1, 3, 5\} \\ \{x, 4, 5\} \\ \{0, 2, 6\} \\ \{1, 3, 6\} \\ \{x, 4, 6\} \end{bmatrix}$$

More generally, *pairing* results in an array, whose rows are obtained by *adding* every element of the second array to every element of the first array. In fact, in the (2) pairing occurs only between arrays whose elements are assignments of disjoint set of parameters: the first $k - 1$ with the $k - th$ one. Please note that in the following will omit the curly brackets for reading convenience.

Recalling that a covering array of strength t is also a covering array of strength $t - 1$, occurrences of the term C_n^{t-1} can be subsumed by C_n^t. Conversely, C_{n-1}^t can be trivially extracted from a C_n^t simply by removing the assignments of the

$n - th$ parameter (that is, the last column). As a consequence, from equation 2 the following secondary result can be derived:

$$\Delta C_n^t = C_n^t - C_n^{t-1} = C_{n-1}^{t-1} \cdot C_n \supset C_n^{t-1} \tag{3}$$

where the $\supset$ symbol stands for coverage subsumption. That is, the additional (ΔC_n^t) test cases, computed in order to increase the coverage strength by one unit, suffice alone to achieve complete coverage of the lower strengths. Another implication, most important for incremental CIT test suite construction algorithms, is that any lower strength test suite that has been seeded to that purpose will be completely redundant in the computed test suite increment. Those seeded tests can in fact be dropped away without losing the lower strength coverage. However, still they are required to achieve the higher strength coverage. Hence, it follows that any t-strength covering array can be partitioned in two *complementary* sub-arrays of identical strength $t - 1$. As a final deduction, whatever the adopted construction technique, a CIT test suite tuples redundancy will at least double at every increase in the strength level.

To help clarifying, Example 2 shows how the (3) can be applied to the computation of a minimal pairwise covering array for the parameter subset $\{P_1, P_2, P_3\}$, that is C_3^2, based on previous knowledge of C_2^1, C_2^2 and the introduced trivial operators.

Example 2

$$C_2^1 = \begin{bmatrix} 0\,2 \\ 1\,3 \\ 0\,4 \end{bmatrix}; \quad C_2^2 = \begin{bmatrix} 0\,2 \\ 0\,3 \\ 0\,4 \\ 1\,2 \\ 1\,3 \\ 1\,4 \end{bmatrix}; \quad C_2^2 - C_2^1 = \begin{bmatrix} 0\,3 \\ 1\,2 \\ 1\,4 \end{bmatrix};$$

$$C_3^2 = C_2^2 + C_2^1 \cdot C_3 = C_2^2 + \begin{bmatrix} C_2^1 \times \{5\} \\ C_2^1 \times \{6\} \end{bmatrix} =$$

$$\begin{bmatrix} C_2^2 \times \{x\} \\ C_2^1 \times \{5\} \\ C_2^1 \times \{6\} \end{bmatrix} = \begin{bmatrix} C_2^1 \times \{5\} \\ C_2^2 - C_2^1 \times \{6\} \end{bmatrix} = \begin{bmatrix} 0\,2\,5 \\ 1\,3\,5 \\ 0\,4\,5 \\ 0\,3\,6 \\ 1\,2\,6 \\ 1\,4\,6 \end{bmatrix}$$

$$\quad (a) \qquad\qquad\qquad (b) \qquad\qquad\qquad (c)$$

In the (a), the array resulting from the pairing operation is appended to the C_2^2 term. Note that this latter can be rewritten as $C_2^2 \times \{x\}$, where x is the unspecified *any* value), in order to equal the column-size of previous array, without changing its coverage. In (b) the eq. 3 has been applied to the lower term of the previous array, that is one instance of C_2^1 has been replaced with its complementary CA with respect to C_2^2. Since now both C_2^1 and its complement $C_2^2 - C_2^1$ are in the array, the term C_2^2 can be removed as it is completely redundant. The resulting array (c) is a correct implementation of C_3^2 and is also minimal by construction.

4.1   Generalization

Starting from equation 2, and recursively applying it to its own first term, this unfolds to the expression shown in Figure 2. After the recursion has been

$$
\begin{aligned}
C_n^t =& \\
& C_{n-1}^t + C_{n-1}^{t-1} \cdot C_n = \\
& C_{n-2}^t + C_{n-2}^{t-1} \cdot C_{n-1} + C_{n-1}^{t-1} \cdot C_n = \\
& \vdots \\
& C_t^t + C_t^{t-1} \cdot C_{t+1} + C_{t+1}^{t-1} \cdot C_{t+2} \cdots + C_{n-2}^{t-1} \cdot C_{n-1} + C_{n-1}^{t-1} \cdot C_n
\end{aligned}
$$

Fig. 2. First-term recursion on C_n^t

applied exactly $n - t$ times, the first term becomes C_t^t, which a trivial array, with exhaustive combinations of the first t factors. At the same time, we observe that computing the *last* array term of the unfolded expression, C_{n-1}^{t-1} suffice to compute all arrays in the expression at once, since these are just sub-arrays:

$$
C_k^{t-1} \subset C_{n-1}^{t-1} \quad \forall k < n - 1
$$

In summary, computing C_n^t relies only on the ability to compute C_{n-1}^{t-1}.

This in turn can be obtained from C_{n-2}^{t-2}, by applying again equation (2). After $n - t$ applications, this other recursion will end up on a trivial last term too, C_{n-t+1}^1, that is, the enumeration of the parameters values without any combination. The overall recursion depth for the starting term C_n^t will be $(n-t)^2$, which is good since allows scalability of the construction technique with respect to increasing the interaction degree.

The middle terms of the recursively unfolded expression, that is the sub-arrays of C_{n-1}^{t-1}, have to be *paired* each with one parameter, and appended to each other: in this operation any redundancy present in the rows of these array terms will add up to the accumulated sum and multiply at a fast rate. Hence, a smarter *merge* operation can here be implemented, instead of the simple *append*, such that redundancy shall be reduced to an acceptable point. The exercise of designing an optimal heuristic algorithm for the efficient implementation of this merge operation is here left outside the scope of this paper. However, one example merge implementation has already been successfully realized in order to empirically test the described approach. The overall efficiency of this approach, both in terms of the test suite final size and of computational complexity, relies mostly on the ability to implement an effective and fast *merge* procedure, that is avoiding redundancy, and keeping its computational complexity low. Note that not a single tuple has to be stored in this process. Note also that the computation will actually proceed on the *return* path from the last recursive unfolding, that is the first computed term will be C_{n-t+1}^1. It is also very easy to mark the rows of the arrays resulting from each level recursion, so that they are always recognizable in the *upper level* recursion. This markings will add up and result by construction in a clear partitioning of the final C_n^t in t incremental sub terms for all lower strengths covering arrays.

5   Early Results

In order to assess the framework and show the scalability improvements achievable with the proposed approach we implemented the framework, testing several possible heuristic strategies with respect to a well know input model, largely used in the literature on combinatorial testing, and compared the execution times achieved. We are here reporting in Figure 3 the results of two such example strategies, contrasted with those of other existing tools, which are the best performing combinatorial testing tools among the few currently available that support also strengths greater than two. These tools are the IPO-G/D algorithms by Y.Lei et al. [12] and the open-source (though undocumented) tool called Jenny [13]. Other existing tools, i.e. ITCH by IBM, have much lower performance than these, and have not been included. The TCAS case study models the input space of a portion of a software system used in aviation to avoid in-flight collisions between aircrafts. The TCAS input space geometry, expressed in exponential notation [14] is $10^2 4^1 3^2 2^7$, that is, it has twelve inputs variables (the exponents) with mixed ranges 10, 4, 3 and 2 (the bases). The results for IPO-G/D are actually those of their publicly available implementation ACTS, which unfortunately does not allow strengths greater than six. For all the other tools, results have been reported only if the computation time was *reasonably* low, which we defined as an upper bound of 900 seconds (15 minutes). Two software tools corresponding to the two variants of our approach have been implemented. These are named INW-7, which implements a recursive computation of the general formula starting from the first formula term, as shown in Figure 2 and INW-a, which implements an alternative recursive computation where the formula terms are combined backwards starting from the rightmost term. Note that these are the two rightmost curves in Figure 3, which reports the computation times (shown in seconds, and smoothed as a cspline) on the vertical axis, at increasing strengths of combinatorial interaction (vertical axis). On Table 1 are contrasted the computation times and computed test suite sizes, instead, for the same set of algorithms. As the experimental results show, our INW-a tool is the only one capable to compute combinatorial suites for TCAS at all possible strengths, that is up to 11, and the trend of its computation times is not to diverge like all other tools do. The scalability of INW-7 equals that of the best of the other compared tools, although their computation times all grow exponentially much faster and, with the exception of INW-7, don't allow to achieve any result in a reasonable time for strengths greater than six.

Table 1. Comparative suite sizes and computation times for task TCAS

t	INW-a		INW-7		IPO-G		IPO-D		Jenny	
	size	time	size	time	size	time	size	time	size	time
3	740	0.051	716	0.055	400	0.55	480	0.13	413	0.71
4	3316	0.12	3364	0.10	1361	4.22	2522	0.34	1536	3.54
5	12114	0.35	12229	0.91	4220	25.39	5306	4.25	4580	43.54
6	35504	1.41	35688	6.11	10918	98.73	14480	47.72	11625	470
7	85956	11.07	87628	37.08	-	-	-	-	-	-
8	169280	22.29	169280	125.21	-	-	-	-	-	-
9	277760	41.22	279896	324.44	-	-	-	-	-	-
10	378656	32.62	378656	630.71	-	-	-	-	-	-
11	441440	13.72	441440	665.53	-	-	-	-	-	-

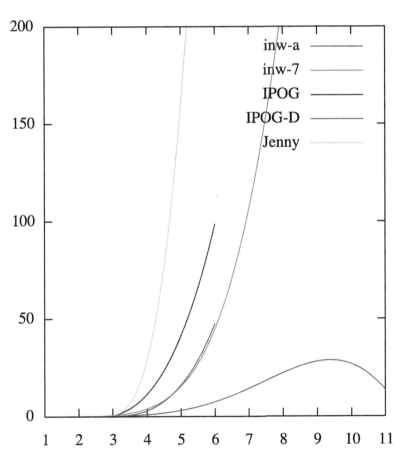

Fig. 3. A visual comparison of the execution times (y axis) with respect to increasing strength (x axis), applied to the TCAS input model

6   Discussion

In which scenario our method can be better? Two measure: time to complete the test suite T_C and time to discover the first fault on average. The time to complete is easily computed as:

$$T_C = T_G + T_t \times n_T$$

where T_G is the time to generate the test suite, T_t is the time to execute a single test and n_T is the number of tests. The time to discover the first faults T_F, assuming that a fault exists and it will be discovered on average after half of the tests have been executes, is

$$T_F = T_G + T_t \times n_T/2$$

For this reason, if the time $T - t$ to execute a single test is negligible, methods with a small test generation time T_G become faster, even if they produce bigger test suites. The situation is depicted in Fig. 4: if T_t is smaller than a certain time limit $\bar{T}_t$, fast methods become preferable.

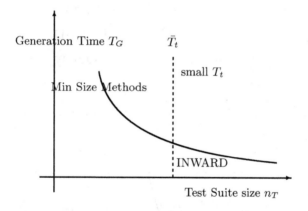

Fig. 4. Generation time vs Suite size

The actual value for $\bar{T}_t$ depends on many factor. Given a method F (Fast) and another M (minum size), we can compute the $\bar{T}_t$ as follows.

$$\bar{T}_t < \frac{T_{GM} - T_{GF}}{n_{TF} - n_{TM}}$$

The value for $\bar{T}_t$ computed for INWARD (which is how we named the presented approach) vs IPO for several test suites is shown in Table 2. If the time to execute a test T_t is less than $\bar{T}_t$ INWARD should be used instead of IPO.

Table 2. The $\bar{T}_t$ in millisecs for INWARD againts IPO

	INWARD		IPO		
t	size	time	size	time	$\bar{T}_t$
3	740	0.051	400	0.55	1.47
4	3316	0.12	1361	4.22	2.10
5	12114	0.35	4220	25.39	3.17
6	35504	1.41	10918	98.73	3.96

7 Conclusions

An original recursive formulation of the combinatorial enumeration problem has been here described formally and has been proposed as the foundation of a new family of stateless construction techniques, which are *inherently* incremental in the coverage strength, allowing incremental delivery of CIT testing at no additional costs. The proposed analytical framework is a starting point for research on more scalable high-strength algorithms, due to the inverse complexity relation with respect to increasing the combinatorial interaction strength and its ability to neatly incapsulate into a delegate *merge* operation any chosen redundancy minimization heuristic strategy, and its additional complexity. The framework has been implemented as the INWARD tool and instantiated with several heuristics, two of which have been compared with state of the art tools available for combinatorial t-wise interaction test suites. Please note that we select available tools that were generally applicable to any task without restrictions of size, strength and ranges, like is the case for ours. The proposed approach shows a significant improvement in the computation times with respect to compared tools, which is more evident at very high strengths of interaction, where the other tools usually diverge or fail. In contrast, INWARD easily allows testing at strengths greater than the usual strength six limit, which may be too low for applications with really strict safety requirements. Finally, we presented an analytical framework to be used to correctly determine when the overall cost of application of a (possibly longer) test suite is however convenient and thus the use of higher strength CIT testing is also economically justified.

References

1. Williams, A.W., Probert, R.L.: Formulation of the interaction test coverage problem as an integer program. In: Proceedings of the 14th International Conference on the Testing of Communicating Systems (TestCom), Berlin, Germany, pp. 283–298 (March 2002)
2. Seroussi, G., Bshouty, N.H.: Vector sets for exhaustive testing of logic circuits. IEEE Transactions on Information Theory 34(3), 513–522 (1988)
3. Kobayashi, N., Tsuchiya, T., Kikuno, T.: Non-specification-based approaches to logic testing for software. Journal of Information and Software Technology 44(2), 113–121 (2002)

4. Cohen, D.M., Dalal, S.R., Fredman, M.L., Patton, G.C.: The AETG system: An approach to testing based on combinatorial design. IEEE Transactions on Software Engineering 23(7), 437–444 (1997)
5. Tai, K.C., Lei, Y.: A test generation strategy for pairwise testing. IEEE Trans. Softw. Eng. 28(1), 109–111 (2002)
6. Bryce, R.C., Colbourn, C.J., Cohen, M.B.: A framework of greedy methods for constructing interaction test suites. In: ICSE 2005: Proceedings of the 27th International Conference on Software Engineering, pp. 146–155. ACM, New York (2005)
7. Cohen, M.B., Colbourn, C.J., Gibbons, P.B., Mugridge, W.B.: Constructing test suites for interaction testing. In: ICSE, pp. 38–48 (2003)
8. Nurmela, K.: Upper bounds for covering arrays by tabu. Discrete Applied Mathematics 138(1-2), 143–152 (2004)
9. Colbourn, C.J., Martirosyan, S.S., Mullen, G.L., Shasha, D., Sherwood, G.B., Yucas, J.L.: Products of mixed covering arrays of strength two. Journal of Combinatorial Designs 14(2), 124–138 (2006)
10. Sherwood, G.B.: Optimal and near-optimal mixed covering arrays by column expansion. Discrete Mathematics 308(24), 6022–6035 (2008), http://www.sciencedirect.com/science/article/B6V00-4RFD3S4-3/2/c6824765b1ecc8edcef07ee1fbc00717
11. Fouché, S., Cohen, M.B., Porter, A.A.: Incremental covering array failure characterization in large configuration spaces. In: ISSTA, pp. 177–188 (2009)
12. Lei, Y., Kacker, R., Kuhn, D.R., Okum, V., Lawrence, J.: IPOG/IPOG-D: efficient test generation for multi-way combinatorial testing. Software Testing Verification and Reliability 18(3), 125–148 (2008)
13. "Jenny combinatorial tool", http://www.burtleburtle.net/bob/math/jenny.html
14. Hartman, A., Raskin, L.: Problems and algorithms for covering arrays. DMATH: Discrete Mathematics 284(1-3), 149–156 (2004)

Web Service Composition Based on a Multi-agent System

Anouer Bennajeh and Hela Hachicha

Stratégies d'Optimisation et Informatique intelligentE (SOIE),
41 Street of liberty Bouchoucha-City CP-2000 Bardo – Tunis, Tunisia
anouer.bennajeh@gmail.com,
hela.hachicha@fsegs.rnu.tn

Abstract. The progression to use web technology has contributed the emergence of a certain number of new approaches. The web services composition is one of the most important approaches ensuring to exploit several services by a single service, called a composite service, in order to reducing time and costs, and increase overall efficiency in businesses. The work presented here is to propose a web services composition model in order to reduce the network problems during the composition and reduce the total duration of composition. In fact, this model is based on software agent technology by exploiting its characteristics, hence eliminate the passivity of web technology and make **it** more responsive. More than the composition of web services, this model ensures the execution of various activities to a composite service, by the integration of a control strategy which offers solutions at real-time in case to execution problem from one of the activities of a composite service. Thus, the control strategy reduces the response time to a web service composition request and reduces the overhead of network traffic.

Keywords: web service, web service composition, multi-agent system, mobile agent.

1 Introduction

The evolution of Internet and the competitiveness between companies based the bulk of their economic activities on the net contributes to the appearance of several new web approaches in order to facilitate the exploitation of this technology. The web service composition is one of the new approaches which enrich the company activities by other external services, in order to meet to the needs of their clients, to reduce time and costs, and to increase overall efficiency in businesses. In fact, the composition works differ according to the adopted criteria by the company. The workflow approach appears as the first work that provides the composition. In fact, workflow approach is based on two modeling. Firstly, the orchestration which is a set of actions to be performed by the intermediate the web services [1]. Moreover, the orchestration is to program a motor from a predefined process by the intermediate web services [5]; this modeling based on a centralized vision. The second modeling is the choreography, where according to [1] the choreography implements a set of services without any

© Springer International Publishing Switzerland 2015
R. Silhavy et al. (eds.), *Software Engineering in Intelligent Systems*,
Advances in Intelligent Systems and Computing 349, DOI: 10.1007/978-3-319-18473-9_29

intermediate in order to accomplish a goal. Then, this modeling allows the design of decentralized coordination applications. Moreover, there are some works which improve the workflow approach, such as, the appearance of the workflow dynamicity notion, among these works; we cite the approach of [4] which uses the semantic approach to make the code understandable to the machines, and then permitting the automation of web services selection.

However, according to [8] the web service composition approach poses many problems coming from web technology itself, such as: the absence of information on the web service environment and the passivity of this technology. Moreover, the web service composition can encounter execution problems of one of its services after the selection, whereas this problem influences on the response time and network traffic overload.

In the first place, the problem proposed by [8] can be treated with the apparition of software agent technology which provides many solutions due to their very innovative characteristics, such as: autonomy, intelligence, reactivity, collaboration, cooperation, mobility, communication, etc. Also, software agent technology can achieve their goals with a flexible manner by using a local interaction and / or remote with other agents on the network. In the second place, the execution problem of a composite service requires following the execution of various activities after the selection of web services.

In this context, we are interested to design and implement a model ensuring the composition of web services based upon a multi-agent system. We propose essentially to treat the two problem of response time and the network traffic overload. Furthermore, this model ensures the integration of a control strategy which offers solutions to real-time.

After the related work presentation in the second section, we pass to present our model for web services composition based upon a multi-agent system with more detail in the third section. Then, in the fourth section we present the evaluation part of our model. Finally, the fifth section concludes this paper.

2   Related Work

Recently the web services composition based on software agent technology has been experiencing much attention. These approaches are classified in two classes.

The first class of approach is based on multi-agent system. In [3] authors have working on a multi-agent system for the composition of web services based on competition between coalitions of services. In this works, each service is represented by a software agent who contacts other agents and offers their services according to their reasoning ability. Thus, each agent forms a coalition of agents capable to solving the goals provided. Then, the different coalitions will make the most competitive offer, where each solution receives a note, and the solution with the highest score will be selected. By contrast, work of [7] ensures the formation of an only coalition by basing to the negotiation approach.

The second class of approach is based on mobile agent. In this context, [9] ensures the composition with the bottom-up approach, that is to say he selects the web services to form an activities plan, unlike to [3], where they select the web services after the preparation of the activities plan. In fact, [9] seeks to reduce the intermediary activities of a composite service, in order to reduce a response time to a composite service.

In [6] authors have integrated a strategy to control the execution of the activities plan, but they did not offer an immediate solution when he meets a problem of execution with any service.

Table 1 summaries the presented works in terms of the fourth characteristics of the web composition: response time, network problems, quality of service and the execution controls.

Table 1. Comparison between existing approaches

	Respo nse Time	Netw ork Problems	Q oS	execu tion control
Muller and Kowalczyk, 2006			✓	
Qian, 2011			✓	
Zhang, 2011	✓	✓		
Qian and al, 2005		✓	✓	✓

With the examination of the Table 1, we can conclude that these approaches have provided solutions at the composition of web services, by taking advantage of features offered by the software agent technology. However, some problems are not treated well by these works, such that, the response time to a web service composition request, especially the queries that require short response time (for example, the requests for stock exchanges and banks). In addition, some approaches do not control the execution to different components of a composite service, where, they can influence on some network problems, such as, network overload, security, etc. Besides, the problems of execution to different components of a composite service, such as, the breakdown of the web service server, the server overload, the update of web service, etc.

In this sense and in order to improve the web service composition, we propose a new model of web service composition based on multi-agent system and mobile agent technology.

3 Proposed Web Service Composition Model

The proposed web service composition model is based on the coordination between three agents: User Agent (UA), Service Agent (SA) and Compositor Agent (CA). In the Figure 1, we present the architecture of our web services composition model.

The proposed model is based on two functionalities. The first concerns the creation of the activities plan for a composite service. This Functionality shows the different selection steps of the best suppliers for each activity. The second functionality is complementary to the first, where it ensures the execution and the control to the various activities of the composite service.

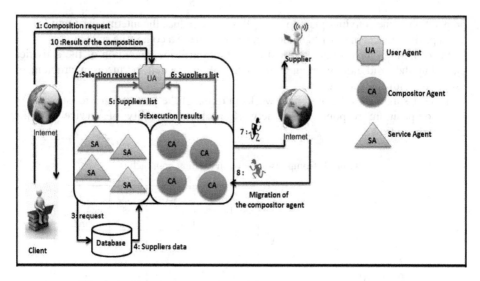

Fig. 1. Architecture proposed of a web services composition model

In the following sub-sections, we will detail the two functionality of the system and the roles of each agent in these two functionalities.

3.1   Creation of the Activity Plan and the Suppliers List

In this first functionality, we find the User Agent (UA) and the Service Agent (SA) ensuring the creation of activities plan of and the suppliers list. More precisely, the UA is responsible for the first step which is the creation of activities plan for a composite service. In facts, it extracts the web services and the data from the composition request of a client, where each service requested and data sent correspond to an activity of a composite service. After the creation of activities plan, the SA creates the suppliers list by selecting the best suppliers matching to each activity in the plan of the composite service.

3.1.1   User Agent

The User Agent (UA) is a stationary agent, where it plays the role to a coordinator component in our model. In fact, its role is like the motor component of the orchestration model in the work of [5]. Hence, the UA communicates with service agents and the compositor agents and receives the composition requests sent by the client.

The behavior of this agent consists of five tasks which are presented in Figure 2. In order to ensure the first functionality, the UA needs only to do the first three tasks which are:

Formation of Activities Plan: This is the first task for the UA, where it starts to verify the composition request sent by the client and extracts the requested services to form the activities plan for a composite service.

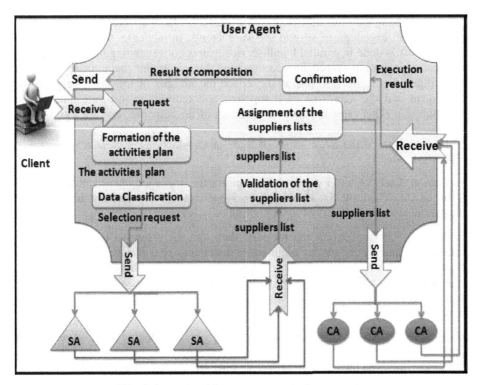

Fig. 2. Internal architecture proposed of user agent

Classification of the Data: The second task is to check the activities plan and to classify the data sent by the client according to their correspondence to the services in the activities plan; these data are the input parameters for each web service requested.

Validation of the Suppliers List: The third task starts after the creation of the suppliers list by the SA, where the UA checks the existence of at least only one supplier in the list for each activity in the plan, in order to ensure the total composition of a composite service.

Figure 2 presents the internal architecture of the user agent, where their tasks and their interactions with the other agents to the first functionality are colored by blue color, while, the green color presents the second functionality.

3.1.2   Service Agent

The Service Agent (SA) is a stationary agent; it plays the role of a representative agent of a set of suppliers having a same web service. Hence, each service agent classifies its suppliers relative to the qualities of service requested by the client. Therefore, the UA is not forced to communicate with all suppliers having the same service, just it communicates with their service agents. The SA has registered all the data about their qualities of service in its database.

In fact, among the most important characteristics of the software agents technology is the parallel execution of several software agents. In this case, the creation of the suppliers list is done in parallel by all service agents corresponding of activities plan. This can contribute to reduce the selection time. Moreover, the goal to using the SA is to facilitate the negotiation operation, reduce the number of messages between the user agent and the agents of each supplier, and reduce the response time of the selection step. This can contribute to reduction of the response time for a composite service.

The behavior of the SA consists of three tasks which are presented in Figure 3. These tasks are:

Validation: Each SA has a standard structure for the messages which will be received from the UA. Then, the SA verifies the message standard structure and its content to the parameters compatibility of input / output of its service.

Extraction: After validating the received data, the SA creates a query taking into account the qualities of service required by the client. After that, it sends this request to the database which contains all information about the suppliers having its service. The SA agent is the only agent which has the right to access in the database of the composition system (suppliers tables which contain the necessary data) to select the best suppliers.

Classification: Finally, the SA classifies the suppliers taking into account their qualities of service (QoS), where it specifies the primary supplier and the reserve suppliers which can replace the primary supplier when it has an execution problem.

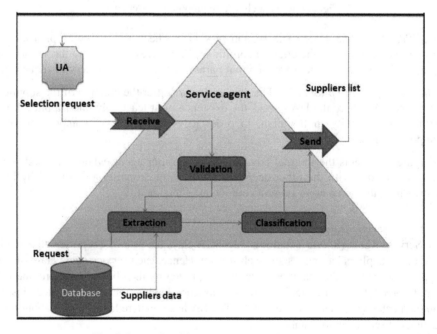

Fig. 3. Internal architecture proposed of service agent

Before finishing this first functionality, there is a question about the database and its role in our model. The proposed of database is inspired from the work of [2], hence the database of the composition system plays the role of the UDDI web service directory, where it contains the data about service quality for each supplier. In this manner, the selection of web services corresponding to the activities plan is done in record time, because the selection of web services is done by local manner. This is help to avoid to discover web services on the UDDI web service directory and to select the best suppliers by using a distance method. In other words, with the discovered and the selection of web services locally, we can avoid several problems such as: the security of each request sent to distance, the overload of traffic on the network and the variation of the response time of each query depending on various factors such the server state to visit, the state of network traffic, etc.

3.2   The Execution Control Strategy

In this second functionality we ensure the execution of different activities of the composite service which was created in the first functionality of our model.

The scenario adopted for this functionality based on the result (suppliers list) to the first functionality. Thus, the user agent assigned each suppliers list to a compositor agent, this later takes the coordinates of the first supplier in the list and it migrates to its platform. Afterward, the compositor agent communicates with the supplier agent in order to execute its service according to the client requirements. If the web service server or the agent platform of the supplier is down, in this case the compositor agent takes the coordinates to the following supplier of the list and migrates toward him. Where, the compositor agent repeats the same steps until it has an execution confirmation by one supplier of its suppliers list.

Therefore, with our execution control strategy, we propose an immediate solution when an execution problem encountered. This is due to the use of mobile agent (compositor agent). Thus, our strategy contributes to reduce the response time of the composite service and to reduce the traffic on the network.

The execution of this functionality is based on the user agent (UA) and compositor agent (CA).

3.2.1   User Agent

In the second functionality, the UA follows the two tasks which are:

Assignment of the Suppliers List: the user agent assigns for each compositor agent a suppliers list.

Confirmation: The user agent checks the result sent by the compositor agents. If this result contents an execution confirmation of one of its suppliers list, then the UA can send the results of the composition to the client.

In the Figure 2, we present the tasks and the interactions with the compositor agents by the green color.

3.2.2   Compositor Agent

The compositor agent is a mobile agent; its role is to ensure the execution of one service in the suppliers list. More precisely, the compositor agent migrates to the supplier platform and ensures the service execution according to the requirements of its client. The internal architecture of compositor agent (Figure 4) consists of two tasks, which are:

Migration: After the reception of a suppliers list, the compositor agent migrates to the first supplier in its list in order to ensure the execution of its service. If the agent compositor encounters an execution problem (the web service server or the supplier agent platform is down), then the compositor agent migrates to the next supplier in its list.

Communication with Supplier Agent: After migrating, the compositor agent communicates with the supplier agent in order to give it the necessary information for the service execution.

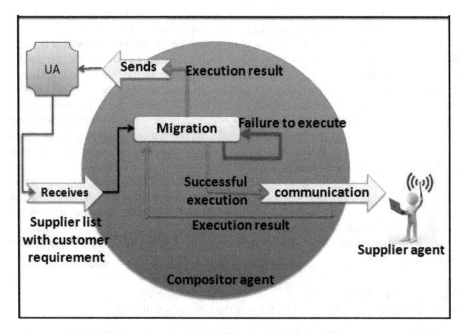

Fig. 4. Internal architecture proposed of compositor agent

4   Evaluations

To simulate our model, we created a composite service based on activities plan containing three services. Consequently, the activities of our composite service which are simulated are: creation to activities plan, selection of the best suppliers and execution of different services selected. Thus, In Figure 5 we represent the durations of different stages to the construction and execution of a composite service based on our model, such as diagram "B" presents the duration of total composition which was done in a local manner (the creation of the activities plan for a composite service and the creation of the suppliers list for each activity) was done in 1916.66 milliseconds, diagram "C" presents the duration of best suppliers' selection, where that was done in 1841.66 milliseconds. For execution time of various activities of a composite service, several factors influence on this step, such as the rate of traffic on the network, the server status of each service requested, etc. In fact, we tested our control strategy on a local server, where the total duration of execution of the composite service is represented by diagram "A" and that was at 7067,66 milliseconds.

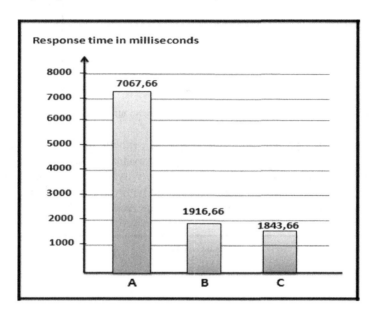

Fig. 5. Durations of different stages to construct a composite service

In fact, the web service composition based in two fashions, which are: a composition in a top-down fashion and a composition in a bottom-up fashion. In our work we adopted a top-down fashion and this is the same thing with the web service composition model of [3]. Thus, with this fashion, we need to start with the creation of a activities plan for each composition request. In this case, [3] adopted the coalition model to ensure the coordination between agents to create the activities plan. However, in our work we adopted the hierarchical model, where, just UA is responsible to create the activities plan.

Consequently, in Figure 6 we present a comparing to the composition duration between our model and the model of [3]. Where, according to the evaluation of the work of [3], the best duration to form a coalition is 3000 milliseconds, where the coalition reflects to a activities plan of a composite service. In contrast, the better duration to composition with our model is 359 milliseconds, where this duration includes the time to form a activities plan and the time to create a suppliers list for each service. In fact, our model presents a very significant improvement relative to the model of [3], where our duration to composition is about the tenth compared to the duration of [3].

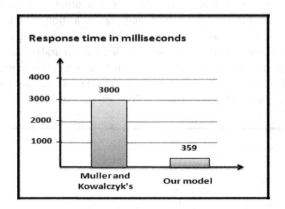

Fig. 6. Comparing to the composition duration between our model and the model of [3]

In Figure 5, we present an evaluation case without taking into account the execution problems. For this reason, in the second evaluation, we varied in each time the number of the execution problems for each activity. In contrast, in Figure 7, we present the execution times of a composite service with variation to the number of the execution problem that the compositor agent encountered.

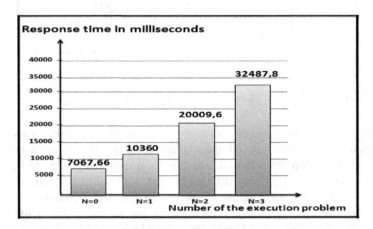

Fig. 7. Execution time of a composite service with variation to number of execution problem

In fact, the variety of number of execution problems influence on the total duration of execution of a composite service, thus the resolution of these problems realized by our control strategy with the proposal of a solution at real-time, where for each time the CA encounters an execution problem with main supplier, it changes its destination by taking the following supplier of the list without returning to its starting platform.

So, the loss average with each execution problem is 4112,18 milliseconds which is an acceptable period, compared to repeating the step of selecting and executing a new supplier.

5   Conclusion

The objective of this paper is to address solutions to the problems of response time, to the overload on the network and especially to the execution problem of different activities of a composite service. We solved this problem by the creation to a composition model based on a multi-agent system. The proposed model is composed by two main functionalities. In the first, we ensure the creation to a activities plan for the composite service and the selection to the best suppliers which can respond to different activities. Actually, these two stages will be realized by a local manner, where on the same server of the composite service. In the second feature, we integrate a control strategy which offers the solutions at real-time in case when where a composite service activity has a execution problem.

References

1. Benatallah, B., Casati, F., Grigori, D., Motahari-Nezhad, H.R., Toumani, F.: Developing Adapters for Web Services Integration. In: Proceedings of the 17th International Conference on Advanced Information System Engineering, Porto, Portugal, pp. 415–429 (2005)
2. Kouki, J., Chainbi, W., Ghedira, K.: Local Registry-Based Approach for Binding Optimization of Web Services. International Journal: The Systemics and Informatics World Network 11, 69–76 (2010)
3. Muller, I., Kowalczyk, R.: Towards Agent-based Coalition Formation for Service Composition. In: Proceedings of the IEEE/WIC/ACM International Conference on Intelligent Agent Technology (2006)
4. Osman, T., Thakker, D., Al-Dabass, D.: Bridging the gap between workflow and semantic-based web services composition. In: Proceedings of WWW Service composition with Semantic Web Services (Compiègne, France), p. 1323 (2005)
5. Peltz, C.: Web Service Orchestration and Choreography. A look at WSCI and BPEL4WS. Web Services – HP Dev Resource Central (2003)
6. Qian, Z., Lu, S., Xie, L.: Mobile-Agent-Based Web Service Composition. This work is partially supported by the National Natural Science Foundation of China under Grant No.60402027 and the National Basic Research Program of China (973) under Grant No.2002CB312002 (2005)
7. Qian, W.: A web services composition model for coalition net construction. In: International Conference on Computer Science and Service System (2011)
8. Tong, H., Cao, J., Zhang, S., Mou, Y.: A fuzzy evaluation system for web services selection using extended QoS model. Library Science: Journal of Systems and Information Technology- Table of Contents 11(1-2) (2009)
9. Zhang, Y.: Web Service Composition Based on Mobile Agent. In: International Conference on Internet Computing and Information Services. IEEE (2011)

Modified Binary FireFly Algorithms with Different Transfer Functions for Solving Set Covering Problems

Broderick Crawford[1,2,3], Ricardo Soto[1,4,5], Marco Riquelme-Leiva[1],
Cristian Peña[1], Claudio Torres-Rojas[1], Franklin Johnson[1,6],
and Fernando Paredes[7]

[1] Pontificia Universidad Católica de Valparaíso, Chile
{broderick.crawford,ricardo.soto}@ucv.cl,
marcoriquelmeleiva@gmail.com,
{cristian.pena.v,claudio.torres.r}@mail.pucv.cl
[2] Universidad Finis Terrae, Chile
[3] Universidad San Sebastián, Chile
[4] Universidad Autónoma de Chile, Chile
[5] Universidad Central de Chile, Chile
[6] Universidad de Playa Ancha, Chile
franklin.johnson@upla.cl
[7] Escuela de Ingeniería Industrial, Universidad Diego Portales, Chile
fernando.paredes@udp.cl

Abstract. In this paper, we propose a set of Modified Binary Firefly Algorithms (MBFF) to solve different instances of the Set Covering Problem (SCP). The algorithms consider eight Transfer Functions and five Discretization Methods in order to solve the binary representation of SCP. The results obtained show that our algorithms are a good and cheap alternative to solve the problem at hand.

Keywords: Modified Binary FireFly Algorithm, Set Covering Problem, Metaheuristics, Combinatorial Optimization.

1 Introduction

The Firefly algorithm is part of the family of algorithms inspired by nature. It was presented by Yang [30] and its operation is based on the social behaviour of fireflies.

The Set Covering Problem (SCP) is considered a classic combinatorial optimization problem, it can be applied in many fields in the real world such as mapping crew buses and airlines [9,27], location emergency services [28], logical analysis of numerical data [7], steel production [29], vehicle routing [20], assigning ships [19]. It belongs to the NP-Complete class [21].

The SCP has been solved using classical optimization algorithms [2,18,11] and Metaheuristics: tabu search [10], simulated annealing [8], genetic algorithms [5],

© Springer International Publishing Switzerland 2015
R. Silhavy et al. (eds.), *Software Engineering in Intelligent Systems*,
Advances in Intelligent Systems and Computing 349, DOI: 10.1007/978-3-319-18473-9_30

ant colony optimization (ACO) [26,13], electromagnetism (unicost SCP) [25], gravitational emulation search [1] and bee colonies [12,16] between others.

In this paper, we propose a discrete Shuffled Frog Leaping Algorithm (SFLA) to solve the SCP. To the best of our knowledge, this is the first work proposing a binary coded SFLA to solve the SCP. The SFLA is a metaheuristic method proposed in [17] that combines the advantages of memetic evolution of genetic-based memetic algorithms [22] and the social behavior of particle swarm optimization (PSO) algorithms [23].

The Set Covering Problem (SCP) can be formally defined as follows. Let $A = (a_{ij})$ be an m-row, n-column, zero-one matrix. We say that a column j covers a row i if $a_{ij} = 1$. Each column j is associated with a non negative real cost a_j. Let $I = 1, 2, ...m$ and $J = 1, 2, ...n$ be the row set and column set, respectively. The SCP calls for a minimum cost subset $S \subseteq J$, such that each row $i \in I$ is covered by at least one column $j \in S$. A mathematical model for the SCP is:

$$Minimize \sum_{j=1}^{n} c_j x_j \tag{1}$$

Subject to:

$$\sum_{j=1}^{n} a_{ij} x_j \geq 1, \qquad \forall i \in I \tag{2}$$

$$x_j \in \{0, 1\}, \qquad \forall j \in J \tag{3}$$

Where x_j is 1 if the column j belongs to the solution, 0 otherwise. Constraints (2) ensures that each row i is covered by at least one column.

2 Binary FireFly Algorithm

Firefly Algorithm is one of many algorithms bio-inspired or inspired in nature, it gets its name because its operation is based on the social behaviour of fireflies using three basic rules [30]:

Rule 1. All fireflies are unisex, one firefly is attracted to other fireflies regard less of gender.

Rule 2. The attractiveness of a firefly is proportional to its brightness, this means that in a pair of fireflies, the less bright will be attracted by the more bright. In the absence of a brighter firefly, the attraction wil be random.

Rule 3. The brightness of a firefly will be determined by the objective function. In a maximization problem the brightness of each firefly is proportional to the value of the objective function. In the case of minimization problema (as is SCP), the brightness of fireflies are inversely proportional to the value of its objective function.

These three basic rules of Firefly algorithm can be explained and represented as follows:

Attractive. The formula of attraction $\beta_{(r)}$ can be any monotonically decreasing function as

$$\beta_{(r)} = \beta_0 e^{-\gamma r^m} \tag{4}$$

Where r is the distance between two fireflies, β_0 is the initial appeal of firefly and γ is a light absorption coefficient.

Distance between Fireflies. The distance r_{ij} between any two fireflies i and j at positions x_i and x_j respectively, can be defined as a Cartesian or Euclidean distance as

$$r_{ij} = ||x_i - x_j|| = \sqrt{\sum_{k=1}^{d}(x_i^k - x_j^k)^2} \tag{5}$$

Where x_i^k is the current value of the k_{th} dimension of the ith firefly (a firefly is a solution of the problem) and d is the number of dimensions or variables of the problem.

Movement of the Fireflies. The movement of a firefly i attracted to another more attractive (brighter) firefly j is determined by

$$x_i^k(t+1) = x_i^k(t) + \beta_0 e^{-\gamma r_{ij}^2}(x_j^k(t) - x_i^k(t)) + \alpha(rand - 1/2) \tag{6}$$

where the first term $x_i^k(t)$ is the current position of the k_{th} dimension of the firefly i at the iteration t. The second term of the equation corresponds to the attraction and the third term introduces a random value to the equation, where α is a randomization parameter and rand is a random number generated uniformly distributed between 0 and 1.

2.1 A Modified Binary FireFly Algorithm

The SCP can not be handled directly by SFLA due to its binary structure. Therefore, to obtain values 0 or 1, a transfer function and a discretization method are performed. The transfer function is applied to the result of the Eq. 6. We tested the eight different functions shown in Table 1, they are separated into two families: S-shape and V-shape. The result of this operation is a real number between 0 and 1, then a binarization method is required to obtain a value 0 or 1. In this paper, we propose a modification to the classical binary algorithm [14,15], this modification consists in a computational optimization, using the ascending order of the population through the evaluation of the objective function. Then, the brightest firefly occupies the zero position into the population matrix. In lines 3 and 6 of the algorithm we can appreciate the computational optimization.

Algorithm 1. MBFF for SCP Algorithm

1: Initialize parameter values of α, β_0, γ, Population size, Number of generations.
2: Evaluate the light intensity I determined by $f(x)$, see Eq. 1
3: Sort the population in ascending order according fitness $1/f(x)$ see Eq. 1
4: **while** $t <$**Number of generations do**
5:    **for** $i = 1 : m$ (m fireflies) **do**
6:       **for** $j = i : m$ (m fireflies) **do**
7:          **if** $I_j < I_i$ **then**
8:             **movement** = calculates value according Table. 1 and Discretization method.
9:          **end if**
10:         Repair solutions using Repair Operator
11:         Update light intensity
12:       **end for**
13:    **end for**
14:    $t = t + 1$
15: **end while**
16: Output result

2.2 Transfer Function

The table 1 shows the eight transfer functions that were used in this work, they was proposed by Mirjalili in [24].

Table 1. Transfer Functions

S-Shape	V-Shape
S1 $T(x) = \frac{1}{1+e^{-2x}}$	**V1** $T(x) = \left\|\text{erf}\left(\frac{\sqrt{\pi}}{2}x\right)\right\|$
S2 $T(x) = \frac{1}{1+e^{-x}}$	**V2** $T(x) = \|tanh(x)\|$
S3 $T(x) = \frac{1}{1+e^{\frac{-x}{2}}}$	**V3** $T(x) = \left\|\frac{x}{\sqrt{1+x^2}}\right\|$
S4 $T(x) = \frac{1}{1+e^{\frac{-x}{3}}}$	**V4** $T(x) = \left\|\frac{2}{\pi}arctan\left(\frac{\pi}{2}x\right)\right\|$

2.3 Discretization Methods

The Discretization methods used in this work were:

Standard

$$x_i^k(t+1) = \begin{cases} 1 \text{ if } rand < T(x_i^k(t+1)) \\ \\ 0 \text{ otherwise} \end{cases} \tag{7}$$

Complement

$$x_i^k(t+1) = \begin{cases} complement(x_i^k(t)) \text{ if } rand < T(x_i^k(t+1)) \\ 0 \qquad\qquad\qquad \text{otherwise} \end{cases} \qquad (8)$$

Static Probability

$$x_i^k(t+1) = \begin{cases} x_i^k(t) \qquad \text{if } 0 < T(x_i^k(t+1)) \le \alpha \\ x_i^k(t+1) \text{ if } \alpha < T(x_i^k(t+1)) \le \frac{1}{2}(1+\alpha) \\ x_{best}^k \qquad \text{if } \frac{1}{2}(1+\alpha) < T(x_i^k(t+1)) \le 1 \end{cases} \qquad (9)$$

Elitist

$$x_i^k(t+1) = \begin{cases} x_{best}^k \text{ if } rand < T(x_i^k(t+1)) \\ 0 \qquad \text{otherwise} \end{cases} \qquad (10)$$

Elitist Roulette. The Discretization method Elitist Roulette (Monte Carlo) is to select randomly between the best fireflies of the population, with a probability proportional its fitness.

2.4   Repair Operator

The Repair Operator used in this work was presented by Beasley's [5]. When the fireflies do not satisfy all constraints, it is necessary to use it. The repair operator acts only on infeasible fireflies and it consists in to identify the rows not covered and changing the required bit values from 0 to 1 in order to assure the coverage of each row.

The second part of repair method consists in to apply a local optimization technique, identifying and eliminating redundant columns. The selection of redundant columns to eliminate is based on the relationship:

$$\frac{\text{cost of the column}}{\text{number of rows that cover the column}}$$

3   Reducing the Problem

In this paper, it was built a method which allows to reduce the problem. This method generates reduced equivalent instances of SCP from OR-Library [4]. We use two techniques to reduce an instance: *Column Domination* and *Column Inclusion*, these were presented by Beasley in [6].

Column Domination. For any column j whose rows $[i|a_{ij} = 1, i = 1, \ldots, m]$ which may be covered by other columns at lower cost than c_j, the column j can be eliminated the problem. Let be

$$d_i = min[c_j|a_{ij} = 1, \ldots, n], i = 1, \ldots, m \tag{11}$$

then any column j for which

$$c_j > \sum_{i=1}^{m} d_i a_{ij} \tag{12}$$

can be eliminated in the problem formulation, because rows covered by column j can be covered by a combination of other columns at lower cost. In this work, we used the reduction of the problem and new instance files were generated to execute our experiments.

4   Experimental Results

The algorithm was implemented using Visual Basic .NET language and conducted on 2.66 GHz with 4 GB RAM running Windows 7 Professional.

The MBFF is executed 30 times for each instance. In this paper are presented the best results solving SCP instances with known optimal value 4.1 to 4.10, 5.1 to 5.10 and 6.1 to 6.5 presented by Balas and Ho [3]; A.1 to A.5, B.1 to B.5, C.1 to C.5 and D.1 to D.5 presented by Beasley [6] and instances with unknown optimal value NRE.1 to NRE.5, NRF.1 to NRF.5, NRG.1 to NRG.5, HNR.1 to HNR.5 presented by [4]. The *T-Function* column represents the Transfer Function used, *D-Method* column represents de Discretization Methods, *Opt* shows the known optimal value, the *min*, *Max* and *Avg*, show the minimum, maximum and average value obtained in our tests. *RPD*, Relative Percentage Deviation, is the distance to the best known value of our best result. The *RPD* formulates is presented in Eq.13.

$$RPD = \frac{100(Z - Z_{opt})}{Z_{opt}} \tag{13}$$

In the experimets were used the following MBFF parameters:

Table 2. MBFF Parameters

Parameter	Value
α	0.5
β_0	1.0
γ	1.0
Population size	50
Number of generations	50
Trials	30

Table 3. Experimental Result of SCP benchmarks

Instance	T-Function	D-Method	Opt	Min	Max	AVG	RPD
4.1	V4	Standard	429	429	436	433,5	0
4.2	S2	Elitist	512	517	557	541,9	0,97
4.3	S3	Elitist	516	519	537	532,36	0,58
4.4	S4	Elitist	494	495	532	519,66	0,20
4.5	V3	Standard	512	514	530	524,16	0,39
4.6	S4	Elitist	560	563	601	583,8	0.53
4.7	V1	Standard	430	430	440	435,56	0
4.8	S1	Elitist	492	497	509	506,6	1,01
4.9	S4	Elitist	641	655	691	675,36	2,18
4.10	V3	Standard	514	519	539	533,76	0,97
5.1	S2	Elitist	253	257	276	268,53	1,58
5.2	S3	Elitist	302	309	317	315	2,31
5.3	S1	Elitist	226	229	247	239,66	1,32
5.4	S2	Elitist	242	242	251	247,73	0
5.5	S4	Elitist	211	211	225	218,83	0
5.6	S2	Static-P	213	213	242	233,96	0
5.7	S1	Elitist	293	298	315	308,66	1,70
5.8	S2	Elitist	288	291	314	303,13	1,04
5.9	S3	Elitist	279	284	308	298,6	1,79
5.10	V2	Elitist	265	268	280	274,46	1,13
6.1	S3	Elitist	138	138	152	148,46	0
6.2	S1	Static-P	146	147	160	154,36	0,68
6.3	V2	Elitist	145	147	160	151,53	1,37
6.4	S4	Elitist	131	131	140	136,43	0
6.5	S2	Elitist	161	164	183	175,53	1,86
A.1	V1	Static-P	253	255	261	259,63	0,79
A.2	S3	Elitist	252	259	277	268,6	2,77
A.3	S3	Elitist	232	238	252	246,36	2,58
A.4	V1	Elitist	234	235	250	246,2	0,42
A.5	S1	Statics-P	236	236	241	240,1	0
B.1	S4	Elitist	69	71	83	78,93	2,89
B.2	V2	Elitist	76	78	88	85,23	2,63
B.3	S1	Elitist	80	80	85	84,43	0
B.4	S3	Elitist	79	80	84	83,36	1,26
B.5	S1	Elitist	72	72	79	75,7	0
C.1	S3	Elitist	227	230	235	234,03	1,32
C.2	S4	Elitist	219	223	236	231,23	1,82
C.3	V1	Static-P	243	253	270	265,2	4,11
C.4	S3	Elitist	219	225	246	238,36	2,73
C.5	S1	Elitist	215	217	223	220,93	0,93
D.1	S2	Elitist	60	60	62	61,33	0
D.2	S1	Elitist	66	68	74	71,33	3,03
D.3	S2	Elitist	72	75	79	78,1	4,16
D.4	V3	Elitist	62	62	68	65,33	0
D.5	S3	Elitist	61	63	66	65,26	3,27
NRE.1	S1	Elitist	29	29	30	29,93	0
NRE.2	S3	Elitist	30	32	35	33,9	6,66
NRE.3	S3	Elitist	27	29	34	31,2	7,40
NRE.4	S3	Elitist	28	29	33	32,36	3,57
NRE.5	S3	Elitist	28	29	30	29,9	3,57
NRF.1	S3	Elitist	14	15	17	16,6	7,14
NRF.2	S3	Elitist	15	16	18	17,4	6,66
NRF.3	S2	Elitist	14	16	17	16,96	14,28
NRF.4	S3	Elitist	14	15	18	16,33	7,14
NRF.5	S3	Elitist	13	15	16	15,63	15,38
NRG.1	S1	Elitist	176	185	196	192,63	5,11
NRG.2	S3	Elitist	154	161	168	165,66	4,54
NRG.3	S3	Elitist	166	175	180	178	5,42
NRG.4	S3	Elitist	168	176	184	180,73	4,76
NRG.5	S3	Elitist	168	177	187	182,7	5,35
NRH.1	S2	Elitist	63	69	73	71,9	9,52
NRH.2	S3	Elitist	63	66	67	66,86	4,76
NRH.3	S1	Elitist	59	65	69	67,83	10,16
NRH.4	S1	Elitist	59	63	68	66,3	6,77
NRH.5	S3	Complement	55	59	61	60,83	7,27

5   Conclusion

In this paper, we presented the algorithm Modified Binary Firefly supported by eight Transfer functions and five Discretization methods. The results presented were validated through experimentation solving SCP benchmarks. The Modified Binary Firefly Algorithm was a great option for solving the Set Covering Problem, because we obtained excellent results and near optimal results with a very low computational cost. Our proposal can reach very good solutions (local optimums or known optimal values) in a few iterations.

Acknowledgments.   Broderick   Crawford   is   supported   by   Grant CONICYT/FONDECYT/REGULA-R/1140897, Ricardo Soto is supported by Grant CONICYT/FONDECYT/INIC-IACION/11130459 and Fernando Paredes is supported by Grant CONICYT/FO-NDECYT/REGULAR/1130455.

References

1. Balachandar, S.R., Kannan, K.: A meta-heuristic algorithm for set covering problem based on gravity 4(7), 944–950 (2010)
2. Balas, E., Carrera, M.C.: A dynamic subgradient-based branch-and-bound procedure for set covering. Operations Research 44(6), 875–890 (1996)
3. Balas, E., Ho, A.: Set Covering Algorithms Using Cutting Planes, Heuristics, and Subgradient Optimization: a Computational Study. In: Padberg, M.W. (ed.) Combinatorial Optimization. Mathematical Programming Studies, vol. 12, pp. 37–60. Elsevier, North-Holland (1980)
4. Beasley, J.: A lagrangian heuristic for set-covering problems. Naval Research Logistics (NRL) 37(1), 151–164 (1990)
5. Beasley, J., Chu, P.: A genetic algorithm for the set covering problem. European Journal of Operational Research 94(2), 392–404 (1996)
6. Beasley, J.E.: An algorithm for set covering problem. European Journal of Operational Research 31(1), 85–93 (1987)
7. Boros, E., Hammer, P.L., Ibaraki, T., Kogan, A.: Logical analysis of numerical data. Mathematical Programming 79(1-3), 163–190 (1997)
8. Brusco, M., Jacobs, L., Thompson, G.: A morphing procedure to supplement a simulated annealing heuristic for cost- and coverage-correlated set-covering problems. Annals of Operations Research 86, 611–627 (1999)
9. Caprara, A., Fischetti, M., Toth, P.: A heuristic method for the set covering problem. Oper. Res. 47(5), 730–743 (1999)
10. Caserta, M.: Tabu search-based metaheuristic algorithm for large-scale set covering problems. In: Doerner, K., Gendreau, M., Greistorfer, P., Gutjahr, W., Hartl, R., Reimann, M. (eds.) Metaheuristics. Operations Research/Computer Science Interfaces Series, vol. 39, pp. 43–63. Springer US (2007)
11. Chvatal, V.: A greedy heuristic for the set-covering problem. Mathematics of Operations Research 4(3), 233–235 (1979)
12. Crawford, B., Soto, R., Cuesta, R., Paredes, F.: Application of the artificial bee colony algorithm for solving the set covering problem. The Scientific World Journal (2014)

13. Crawford, B., Soto, R., Monfroy, E., Castro, C., Palma, W., Paredes, F.: A hybrid soft computing approach for subset problems. Mathematical Problems in Engineering, Article ID 716069, 1–12 (2013)

14. Crawford, B., Soto, R., Olivares-Suárez, M., Paredes, F.: A binary firefly algorithm for the set covering problem. In: Silhavy, R., Senkerik, R., Oplatkova, Z.K., Silhavy, P., Prokopova, Z. (eds.) Modern Trends and Techniques in Computer Science. AISC, vol. 285, pp. 65–73. Springer, Heidelberg (2014)

15. Crawford, B., Soto, R., Olivares-Suárez, M., Paredes, F.: Using the firefly optimization method to solve the weighted set covering problem. In: Stephanidis, C. (ed.) HCII 2014 Posters, Part I. CCIS, vol. 434, pp. 509–514. Springer, Heidelberg (2014)

16. Cuesta, R., Crawford, B., Soto, R., Paredes, F.: An artificial bee colony algorithm for the set covering problem. In: Silhavy, R., Senkerik, R., Oplatkova, Z.K., Silhavy, P., Prokopova, Z. (eds.) Modern Trends and Techniques in Computer Science. AISC, vol. 285, pp. 53–63. Springer, Heidelberg (2014)

17. Eusuff, M.M., Lansey, K.E.: Optimization of Water Distribution Network Design Using the Shuffled Frog Leaping Algorithm. Journal of Water Resources Planning and Management 129(3), 210–225 (2003)

18. Fisher, M.L., Kedia, P.: Optimal solution of set covering/partitioning problems using dual heuristics. Management Science 36(6), 674–688 (1990)

19. Fisher, M.L., Rosenwein, M.B.: An interactive optimization system for bulk-cargo ship scheduling. Naval Research Logistics (NRL) 36(1), 27–42 (1989)

20. Foster, B.A., Ryan, D.M.: An integer programming approach to the vehicle scheduling problem. Operational Research Quarterly, 367–384 (1976)

21. Garey, M.R., Johnson, D.S.: Computers and Intractability; A Guide to the Theory of NP-Completeness. W. H. Freeman & Co., New York (1990)

22. Glover, F.W., Kochenberger, G.A.: Handbook of Metaheuristics (International Series in Operations Research & Management Science). Springer (January 2003)

23. Kennedy, J., Eberhart, R.: Particle swarm optimization. In: IEEE International Conference on Neural Networks, pp. 1942–1948 (1995)

24. Mirjalili, S., Hashim, S.M., Taherzadeh, G., Mirjalili, S., Salehi, S.: A study of different transfer functions for binary version of particle swarm optimization. In: GEM 2011. CSREA Press (2011)

25. Naji-Azimi, Z., Toth, P., Galli, L.: An electromagnetism metaheuristic for the unicost set covering problem. European Journal of Operational Research 205(2), 290–300 (2010)

26. Ren, Z.-G., Feng, Z.-R., Ke, L.-J., Zhang, Z.-J.: New ideas for applying ant colony optimization to the set covering problem. Computers & Industrial Engineering 58(4), 774–784 (2010)

27. Smith, B.M.: Impacs a bus crew scheduling system using integer programming. Mathematical Programming 42(1), 181–187 (1988)

28. Toregas, C., Swain, R., ReVelle, C., Bergman, L.: The location of emergency service facilities. Operations Research 19(6), 1363–1373 (1971)

29. Vasko, F.J., Wolf, F.E., Stott Jr., K.L.: A set covering approach to metallurgical grade assignment. European Journal of Operational Research 38(1), 27–34 (1989)

30. Yang, X.-S.: Nature-inspired metaheuristic algorithms. Luniver Press (2010)

Binarization Methods for Shuffled Frog Leaping Algorithms That Solve Set Covering Problems

Broderick Crawford[1,2,3], Ricardo Soto[1,4,5], Cristian Peña[1],
Marco Riquelme-Leiva[1], Claudio Torres-Rojas[1],
Franklin Johnson[1,6], and Fernando Paredes[7]

[1] Pontificia Universidad Católica de Valparaíso, Chile
{broderick.crawford,ricardo.soto}@ucv.cl,
{cristian.pena.v,claudio.torres.r}@mail.pucv.cl,
marcoriquelmeleiva@gmail.com
[2] Universidad Finis Terrae, Chile
[3] Universidad San Sebastián, Chile
[4] Universidad Autónoma de Chile, Chile
[5] Universidad Central de Chile, Chile
[6] Universidad de Playa Ancha, Chile
franklin.johnson@upla.cl
[7] Escuela de Ingeniería Industrial, Universidad Diego Portales, Chile
fernando.paredes@udp.cl

Abstract. This work proposes Shuffled Frog Leaping Algorithms (SFLAs) to solve Set Covering Problems (SCPs). The proposed algorithms include eight transfer function and five discretization methods in order to solve the binary representation of SCP. Different instances of the Set Covering Problem are solved to test our algorithm showing very promising results.

Keywords: Shuffled Frog Leaping Algorithm, Set Covering Problem, Metaheuristics, Combinatiorial Optimization.

1 Introduction

The Set Covering Problem (SCP) is defined as follows, let $A = (a_{ij})$ be an m-row, n-column, zero-one matrix. We say that a column j covers a row i if $a_{ij} = 1$. Each column j is associated with a non negative real cost c_j. Let $I = \{1, 2, \ldots, m\}$ and $J = \{1, 2, \ldots, n\}$ be the row set and column set, respectively. The SCP calls for a minimum cost subset $S \subseteq J$, such that each row $i \in I$ is covered by at least one column $j \in S$. A mathematical model for the SCP is:

$$Minimize \ \ f(x) = \sum_{j=1}^{n} c_j x_j \qquad (1)$$

subject to:

$$\sum_{j=1}^{n} a_{ij} x_j \geq 1, \qquad \forall i \in I \qquad (2)$$

$$x_j \in \{0, 1\}, \qquad \forall j \in J \qquad (3)$$

© Springer International Publishing Switzerland 2015
R. Silhavy et al. (eds.), *Software Engineering in Intelligent Systems*,
Advances in Intelligent Systems and Computing 349, DOI: 10.1007/978-3-319-18473-9_31

This work proposes to solve the SCP with the recent metaheuristic Shuffled Frog Leaping Algorithm (SFLA), a method of optimization that is based on the observation, the imitation and the modeling of the behavior of a group of frogs searching a location that has the maximum available quantity of food [12]. The most distinguished benefit of SFLA is its fast convergence speed [10]. SFLA combines the benefits of a genetic-based memetic algorithm and the social behavior-based of particle swarm optimization [14]. The SCP has many practical applications like location of emergency facilities [5], airline and bus crew scheduling [4,18], steel production [13], logical analysis of numerical data [3], ship scheduling [15], vehicle routing [1]. The SCP has been solved using complete techniques and different metaheuristics [20,7,6].

SFLA has been applied to multi-mode resource-constrained project scheduling problem [21], bridge deck repairs [9], water resource distribution [11], unit commitment problem [8], traveling salesman problem (TSP) [17] and job-shop scheduling arrangement [19].

This work has been organized of the following way: in the section 2, a description of SFLA to solve the SCP is given. In the section 3 are presented binarization issues, transfer functions, methods of discretization and the repairing operator for the treatment of infeasibility. In section 4, the experimental results are presented and finally we conclude.

2   Shuffled Frog Leaping Algorithm

In the SFLA a set of P frogs are generated randomly. Then, the population is divided in frog subgroups named *memeplexes*. For each subgroup, a local search is realized to improve the position of the worst frog, which in turn can be influenced by other frogs since each frog has ideas affecting others. This process is named *evolution memetica*, which can repeat up to a certain number of iterations. The ideas generated by every memeplexe are transmitted to other memeplexes in a process of redistribution [16]. The local search and the redistribution continue until the criterion of stop is reached [11].

For a problem of n variables, a frog i is represented as a vector $X_i = (x_i^1, x_i^2, \ldots, x_i^n)$. After the initial population is created, the fitness is calculated for each frog and they are arranged in descending order according to the obtained fitness. m memeplexes are generated from the division of the population, each subgroup contains f frogs (i.e. $P = m \times f$). In this process, the first frog goes to the first memeplexe, the second frog goes to the second memeplexe, ..., frog f goes to the memeplexe m, and the frog $f + 1$ goes back to the first memeplexe, ...

In each memeplexe the best frog is identified as x_b, the frog with the worst fitness as x_w and the best global frog as x_g. In the local search applied in each memeplexe an adjustment is applied to the worst frog in the following way:

$$d_w^j = (x_b^j - x_w^j)\ rand, \qquad 1 \le j \le n \tag{4}$$

$$x_{new}^j = x_w^j + d_w^j, \qquad d_{min}^j \le d_w^j \le d_{max}^j \tag{5}$$

Where $rand$ is a random number ($rand \sim U(0,1)$) and d^j_{max} is the maximum change allowed in the position of a frog. The result of this process is compared with the fitness of the worst frog, if this one is better than the worst frog, the worst frog is replaced. Otherwise, Eqs. 4 and 5 are applied with the best global frog (i.e. x_g replaces to x_b). If the new fitness is better than the fitness of the worst frog, the worst frog is replaced. Otherwise, a frog is generated randomly to replace the worst frog. The process is realized by a certain number of iterations.

3   A Binary Coded Shuffled Frog Leaping Algorithm

The SCP can not be handled directly by SFLA due to its binary structure. Therefore, to obtain values 0 or 1, a transfer function and a discretization method are performed. The transfer function is applied to the result of the Eqs. 4 and 5. We tested the eight different functions shown in Table 1. They are separated into two families: S-shape and V-shape (Fig. 1). The result of this operation is a real number between 0 and 1, then a binarization method is required to obtain a value 0 or 1.

3.1   Transfer Functions

Table 1. Transfer functions

S-Shape	V-Shape
S1 $T(d^j_w) = \dfrac{1}{1+e^{-2d^j_w}}$	**V1** $T(d^j_w) = \left\| erf\left(\dfrac{\sqrt{\pi}}{2}d^j_w\right)\right\|$
S2 $T(d^j_w) = \dfrac{1}{1+e^{-d^j_w}}$	**V2** $T(d^j_w) = \left\| tanh(d^j_w)\right\|$
S3 $T(d^j_w) = \dfrac{1}{1+e^{\frac{-d^j_w}{2}}}$	**V3** $T(d^j_w) = \left\| \dfrac{x}{\sqrt{1+d^j_w{}^2}}\right\|$
S4 $T(d^j_w) = \dfrac{1}{1+e^{\frac{-d^j_w}{3}}}$	**V4** $T(d^j_w) = \left\|\dfrac{2}{\pi}arctan\left(\dfrac{\pi}{2}d^j_w\right)\right\|$

3.2   Binarization Methods

- Standard.

$$x^j_{new} = \begin{cases} 1 \text{ if } rand \leq T(d^j_w) \\ 0 \text{ } otherwise \end{cases} \qquad (6)$$

- Complement.

$$x^j_{new} = \begin{cases} \overline{x^j_w} \text{ if } rand \leq T(d^j_w) \\ 0 \quad otherwise \end{cases} \qquad (7)$$

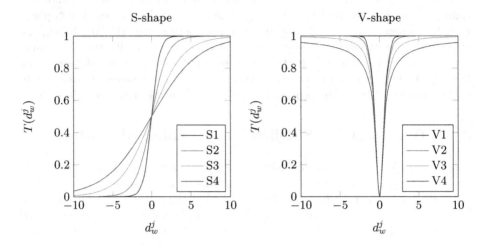

Fig. 1. S-shape and V-shape Transfer functions

- Static probability.

$$
x_{new}^j = \begin{cases} 0 & \text{if } T(d_w^j) \le \alpha \\ x_w^j & \text{if } \alpha < T(d_w^j) \le \frac{1}{2}(1 + \alpha) \\ 1 & \text{if } T(d_w^j) \ge \frac{1}{2}(1 + \alpha) \end{cases} \tag{8}
$$

- Elitist.

$$
x_{new}^j = \begin{cases} x_b^j & \text{if } rand < T(d_w^j) \\ 0 & otherwise \end{cases} \tag{9}
$$

- Elitist roulette. Each x_{new}^j value is assigned from a frog selected via a roulette wheel selection.

3.3   Avoiding Infeasible Solutions

When a solution does not satisfy the constraints of the problem this is called infeasible. Then a repair operator should be applied to the solution becomes feasible. The steps are: identify rows that are not covered by at least one column.

Then, to add at least a column so that all rows are covered, the search for these columns are based on the ratio [2]:

$$\frac{\text{cost of a column}}{\text{number of uncovered rows that it covers}}$$

As soon as columns are added converting infeasible solutions to feasibles, it is applied a local optimization phase, which eliminates a set of redundant columns that are in the solution. A redundant column is one that can be removed from the solution without altering its feasibility [2]. The pseudo-code of this process is the following one:

Algorithm 1. Repair Operator.

1: I = the set of all rows,
2: J = the set of all columns,
3: α_i = the set of columns that cover row i, $i \in I$,
4: β_j = the set of rows covered by column j, $j \in J$,
5: S = the set of columns in a solution,
6: U = the set of uncovered rows,
7: w_i = the number of columns that cover row i, $i \in I$ in S.
8: Initialise $w_i := \mid S \cap \alpha_i \mid, \forall_i \in I$.
9: Initialise $U := \{i \mid w_i = 0, \forall_i \in I\}$.
10: **for all** row i in U **do** {in increasing order of i}
11:    find the first column j (in increasing order of j) in α_i that minimises $c_j / \mid U \cap \beta_j \mid$
12:    $S := S + j$
13:    $w_i := w_i + 1, \forall_i \in \beta_j$
14:    $U := U - \beta_j$
15: **end for**
16: **for all** column j in S **do** {in decreasing order of j}
17:    **if** $w_i \geq 2, \forall_i \in \beta_j$ **then**
18:        $S := S - j;$
19:        $w_i := w_i - 1, \forall_i \in \beta_j;$
20:    **end if**
21: **end for**
22: S is now a feasible solution for the SCP that contains no redundant columns.

3.4 Flowchart of SFLA

The flowchart explaining SFLA solving the SCP is:

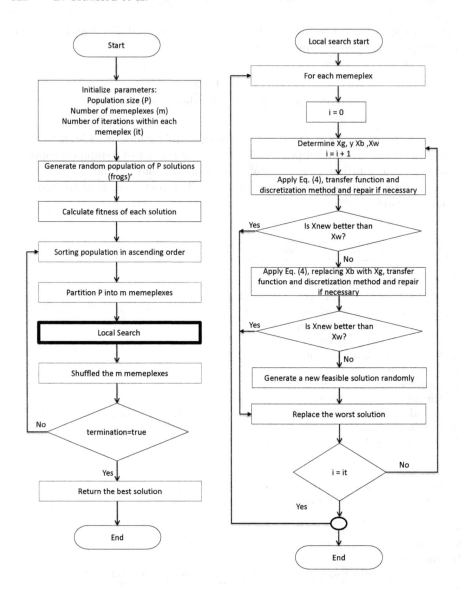

Fig. 2. Flowchart of the proposed coded SFLA

4   Experimental Evaluation

The table 2 shows the attributes of the SCP instances: number of the file set, m (number of rows), n (number of column), range of costs, the percentage of non-zeros in matrix A and if the optimal solution is known or unknown.

Table 2. SCP instances

Instance set	No. of instance	m	n	Cost range	Density (%)	Optimal solution
4	10	200	1000	[1, 100]	2	Known
5	10	200	2000	[1, 100]	2	Known
6	5	200	1000	[1, 100]	5	Known
A	5	300	3000	[1, 100]	2	Known
B	5	300	3000	[1, 100]	5	Known
C	5	400	4000	[1, 100]	2	Known
D	5	400	4000	[1, 100]	5	Known
NRE	5	500	5000	[1, 100]	10	Unknown
NRF	5	500	5000	[1, 100]	20	Unknown
NRG	5	1000	10000	[1, 100]	2	Unknown
NRH	5	1000	10000	[1, 100]	5	Unknown

The results are evaluated using the relative percentage deviation (RPD). The value RPD quantifies the deviation of the objective value Z from Z_{opt} that in our case is the best known value for each instance (Opt in Tables 4 and 5). The minimum (Min), maximum (Max) and average (Avg) of the solutions obtained using the 40 instances of SFLA (different combinations of 8 transfer functions with 5 discretization methods) are shown. To calculate RPD we use $Z = Min$. This measure is calculated as:

$$RPD = \frac{100(Z - Z_{opt})}{Z_{opt}} \tag{10}$$

In all experiments, each SFLA isntance is executed 30 times over each one of the 65 SCP test instances. We test all the combinations of transfer functions and discretization methods over all these instances. We used a population of 200 frogs ($P = 200$), 10 memeplexes ($m = 10$), 20 iterations within each memeplex ($it = 20$) and number of iterations as termination condition ($iMax = 100$). In the experiments using the static probability discretization method, the parameter $\propto$ was set to 0.5.

Table 3. Experimental Results

Instance	Opt	Min	Max	Avg	RPD
4.1	429	430	437	432,13	0,23
4.2	512	516	542	524,47	0,78
4.3	516	520	531	525,93	0,78
4.4	494	501	512	507,23	1,42
4.5	512	514	525	518,27	0,39
4.6	560	563	579	571,23	0,54

Table 4. Experimental Results

Instance	Opt	Min	Max	Avg.	RPD
4.7	430	431	434	433,77	0,23
4.8	492	497	499	498,67	1,02
4.9	641	656	671	663,40	2,34
4.10	514	518	536	527,83	0,78
5.1	253	254	264	261,03	0,40
5.2	302	307	312	309,57	1,66
5.3	226	228	232	229,73	0,88
5.4	242	242	244	243,13	0
5.5	211	211	225	218,87	0
5.6	213	213	224	218,93	0
5.7	293	297	303	299,77	1,37
5.8	288	291	298	296,60	1,04
5.9	279	281	292	285,87	0,72
5.10	265	265	271	267,70	0
6.1	138	140	146	144,03	1,45
6.2	146	147	153	150,93	0,68
6.3	145	147	150	148,00	1,38
6.4	131	131	133	131,47	0
6.5	161	166	174	169,17	3,11
A.1	253	255	258	255,60	0,79
A.2	252	260	265	262,20	3,17
A.3	232	237	245	242,70	2,16
A.4	234	235	240	237,77	0,43
A.5	236	236	240	238,23	0
B.1	69	70	76	72,00	1,45
B.2	76	76	81	79,67	0
B.3	80	80	82	81,50	0
B.4	79	79	84	80,70	0
B.5	72	72	72	72,00	0
C.1	227	229	234	233,07	0,88
C.2	219	223	227	224,17	1,83
C.3	243	253	262	257,50	4,12
C.4	219	227	234	230,83	3,65
C.5	215	217	218	217,80	0,93
D.1	60	60	61	60,53	0
D.2	66	67	70	68,83	1,52
D.3	72	75	78	76,53	4,17
D.4	62	63	65	63,90	1,61
D.5	61	63	64	63,43	3,28
NRE.1	29	29	30	29,50	0
NRE.2	30	31	34	32,80	3,33
NRE.3	27	28	30	29,03	3,70
NRE.4	28	29	32	31,87	3,57
NRE.5	28	28	30	29,10	0
NRF.1	14	15	17	15,77	7,14
NRF.2	15	15	18	16,93	0

Table 5. Experimental Results

Instance	Opt	Min	Max	Avg.	RPD
NRF.3	14	16	17	16,83	14,29
NRF.4	14	15	16	15,47	7,14
NRF.5	13	15	16	15,1	15,38
NRG.1	176	182	190	186,17	3,41
NRG.2	151	161	165	163,00	4,55
NRG.3	166	173	177	175,50	4,22
NRG.4	168	173	177	175,03	2,98
NRG.5	168	174	180	176,67	3,57
NRH.1	63	68	72	70,00	7,94
NRH.2	63	66	67	66,43	4,76
NRH.3	59	62	67	65,43	5,08
NRH.4	59	63	65	64,23	8,62
NRH.5	55	59	61	59,70	7,27

5 Conclusion

In this paper, SFLA metaheuristic was presented as a very good alternative to solve the SCP. We used eight transfer function and five methods of discretization with the aim to solve the binary problem at hand. We have performed experiments through several instances. Our proposal has demonstrated to be very effective. Solving the 65 instances we obtained an important number of optimums (14). Furthermore, 29 RPDs are under the one percent.

Acknowledgments. Broderick Crawford is supported by Grant CONICYT/FONDECYT/REGULA-R/1140897, Ricardo Soto is supported by Grant CONICYT/FONDECYT/INIC-IACION/11130459 and Fernando Paredes is supported by Grant CONICYT/FO-NDECYT/REGULAR/1130455.

References

1. Foster, B.A., Ryan, D.M.: An integer programming approach to the vehicle scheduling problem. Operational Research Quarterly 27, 367–384 (1976)
2. Beasley, J., Chu, P.: A genetic algorithm for the set covering problem. European Journal of Operational Research 94(2), 392–404 (1996)
3. Boros, E., Hammer, P.L., Ibaraki, T., Kogan, A.: Logical analysis of numerical data. Math. Program. 79, 163–190 (1997)
4. Caprara, A., Fischetti, M., Toth, P.: A heuristic method for the set covering problem. Operations Research 47(5), 730–743 (1999)
5. Constantine, T., Ralph, S., Charles, R., Lawrence, B.: The location of emergency service facilities. Operations Research 19, 1363–1373 (1971)
6. Crawford, B., Soto, R., Cuesta, R., Paredes, F.: Application of the Artificial Bee Colony Algorithm for Solving the Set Covering Problem. The Scientific World Journal 2014, 8 (2014)

7. Crawford, B., Soto, R., Olivares-Suárez, M., Paredes, F.: A Binary Firefly Algorithm for the Set Covering Problem. In: Silhavy, R., Senkerik, R., Oplatkova, Z.K., Silhavy, P., Prokopova, Z. (eds.) Modern Trends and Techniques in Computer Science. AISC, vol. 285, pp. 65–73. Springer, Heidelberg (2014)

8. Ebrahimi, J., Hosseinian, S., Gharehpetian, G.: G.b. gharehpetian, unit commitment problem solution using shuffled frog leaping algorithm. IEEE Transaction on Power Systems 26, 573–581 (2011)

9. Elbehairy, H., Elbeltagi, E., Hegazy, T.: Comparison of two evolutionary algorithms for optimization of bridge deck repairs. Computer-Aided Civil and Infrastructure Engineering 21, 561–572 (2006)

10. Elbeltagi, E., Hegazy, T., Grierson, D.: Comparison among five evolutionary-based optimization algorithms. Advanced Engineering Informatics 19, 43–53 (2005)

11. Eusuff, M., Lansey, K.: Optimization of water distribution network design usingthe shuffled frog leaping algorithm. Journal of Water Resource Plan Management 129, 210–225 (2003)

12. Eusuff, M., Lansey, K., Pasha, F.: Shuffled frog-leaping algorithm: a memeticmetaheuristic for discrete optimization. Engineering Optimization 38, 129–154 (2006)

13. Vasko, F.J., Wolf, F.E., Stott, K.L.: A set covering approach to metallurgical grade assignment. European Journal of Operational Research 38(1), 27–34 (1989)

14. Kennedy, J., Eberhart, R.: Particle swarm optimization. In: Proc. IEEE Conf. Neural Networks, vol. 4, pp. 1942–1948 (1995)

15. Fisher, M.L., Rosenwein, M.B.: An interactive optimization system for bulk-cargo ship scheduling. Naval Research Logistics 36, 27–42 (1989)

16. Liong, S., Atiquzzaman, M.: Optimal design of water distribution network usingshuffled complex evolution. Journal of Instrumentation Engineering 44, 93–107 (2004)

17. Luo, X., Yang, Y., Li, X.: Solving tsp with shuffled frog-leaping algorithm. In: Proc. ISDA, vol. 3, pp. 228–232 (2008)

18. Smith, B.M.: Impacs - a bus crew scheduling system using integer programming. Mathematical Programming 42(1-3), 181–187 (1998)

19. Rahimi-Vahed, A., Mirzaei, A.: Solving a bi-criteria permutation flow-shop problem using shuffled frog-leaping algorithm. Soft Computing (2007)

20. Valenzuela, C., Crawford, B., Soto, R., Monfroy, E., Paredes, F.: A 2-level metaheuristic for the set covering problem. International Journal of Computers Communications and Control 7(2), 377–387 (2012)

21. Wang, L., Fang, C.: An effective shuffled frog-leaping algorithm for multi-modre source-constrained project scheduling problem. Information Sciences 181, 4804–4822 (2011)

Monitoring the Effectiveness of Security Countermeasures in a Security Risk Management Model

Neila Rjaibi and Latifa Ben Arfa Rabai

Department of Computer Science, Institut Supérieur de Gestion, University of Tunis, Tunisia
Rjaibi_neila@yahoo.fr, latifa.rabai@gmail.com

Abstract. Through our studies, successful and relevant security risk management models help to choose the right security measures which are vital in business analysis. In earlier works, an interesting value based cybersecurity metric namely the Mean failure Cost (MFC) has been presented. It computes for each system's stakeholder his loss of operation in monetary term taking into consideration the security requirements, the architectural components and threats of such a system. In this paper, our intention is to extend this measure into a security risk management process in order to highlight the security priorities, implement controls and countermeasures then monitor the effectiveness of the chosen security solution by using the return on investment (ROI). Our attempt is to maximize the security management performance and business decisions by saving both time and money. The practical investigation is conducted thought the context of e-learning systems.

Keywords: Cyber Security Metrics, E-Learning, Mean Failure Cost, Risk Management, Security Measures, the Return on Investment, Decision Making.

1 Introduction

Information security management is a critically important and challenging task, given the increase of threats and vulnerabilities, effective security risk management models help to deploy the suitable security resources such including attack prevention, vulnerability reduction, and threat deterrence [3], [10]. A variety of security risk management models and approaches have emerged to assess and quantify security through a financial measure [4], [10], [11], [12]. But among the research challenges found in security we can highlight decision concerns. So, the security managers need to be helped to set the relevant security decisions and make an appropriate risk versus cost.

In previous works, we defined and computed a value based cyber security metric which is the Mean failure Cost (MFC) measure by taking into consideration the system's stakeholders, security requirements, architectural components and threats. This cybersecurity metric was applied in a practical case study to the quantification of security threats of e-learning systems [3]. A review of the literature identified deficiencies in the practice of information security risk assessment and decision-making involving the system's stakeholders, security requirements, architectural components and threats. For these reason, this paper emerges new ideas to take advantage of the

© Springer International Publishing Switzerland 2015
R. Silhavy et al. (eds.), *Software Engineering in Intelligent Systems*,
Advances in Intelligent Systems and Computing 349, DOI: 10.1007/978-3-319-18473-9_32

MFC's features in security business decision in a structured way and discusses the use of this cyber security measure for quantitative decision-making in the practice. To achieve this goal an extension of the MFC measure to a dynamic and quantitative security risk management model will be done, our attempt is to incorporate three new components in the risk management process for decision making:

1. Diagnosing and setting the security priorities in the MFC matrices
2. Choosing the suitable security solutions
3. Ensuring a better security solution.

The Mean failure Cost computes each stakeholder's loss of operation in the given system ($/H). This quantitative model is a cascade of linear models to quantify the system's security in terms of the loss resulting from system's vulnerabilities as: [2], [4], [5] [16].

$$MFC = ST \circ DP \circ IM \circ PT \qquad (1)$$

The MFC cybersecurity metric is the product of several factors (the stakes matrix ST, the dependency matrix DP, the impact matrix IM, and the threat vector PT) [6-8]. Where:

- The stakes matrix (ST): This matrix is composed of the list related to stakeholders and the list of security requirements. Each row is filled by relevant stakeholders who have internal or external usage of the platform. Each cell is expressed in dollars and represents the loss incurred and/or premium placed on requirement.
- The dependency matrix (DP): Each row for this matrix is filled by system architects; each cell represents probability of failure with respect to the requirement that a component has failed. DP (Rj, Ck): The probability that the system fails to meet requirement Rj if component Ck is compromise
- The impact matrix (IM): Each row for this matrix is filled by V&V Team; each cell represents probability of compromising a component given that a threat has materialized, it depends on the target of each threat, likelihood of success of the threat. IM (Ck, Th): The probability that Component Ck is compromised if Threat Th has materialized
- The threat vector (PT): Each row for this vector is filled by a security team; each cell represents to which extent each threat is likely to be successful, it depends on perpetrator models, empirical data, known vulnerabilities, known countermeasures, etc. PT (Ti): The probability that threat Ti is materialized for a unit of operational time (one hour of operation).

After estimating and calculating the assets for each security risk management process it is evident that we need to choose the right security measures. This paper focuses on the MFC cybersecurity measure and aims at spreading it to a cybersecurity risk management model: once the assets are estimated and calculated the objective is to highlight the three remaining steps namely: setting priorities, implementing controls and countermeasures, monitoring risks and the effectiveness of countermeasures. Our practical case study is for current and standard e-learning systems [13-15]. Therefore we propose a theoretical and practical idea in monitoring the effectiveness of security countermeasures and ensuring its best choice based on the MFC cyber-security measure's values. As a first step, an idea to diagnose problems in the MFC matrix is to be conducted in order to set the highest security priorities in a practical case study of current e-learning systems as an example [1- 2]. The results show that we are facing a critical impact matrix (IM). Possible security measures to control the impact matrix are a set of reinforcement measures which aim to reduce the probability of failure of one or more components if the threats occur. To implement the proposed counter-measures we must invest through software and / or hardware solutions. This can be done for example by duplicating components [9]. The next step is to focus on the relevance and pertinence of the redundancy of e-learning architectural components and how it can be considered as a solution to countermeasure the security diagnostic results: Knowing that the problem to be confronted is how we can ensure the proper choice of solutions? This is done through the computing of the return on investment (ROI) of the proposed security solution which is a measure to evaluate the efficiency of an investment. The ROI is the benefit (return) of an investment divided by the cost of the investment, in other words how can we calculate and evaluate the gain periodically (the gain / period)? It is the main problem of decision making. In fact, our contribution is to be done using the MFC's values.

The main concern of this paper is to calculate the profitability of the solution through the return on investment (ROI) based on the MFC's values for each system's stakeholders. Our theoretical contribution is to help monitoring the effectiveness of the proposed countermeasures and maximizing management performance by taking the most suitable decisions. Having reached this point we can answer and justify the question: how can we ensure the proper choice of security solution?

This paper is organized as follows. In section 2, the focus is on presenting the security problems diagnostic in the MFC matrix in order to find the critical level problems. In section 3, the objective is to propose a solution aiming at reducing the MFC values when the IM matrix is critical. Section 4, is a presentation of the remedy approach to make choice in other words the calculation of the profitability through the return on investment of the proposed solution based on the MFC model. Our next focus in section 5 is to present in details the computational steps in order to monitor the effectiveness of the proposed solution. In section 6, we will provide an answer to the question: Is the duplication of the application server profitable and what the system stakeholders' gain from it? Finally, in section 7 we will conclude by summarizing our results and sketching directions for further research.

2   PROBLEMS Diagnostic in the MFC Matrix: E-learning Application Case Study

To find and prove quantitatively the critical level problems of the MFC cybersecurity measure, we need to focus on diagnosing security problems in relationship with these factors (the stakes matrix ST, the dependency matrix DP, the impact matrix IM, and the threat vector PT). The computing steps are as follow:

- Assuming the DP matrix is perfect (0 in each line except the last column of 1) the same for IM matrix, we assume that it is perfect (0 in each line except the last row of 1)
- Then the calculation of the MFC for the two hypothesis, the first (assuming the DP matrix is perfect), the second (assuming the IM matrix is perfect): The large difference represents the matrix of the problem
- Next, we try to find the most suitable security solutions and countermeasures for the critical matrix.

Table 1. Security problems diagnostic of the MFC'for e-learning systems case study

Stakeholders	Initial MFC' $ /hour	MFC' $ /hour DP matrix is perfect	MFC' $ /hour IM matrix is perfect
System administrator	643,457	667,144	647,666
Teacher	455,374	473,425	458,345
Student	81,768	83,022	82,308
Technician	208,878	216,451	210,244

We note that the highest values of the MFC application are when the DP matrix is perfect; we conclude that in this practical case study of the security risk management process of e-learning systems, the critical level's problem resides in the impact matrix (IM). Possible security measures to control the impact matrix are a set of reinforcement measures which aim to reduce the probability of failure of one or more components if the threats occur. This can be done for example by duplicating architectural components.

3   Security Measures to Reduce the MFC Values When the IM Matrix Is Critical

In order to reduce the MFC values to the half we opted for a solution: the duplication of the architectural hardware components. This technique is known as the redundancy technique, it is the duplication of critical components or functions of a system, in order to increase the system's reliability. In general, it takes the forms of backup. The redundant elements work in parallel. This is recommended for complex computer systems and for ultra large systems with a great number of stakeholders. The problem that we confront is how can we ensure the proper choice of this solution? The only remedy to the judgment of the good choice is the calculation of the profitability through the return on investment (RSI French). The ROI is defined, for a given period as the sum of discounted profits of the project, that is to say the lowest income costs divided by the funds invested in the project.

4   Computing the Return on Investment (ROI) of the Proposed Solution Based on the MFC Measure: The Approach

The return on investment (ROI) is the measure to evaluate the efficiency of an investment. The ROI is the benefit (return) of an investment divided by the cost of the investment; the result is expressed as a percentage or a ratio. The return on investment formula (ROI):

$$\frac{\text{Gain from Investment} - \text{Cost of Investment}}{\text{Cost of Investment}} \tag{2}$$

However, this equation is not suitable because we need to calculate the gain per period, to define the sum of discounted profits (gain) of the project for a given period. The calculation method is as follows [9]. The return on investment formula (ROI):

$$\frac{\sum_{W=1}^{w=W} \text{regular updated income} - \text{Periodic discounted costs}}{\text{Amount originally invested}} \tag{3}$$

If we consider the factor money value in time, the equation of the ROI is:

$$\frac{\sum_{W=1}^{w=W} B(w)/(1+d)}{C(0)} \tag{4}$$

With:

- **W**: The total number of discount periods.
- **w**: The number of the period
- **B(w)**: (Revenue - Cost) during the period w.
- **d**: The   amortization period (discount rate)
- **C(0)**: Amount originally invested

The main problem of decision making is how to calculate periodically (period w) the gain: which is the gain of the proposed solution for a stakeholder: B (w)? The Mean failure cost is a solution to the problem, the MFC is the monetary value of a failure during a period (generally 1 hour). If we implement a solution for a period w, the income generated by this solution is:

$$B_i(w) = (MFC_i(w_j) - MFC_i(w_{j+1}))*Nbh \tag{5}$$

With:

- **Bi (w)**: the benefit of the stakeholder i in period w.
- **Nbh**: the number of hours when the system is functional.
- **W $_j$**: this is the period number j.
- **W $_{j+1}$**: this is the period number j+1.
- **MFC$_i$ (w$_j$)**: is the mean failure cost of the stakeholder i occurring during period wj.

5   Monitoring the Effectiveness of Countermeasures with the MFC Cybersecurity Measure: Steps in Computations

Steps in computations are:

1. Define and justify the security solutions (what are the components that we intend to duplicate in order to strengthen the security of the considered system).
2. Fix the architectural components price, the discount rate and the number of periods.
3. What remains is the B (w): (Revenue - Cost) during the period w: We need to determine the gain / period, how? Solution and contribution: We use the difference between two MFC's values (initial MFC and MFC / period after applying the solution)
4. We have the initial MFC values (calculated in the quantitative risk management process initially)
5. Computing the MFC S_i for the architectural component / period
   MFC'Si (architectural component) = ST' * DP'* $\textbf{IM}_{\textbf{si}}$ * PT
   - Only the impact matrix varies because it is the critical element according to the MFC security problem's diagnostic.
   - The empirical information in the impact matrix is the probability that a component C_k fails once threat T_q has materialized.
   — When we duplicate the architectural components of our system, the probability that each one of these threats may materialize within a unitary operation time decreases to half every period of time.
   — We need to identify the evolution of the probabilities for one architectural component of the impact matrix during a number of periods.

   - We Compute the MFC S_i (MFC/ period) for the architectural component / period.

6. Computing the gain for the architectural component/ period

   gain $_{Si}$ = **the initial MFC - MFC'$_{Si}$** (architectural component) * Nbh
   Gain = Somme (gain)

7. Computing the return on investment of the architectural component

$$\frac{\sum_{W=1}^{w=W} B(w)/(1 + d)}{C(0)} \qquad (6)$$

8. Discussions and decisions: Depending on the values of the ROI (Positive or negative for all stakeholders) Decisions can be made.

6   The Decision Problem: The Duplication of the Application Server

It is mainly concerned with duplication of the application server and its profitability to the current e-learning system's case study and what gain it offers to the system's stakeholders?

6.1   The Architectural Components Prices and the Amortization Period

The standard architecture of an e-learning system includes six linked components namely: the browser, the web server, the application server, the Db server, the firewall server and the mail server. This table shows details about the architectural components prices and the amortization period of current and standard e-learning systems.

Table 2. The architectural components prices and the amortization period for an e-learning system case study

	The Architectural Components of E-learning Systems				
	Web server	Application Server	DB server	Firewall server	Mail server
Max prices	3 109€	1 028,99€	3080,00 €	5 379€95	149 € / month
Min prices	861€	420,40€	516,00 €	159€	0€ Open source server
Average price	3539,5€	1239,19€	3338€	5459,45€	149€
conversion $	4884,191$	1709,971$	4606,139$	7533,549$	205,607$
The amortization period	3 years	3 years	3 years	3 years	

Hypotheses of the Decision Problem: The price of the application server (= 1239,19€ / 1709,971$), Discount rate = 3 years, Period (T) = 6 month (semester), Six computations.

6.2   Defining the Evolution of the Impact Matrix for 3 Years

The impact matrix IM of the MFC can be filled by analyzing which threats affect which components, and assessing the likelihood of success of each threat. The empirical information in the impact matrix is the probability that component C_k fails once threat T_q has materialized. When we duplicate the architectural components of our system, the probability that each one of these threats may materialize within a unitary operational time decreases to the half every period of time ($S1$ = one semester).

Table 3. Probability evolution of the application server during 3 years

Application server	0,271	0,135	0,007	0	0	0	0,338	0,007	0	0,225	0,014	0,003	0
S1		0,1355	0,0675	0,0035	0,0000	0,0000	0,0000	0,1690	0,0035	0,0000	0,1125	0,0070	0,0015	0,0000
S2		0,0678	0,0338	0,0018	0,0000	0,0000	0,0000	0,0845	0,0018	0,0000	0,0563	0,0035	0,0008	0,0000
S3		0,0339	0,0169	0,0009	0,0000	0,0000	0,0000	0,0423	0,0009	0,0000	0,0281	0,0018	0,0004	0,0000
S4		0,0169	0,0084	0,0004	0,0000	0,0000	0,0000	0,0211	0,0004	0,0000	0,0141	0,0009	0,0002	0,0000
S5		0,0085	0,0042	0,0002	0,0000	0,0000	0,0000	0,0106	0,0002	0,0000	0,0070	0,0004	0,0001	0,0000
S6		0,0042	0,0021	0,0001	0,0000	0,0000	0,0000	0,0053	0,0001	0,0000	0,0035	0,0002	0,0000	0,0000

6.3   Computing the MFC Si

Given the duplication of the application server and the probability evolution of the application server during 3 years, we compute the Mean Failure Cost for e-learning Systems/semester.

MFC'Si (application server) = ST' * DP'* IM Si * PT

Table 4. MFC'Si application for e-learning systems for the application server

Stakeholders	MFC'0 $ /hour	MFC'S1 $ /hour	MFC'S2 $ /hour	MFC'S3 $ /hour	MFC'S4 $ /hour	MFC'S5 $/hour	MFC'S6 $ /hour
System administrator	643,457	643.434	643.430	643.428	643.427	643.427	643.426
Teacher	455,374	455.355	455.352	455.350	455.349	455.349	455.349
Student	81,768	81.767	81.766	81.766	81.766	81.766	81.766
Technician	208,878	208.870	208.869	208.868	208.868	208.867	208.867

6.4   Computing Gain for the Application Server Duplication

According to table 5, we compute gain for the application server duplication
Gain= The initial MFC - MFC'Si (application server) * Nbh

Table 5. Gain for the application server

Stakeholders	MFC' $ /hour	Gain S1	Gain S2	Gain S3	Gain S4	Gain S5	Gain S6
System administrator	643,457	0,552	0,648	0,696	0,72	0,72	0,744
Teacher	455,374	0,456	0,528	0,576	0,6	0,6	0,6
Student	81,768	0,024	0,048	0,048	0,048	0,048	0,048
Technician	208,878	0,192	0,216	0,24	0,24	0,264	0,264

6.5   The Return on Investment of the Application Server Duplication

Table 6. The return on investment of the application server duplication

Stakeholders	ROI
System administrator	0,001
Teacher	0,000
Student	0,001
Technician	0,000

6.6   Discussion

Depending on the ROI values (positive for all stakeholders) we conclude that it is a good solution to adopt the duplication of the application server components because all of the stakeholders are winners. Given the positive values of the ROI, they are significant because only the system administrator and the student in the e-learning case study systems could extremely gain from the duplication solution. But from another side it seems not very interesting to adopt the duplication solution of the application server because the ROI values are not high. Therefore, we decide not to adopt this solution and not to waste the budget because there is not really an important gain. It is not really a profitable solution. In order to maximize the security management performance and decisions of the considered system we intend to focus on the duplication of the other remaining architectural components like the firewall and to focus on the other suitable security solutions and countermeasures to reduce the MFC matrix values and control it in the most adequate way.

7   Conclusions

The MFC cyber security metric is the product of several factors (the stakes matrix ST, the dependency matrix DP, the impact matrix IM, and the threat vector PT). It is possible to control the MFC through its factors in order to minimize and to reduce its values. This leads to maximize the security management performance and decisions of the considered system in the appropriate time and by maintaining the budget. To implement the proposed countermeasures we must invest through software and / or hardware solutions and ensure the proper choice of solution through the return on investment (ROI). The main problem of decision making is: How to calculate periodically (period w) the gain of the proposed solution for the entire systems' stakeholder? The most important elements of our study reside in using the Mean failure cost measure's values to evaluate periodically the gain. It is the best solution that helps in monitoring quantitatively the effectiveness of the security countermeasures for the most suitable decisions. The calculation of the return on investment based on the MFC measure's values is a good solution for decision making for all the stakeholders of the considered system. It is an optimal solution for both simple and complex systems such as systems in which users/ stakeholders have a variety of benefits from the proposed practical solution. In these cases, security management decisions can be easily and quickly managed. Since our approach provided encouraging results, this work can be extended and completed to study and monitor the effectiveness of security countermeasures of the other remaining architectural components like the firewall or to study the combination redundancy possibility such as the two architectural components: The DB server and The Web server. This is the only way to ensure the best solution without wasting the budget. Our future plans are to explore such opportunities to control other factors of the MFC measure in order to minimize its values which lead to more secure and safe e-systems.

References

1. Rabai, L.B.A., Rjaibi, N.: Assessing Quality in E-learning including learner with Special Needs. In: Proceedings of The Fourth National Symposium on Informatics, Technologies for Special Needs, April 23-25. King Saud University, Riyadh (2013), http://nsi.ksu.edu.sa/node/2
2. Rjaibi, N., Rabai, L.B.A.: Toward A New Model For Assessing Quality Teaching Processes In E-learning. In: Proceedings of 3rd International Conference on Computer Supported Education, CSEDU 2011, Noordwijkerhout, The Netherlands, May 6-9, vol. 2, pp. 468–472. SciTePress (2011), http://www.csedu.org, ISBN: 978-989-8425-50-8
3. Rabai, L.B.A., Rjaibi, N., Aissa, A.B.: Quantifying Security Threats for E-learning Systems. In: IEEE Proceedings of International Conference on Education & E-Learning Innovations- Infrastructural Development in Education (ICEELI 2012), Sousse,Tunisia, July 1-3 (2012), http://www.iceeli.org/index.htm, doi:10.1109/ICEELI.2012.6360592, Print ISBN: 978-1-4673-2226-3
4. Rjaibi, N., Rabai, L.B.A., Omrani, H., Aissa, A.B.: Mean failure cost as a measure of critical security requirements: E-learning case study. In: Proceedings of the 2012 World Congress in Computer Science, Computer Engineering, and Applied Computing (WORLDCOMP 2012, Las Vegas, Nevada, USA), July 16-19, pp. 520–526. CSREA Press, U. S. A. (2012), The 11 th International Conference on e-Learning, e-Business, Enterprise Information Systems, and e-Government (EEE 2012: July 16-19, USA), ISBN: 1-60132-209-7
5. Rjaibi, N., Rabai, L.B.A., Aissa, B.A., Louadi, M.: Cyber security measurement in depth for e-learning systems. International Journal of Advanced Research in Computer Science and Software Engineering (IJARCSSE) 2(11), 107–120 (2012), http://www.ijarcsse.com, ISSN (Online): 2277 128X, ISSN (Print): 2277 6451
6. Rjaibi, N., Rabai, L.B.A., Aissa, B.A., Mili, A.: Mean failure Cost as a Measurable Value and Evidence of Cybersecurity: E-learning Case Study. International Journal of Secure Software Engineering (IJSSE) 4(3), 64–81 (2013), http://www.igi-global.com/ijsse, doi:10.4018/jsse.2013070104
7. Rjaibi, N., Rabai, L.B.A., Aissa, B.A.: A basic security requirements taxonomy to quantify security threats: an e-learning application. In: Proceedings of the Third International Conference on Digital Information Processing and Communications (ICDIPC 2013), Session: Information security, Islamic Azad University (IAU), Dubai, United Arab Emirates (UAE), January 30-February 1, pp. 96–105 (2013), http://www.sdiwc.net/conferences/2013/Dubai/, ISBN: 978-0-9853483-3-5 ©2013 SDIWC

8. Rjaibi, N., Rabai, L.B.A., Aissa, A.B.: The Mean Failure Cost Cybersecurity Model toward Security Measures And Associated Mechanisms. International Journal of Cyber-Security and Digital Forensics (IJCSDF) 2(2), 23–35 (2013)

9. Aissa, A.B.: Vers une mesure économétrique de la sécurité des systèmes informatiques, Doctoral dissertation, Faculty of Sciences of Tunis, submitted (Spring 2012)

10. Nazareth, D.L., Choi, J.A.: System Dynamics Model for Information Security Management. Information & Management (2014)

11. Fenz, S., Heurix, J., Neubauer, T., Pechstein, F.: Current challenges in information security risk management. Information Management & Computer Security 22(5), 410–430 (2014)

12. Feng, N., Wang, H.J., Li, M.: A security risk analysis model for information systems: causal relationships of risk factors and vulnerability propagation analysis. Information Sciences 256, 57–73 (2014)

13. Weippl, E.R., Ebner, M.: Security Privacy Challenges in E-Learning 2.0. In: World Conference on E-Learning in Corporate, Government, Healthcare, and Higher Education Healthcare, and Higher Education, pp. 4001–4007 (2008)

14. Raitman, R., Ngo, L., Augar, N., Zhou, W.: Security in the online e-learning environment. In: Fifth IEEE International Conference on Advanced Learning Technologies, ICALT 2005, pp. 702–706. IEEE (2005)

15. Weippl, E.: Security In E-Learning, eLearn Magazine. Association for Computing Machinery (ACM), article from 16, 03–05 (2005)

16. Aissa, A.B., Abercrombie, R.K., Sheldon, F.T., Mili, A.: Defining and Computing a Value Based Cyber-Security Measure. Information Systems and e-Business Management 10(4), 433–453 (2012), doi:10.1007/s10257-011-0177-1

Proposal of an Expert System for Predicting Warehouse Stock

Bogdan Walek and Radim Farana

Department of Informatics and Computers, University of Ostrava, 30. dubna 22,
701 03 Ostrava, Czech Republic
{bogdan.walek,radim.farana}@osu.cz

Abstract. Currently, the area of stone-store sale requires a suitable solution how to solve the issue of warehouse stock as well as the process of purchasing. This paper proposes an expert system which takes into consideration past periods and various impacts on sales and subsequently proposes how much goods should be ordered for the next period. Individual steps and parts of the expert system are described in the paper. The proposed expert system is verified on a particular example.

Keywords: prediction, warehouse stock, expert system, shopkeeper, goods.

1 Introduction

Currently, the area of stone-shop sale is an everyday topic for solving warehouse stock (1). Every shopkeeper frequently copes with a question how much goods should be ordered in order to store it with no problems and sell it in a reasonable time. If a shopkeeper does not order enough goods, he will risk its unavailability. This can result in either short-term or long-term loss of customers. The shopkeeper loses profit not only for the unavailable goods, but for other goods that a potential customer would buy as well. On the other hand, if the shopkeeper orders too many goods, there is a risk of insufficient warehouse area. In some cases, big stock can be economically disadvantageous as the price of the good might go down, thus the margin goes down as well. Be worse, it might be sold at a loss.

Currently, there are various information systems which more or less successfully anticipate and predict the quantity of goods that a shopkeeper should order (2). These systems usually use prediction based on sales in the previous period. The shopkeeper can adjust the prediction based on his knowledge and experience with the given type of goods.

This paper will introduce a fuzzy logic expert system (3, 4) which will use a knowledge base created based on shopkeeper's knowledge and other influences affecting the prediction in order to predict and propose the quantity of goods which the shopkeeper should order. Expert systems, especially based on fuzzy logic, are often used for system behaviour prediction in many application areas (e.g. 5, 6, 7), but applications in predicting warehouse stock domain are very rare in literary sources. Authors usually use data mining (e.g. 8, 9), statistical methods or neural networks (e.g. 10, 11) to solve this problem. This paper will present possibility of fuzzy logic expert systems in this problem domain.

© Springer International Publishing Switzerland 2015
R. Silhavy et al. (eds.), *Software Engineering in Intelligent Systems*,
Advances in Intelligent Systems and Computing 349, DOI: 10.1007/978-3-319-18473-9_33

2   Problem Formulation

As said above, there are several issues to solve in the area of warehouse stock.

The first issue is inaccurate prediction of an order for next period. Current information systems perform this prediction based on sales in previous periods. This step, however, takes into consideration only statistics on the given goods. But prediction of an order is affected by many other factors, which have a considerable influence on the final sale. These influences include possible discounts, correlation with other goods, occurrence of the product on a leaflet, or placement of the product in the shop. Such factors can significantly change sales of the goods in the next period. Current information systems do not take those into consideration and leave the order adjustment on the shopkeeper. The shopkeeper then uses his knowledge and experience to manually adjust the order. This makes the prediction more accurate, but some factors cannot be accurately considered by the shopkeeper and the final quality is then highly dependent on his ability to include them in the order adjustment.

Another issue is the fact that the shopkeeper has an idea of a long-term strategy of selling a given product (discounts, supportive actions, making the product visible, etc.), but there is no tool to grasp this strategy and include it into future order predictions.

3   Problem Solution

Based on the above-stated definition of the main issues, we are proposing an expert system which will use a knowledge base and input information to propose (predict) an order of given goods for the next period.

Firstly, the main influences entering the process of prediction must be defined. Such influences can be divided into several categories:

Direct influences:
- Warehouse area (in m2 or m3)
- Quantity of goods in the shop
- Quantity of goods in the warehouse
- Average delivery time from supplier (hours, days or weeks)

Internal influences:
- Discount level (in percent for the given period)
- Correlation with other products, we distinguish two types:
  - Mutual correlation – if a customer buys ham, he usually buys bakery products
  - Competition – if a customer buys a certain amount of rolls, he might not buy baguettes or buns
- Occurrence of the product in a regular leaflet (if the goods appears there, there is a possibility of its higher sale than of other products from the same category)
- Placement of the product in a regular leaflet (if the goods appears on the first page or among top products, the possibility of its sale is very high)
- Placement of the product in the shop (it presumes that a product placed right next to the shop entry or on a shelf within eye is likely to sell more than a product in back corners or lower shelves)

External influences regular:
- Public holidays, holidays, period of vacation – in case the following period includes a public holiday or a day with school holiday, this can significantly influence sale of given types of products
- Seasonal products (only summer, only winter, etc.)
- Competitors' actions (discounts of the same product)

• Competitors' product is permanently cheaper or their product is the cheapest in the city

External influences irregular:

• Ceremonial opening of a new shop (significant impact on sale)

• Worse access to the given shop (road works, reconstruction of surrounding buildings, etc.)

• Important events – world championships in ice-hockey or football (more beer will be sold), elections (possibly less customers)

The proposed expert system is shown in the following figure:

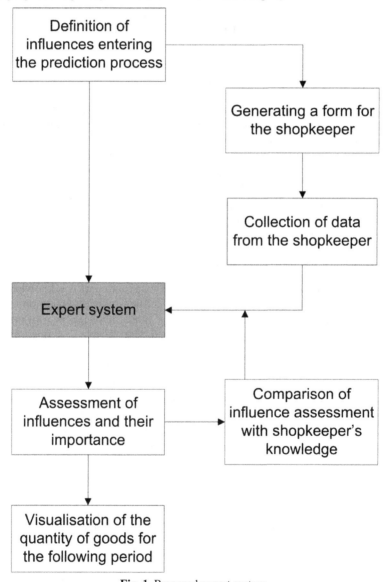

Fig. 1. Proposed expert system

The following sub-chapters will define the main steps of the expert system:

3.1   Definition of Influences Entering the Prediction Process

First of all, it is necessary to recognize and define all influences entering the prediction process of a given type of goods. These influences will then be included into the knowledge base and their definition will have an impact on assessing and predicting the quantity of orders. The defined influence will also be a base for generating a form for the shopkeeper. Using the form, the shopkeeper can set values of individual influences and their importance. The *influence* category is automatically read based on selection of a given influence and its classification according to the categorization stated above.

3.2   Generating a Form for the Shopkeeper

Next step consists in generating a form based on selected influences. Using this form, the shopkeeper sets values for the selected influences and their importance for the final prediction.

3.3   Collection of Data from the Shopkeeper

Once the shopkeeper submits the form, the values are stored in a database and ready to be processed by the expert system.

3.4   Creation of the Expert System

This step consists in creating an expert system whose main part is a knowledge base containing IF-THEN rules. The knowledge base is filled-in by an expert on sales of a given product range, which might include one or more shopkeepers who have wide experience with selling the given product range and who are able to define importance of individual influences.

The input linguistic values of the IF-THEN rules are influence defined in the first step together with sales statistics of the given type of goods in the past, preferably in the same period when the prediction is done. For instance, if the prediction is done for December, it would be best to get data from last December (or more years back).

The output variable is assessment of the quantity of the goods which should be sold in the following period.

3.5   Assessment of Influences and Their Importance

The expert system performs assessment of influences and their importance. Then it proposes the quantity of goods which should be ordered for the given period.

3.6   Comparison of Influence Assessment with Shopkeeper's Knowledge

This step compares the assessed influences and outputs from the expert system with shopkeeper's knowledge. In case the interpretation differs, there is a possibility to refine the influences and perform the assessment using the expert system once again.

3.7   Visualisation of the Quantity of Goods for the Following Period

The last step consists in visualising the quantity of goods which should be ordered for the following period. The quantity is approved or adjusted by the shopkeeper in case any influence has been ignored or a specific situation in the given period occurs.

4   Verification

Verification of the proposed expert system has been performed on real data from a shop primarily selling food and drinks. We have randomly selected five products. The data sample included sales statistics of the products for the period June 2012 – August 2014.

The selected products are the following:

- Camembert-type cheese 120g
- Bread 700g
- Kofola (soft drink) 2L
- Ostravar (beer) 0.5l
- Roll 50g

The graph below represents monthly statistics for the monitored period:

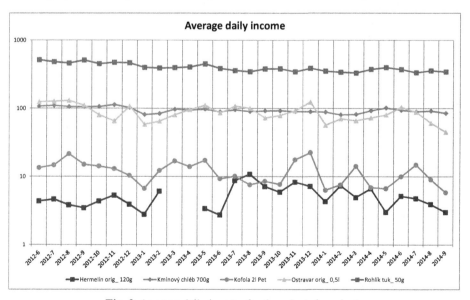

Fig. 2. Average daily income for the selected products

4.1   Definition of Influences Entering the Prediction Process

Before predicting the sales in the following period, the following influences have been identified:

- Discount –(discount x% for everything)
- Competition – competitors in the vicinity will have higher prices due to introduction of the discount for the following period
- Worse accessibility of the shop – the following period will be influenced by road works next to the shop, which might cause accessibility problems. The shop will be open as usual.

4.2   Generation of the Form for the Shopkeeper

Based on selected influenced, a form will be generated for the shopkeeper. The influence *Discount* needs a specific figure in percent. Each influence can be assigned with its impact extent, which defines what impact each influence will have on the final prediction – either positive or negative (increase or decrease of sales). The extent of impact is expressed by the following linguistic expressions:

- Very small negative
- Small negative
- Medium negative
- Big negative
- Very big negative
- Very small positive
- Small positive
- Medium positive
- Big positive
- Very big positive

The following figure shows the form for the shopkeeper:

4.3   Collecting Data from the Shopkeeper

Based on the filled-in form, the database stores influence specification and their importance:

V1 – 5% - discount 5% on everything
V1_IMP – big positive
V2_IMP – medium positive
V3_IMP – medium negative

4.4   Creation of the Expert System

Next, the expert system and its knowledge base will be created. The knowledge base will be created for each type of product separately as the IF-THEN rules for those can differ:

The input linguistic variables of the IF-THEN rules are as follows:

- P1 – sales of the product in the previous period (values: very small, small, medium, big, very big)
- V1_IMP
- V2_IMP
- V3_IMP

The output linguistic variable is assessment of the quantity of goods to be sold in the following period:

- OUT (values: very small, small, medium, big, very big)

Below are presented examples of the IF-THEN rules of the expert system:

```
IF(P1 IS VERY SMALL) AND
(V1_IMP IS BIG POSITIVE) AND
(V2_IMP IS MEDIUM POSITIVE) AND
(V3_IMP IS MEDIUM NEGATIVE) THEN
(OUT IS SMALL)

IF(P1 IS BIG) AND
(V1_IMP IS BIG POSITIVE) AND
(V2_IMP IS MEDIUM POSITIVE) AND
(V3_IMP IS MEDIUM NEGATIVE) THEN
(OUT IS VERY BIG)
```

4.5   Assessment of the Influences and Their Importance

The expert system has assessed individual influences as well as sales of the given goods in the previous period (34th week of 2014)

- Camembert-type cheese 120g - **3**
- Bread 700g - **86**
- Kofola (soft drink) 2L - **9**
- Ostravar (beer) 0.5l - **54**
- Roll 50g - **362**

4.6   Comparison of Influence Assessment with Shopkeeper's Knowledge

The assessed influences and outputs of the expert system correspond to shopkeeper's knowledge. There is no need perform adjustment of the impact of the influences and subsequent re-assessment using the expert system.

4.7   Visualisation of the Quantity of Goods for the Following Period

Finally, the final prediction of sales in the following period is visualised. The following table shows individual items with their prediction for the 35th week of 2015.

WeekNumber	WeekDay	Hermelín orig_ 120g	Kmínový chléb 700g	Kofola 2l Pet	Ostravar orig_ 0,5l	Rohlík tuk_ 50g
35	1	2	73	9	69	364
35	2	7	85	21	121	328
35	3	6	74	6	80	332
35	4	3	79	13	79	282
35	5	5	186	10	144	526
35	6	2	65	14	84	443
35	Avg	4	94	12	96	379
35	Real data	3	92	7	69	356
35	Difference	1	2	5	27	24
35	Rel.diff.	30%	2%	66%	39%	7%

We can see that the prediction for the 35th week differs from real sales to various extent. For sales in order of units, the difference is relatively high, but for sales in dozens or hundreds of units, the difference significantly decreases. For low sales, the number of influencing parameters is not sufficient and random influences prevail. The following figure shows prediction of sales for individual days.

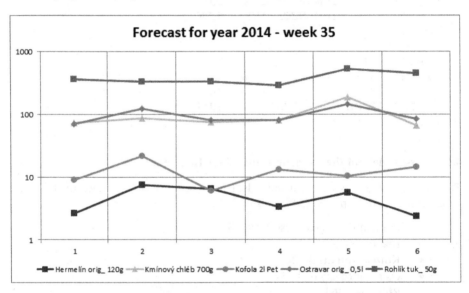

Fig. 3. Prediction of sales for individual week days

5   Conclusion

The paper has presented a proposal of an expert system for prediction of sales of goods. First of all, the main issues affecting the prediction process were identified as well as impacts of an inaccurate prediction. Secondly, an expert system was created which, based on expert knowledge of shopkeepers', statistical data on sales for past periods, and other influences, proposes prediction of sales for the following period. The expert system is only a supportive tool for the shopkeeper. The shopkeeper can adjust the proposed prediction. The proposed expert system was verified on a real shop and its sales statistics.

Acknowledgment. This paper was supported by the European Regional Development Fund in the IT4Innovations Centre of Excellence project (CZ.1.05/1.1.00/02.0070) and by the internal grant SGS15/PřF/2014, called Fuzzy modeling tools for securing intelligent systems and adaptive search burdened with indeterminacy, at Department of Informatics and Computers, University of Ostrava.

References

1. Brown, S.A.: Customer Relationship Management: A Strategic Imperative in the World of E-Business. John Wiley, New York (2000)
2. Swift, R.S.: Accelerating Customer Relationships: Using CRM and Relationship Technologies. Prentice Hall, Upper Saddle River (2001)
3. Novák, V.: Linguistically Oriented Fuzzy Logic Control and Its Design. Int. Journal of Approximate Reasoning 12, 263–277 (1995)
4. Pokorny, M.: Artificial Intelligence in modelling and control (in Czech), 189 p. BEN - technická literatura, Praha (1996) ISBN: 80-901984-4-9
5. Xu, B., Liu, Z.-T., Nan, F.-Q., Liao, X.: Research on energy characteristic prediction expert system for gun propellant. In: IEEE International Conference on Intelligent Computing and Intelligent Systems (ICIS), vol. 2, pp. 732–736 (2010) ISBN: 978-1-4244-6582-8
6. Zhang, B., Wang, N., Wu, G., Li, S.: Research on a personalized expert system explanation method based on fuzzy user model. In: Fifth World Congress on Intelligent Control and Automation, WCICA 2004, vol. 5, pp. 3996–4000 (2004) ISBN 0-7803-8273-0
7. Zhang, B., Liu, Y.: Customized explanation in expert system for earthquake prediction. In: 17th IEEE International Conference on Tools with Artificial Intelligence, ICTAI 2005, pp. 367–371 (2005) ISBN: 0-7695-2488-5
8. John, W.: Data warehousing and mining: concepts, methodologies, tools, and applications, 6 v. (lxxi, 3699, 20 p.). Information Science Reference, Hershey (2008) ISBN 978-1-59904-951-9
9. Khosrow-Pour, M.: Encyclopedia of information science and technology, 3rd edn., 10384 p. IGI Global (2014) ISBN 978-1-46665-889-9
10. Vaisla, K.S., Bhatt, A.K., Kumar, S.: Stock Market Forecasting using Artificial Neural Network and Statistical Technique: A Comparison Report. (IJCNS) International Journal of Computer and Network Security 2(8) (August 2010)
11. Vaisla, K.S., Bhatt, A.K.: An Analysis of the Performance of Artificial Neural Network Technique for Stock Market Forecasting. (IJCSE) International Journal on Computer Science and Engineering 02(06), 2104–2109 (2010)

A Multiplatform Comparison of a Dynamic Compilation Using Roslyn and Mathematical Parser Libraries in .NET for Expression Evaluation

Petr Capek, Erik Kral, and Roman Senkerik

Tomas Bata University in Zlin, Faculty of Applied Informatics, Nam T.G. Masaryka 5555,
760 01 Zlin, Czech Republic
capek@fai.utb.cz

Abstract. This work aims to provide a multiplatform comparison of a dynamic compilation using Roslyn and mathematical parser libraries in .NET for expression evaluation. In the first part, our own implementation of a mathematical parser, based on a new .NET compiler platform called Roslyn to implement on-fly compiling for mathematical expressions, is introduced. We then define the benchmark and perform performance measurements of our solution against selected mathematical parsers. At the end of the work, we discuss the results of the benchmark on different platforms.

Keywords: .NET, Roslyn, mathematical parser, dynamic compilation.

1 Introduction

In the world of science, people very often complain about the evaluation of some mathematical formulas. They have some data and they need to apply functions to this data. Small amounts of data can be calculated by hand, but it is necessary to use computer power for large amounts of data. For simple calculations, it is possible to use spreadsheet software which allows one to easily modify the functions expression, if so required. It is often need to create and compile one´s own program for complex data processing.

In some cases, there is a requirement to give users abstract control in order to change the mathematical expression in a program without recompiling or reinstalling the program. If the program uses a data processing method with formulas that can be changed, one needs to choose the right techniques to allow users to do that.

One of the solutions is to provide a predefined set of functions for users so that a user can choose a function formula from them; while another is to provide users with the ability to design and use their own formulas. In the latter case, it is necessary to implement some sort of mathematical parser engine which allows users to enter new formulas into the software.

A Parser engine is a complex system which has specific phases [3] [4]. In general, the principle of parsing is described in Fig. 1.

© Springer International Publishing Switzerland 2015
R. Silhavy et al. (eds.), *Software Engineering in Intelligent Systems*,
Advances in Intelligent Systems and Computing 349, DOI: 10.1007/978-3-319-18473-9_34

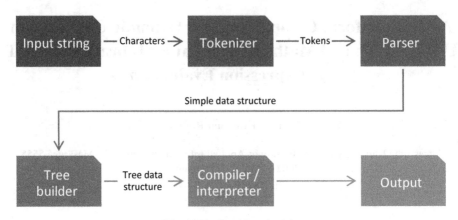

Fig. 1. The Parsing principle

To implement one´s own parser system, it is essential that one knows each component of the parser system. It can be hard to implement it without bugs and to achieve a comparable performance in relation to the native code [1] [5].

In this paper, we compared the existing mathematical expression parsers and also compared them with our solution. We chose an alternative approach to the implementation of a mathematical parser system to simplify the complexity of the parser engine. There is a similarity between the C# compilation code and the expression parsing process. Our solution is a string replacement engine based on processing Regex expressions, which translate the mathematical expression into C# code. Finally, the .NET dynamic code compilation [2] is used to "revive" this code.

2   Implementation Details

Our engine has three basic phases. The first is the expression preparation phase, in which mathematical language tokens are replaced with tokens accepted by the C# language. This phase consists of the following steps:

- String token replacements for mathematical functions
- String token replacement using Regex for replacing mathematic constants
- Replacing input parameters tokens

The second phase is the compilation phase. In this phase, we need to input our expression into a prepared class template and perform an on-fly compilation using Roslyn services [16]. This phase involves several steps, as follows:

- Template class preparation
- Assembly of a template class with a prepared string expression
- Initialization of Roslyn services
- Parsing the class into the syntax tree
- Setting the compilation parameters and loading the dependent assemblies
- Starting memory compilation

The last phase is called code revival. In this phase, we try to load the compiled code, find the target function and cache it. This phase includes the following steps:

- Loading the compiled assembly into an application domain
- Searching the domain for the target class
- Creating an instance of a target class
- Searching for the target function in the class instance
- Translating the function from a general type into a generic type, and then caching it

Finally, everything is ready for the evaluation of the input data against the compiled function. Originally, we used an old C# compilation engine in the previous version of our mathematic parser, which had one big disadvantage. This involved bugs [15] in the *CSharpCodeProvider* class, which did not allow in-memory compilation. This caused serious time drawbacks for the compilation process. With the new Roslyn compiler platform, this handicap is eliminated (See Figure 2).

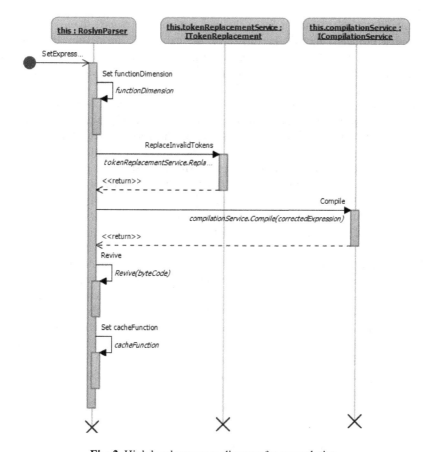

Fig. 2. High level sequence diagram for our solution

3 Benchmarking

So as to measure the performance of our mathematical parser engine, we defined a list of functions and we tried to parse these function and evaluate this function against different input values in two different scenarios.

Due to the varied complexity of expressions, we had to categorize the expressions depending on the complexity of their expressions. For instance, there are categories based on expression complexity in which the complexity is defined by the number of operators, operands and variables:

- Simple expressions – up to 5 operands and 5 operators
- Medium expressions – up to 10 operands and 10 operators, up to 3 function nesting
- Complex expressions – more than 10 operands and operators, more than 3 function nesting

A list of the functions used in the benchmarking process and its categories is shown in Table 1.

There are two test scenarios for evaluating expressions because there are two main factors that influence the test performance, the expression processing time and the expression evaluation time. N represents the number of different expressions which are used in the test, and M represents the number of expression evaluations with given input variables.

The first scenario is focused on measuring the performance of processing different expressions (N >> M). In this case, this involves the measured time of the evaluation.

The second scenario is focused on measuring the performance of evaluating the same expression against different input variable values (N << M). In this case, the measured time represents the expression processing process.

The high level scheme of the benchmarking process is described in Figure 3.

The main class in the benchmark process is the Benchmark class which contains the two test scenarios. Also, this class has a constructor which searches and registers all mathematical parsers (i.e. a class implemention IEvaluator interface) as well as all testing functions (i.e. class derived from ExpressionBase) via reflection.

The Interface IEvaluator assures the abstraction for the mathematical parsers; where a parser engine is usually loaded into the constructor and then a parsing string expression into the tree structure is carried out in the SetExpression mode. Evaluation of the tree structure against the input data is carried out in the Evaluate mode. This class also contains the PreferSecondExpression property which indicates which string expression should be used for the current parser. This is because some parsers do not allow one to define operators like '^' power or '%' modulo, and those formulas which have these operators must be rewritten without using these operators and stored as SecondaryExpressions in the Expression class.

The ExpressionBase Abstract class provides a base class for all mathematical formulas tested. It provides the following attributes:

- *Dimension* – the number of input arguments for a function
- *Expression* – the string expression of a mathematical formula

- *Level* – the number which indicates the complexity of an expression
- *SecondaryExpression* – the string expression of a mathematical formula which has replaced any incompatible operators

Additionally, this contains the NativeCall method that contains all mathematical formulas expressed in C# code. This method thus ensures the maximum theoretical performance when evaluating a given function on a .NET platform.

We also wanted to know how fast our mathematica parser is as compared to other existing parsers. We chose some existing parsers and included them in our benchmark. The parsers used are:

- NCalc [6]
- Sprache.Calc [7]
- Flee [8]
- Jace.NET [9]
- Mathos Parser [10]
- xFunc [11]
- Expression Evaluator (in result tables as EE) [12]
- Dynamic Expresso (in result tables as DE) [13]

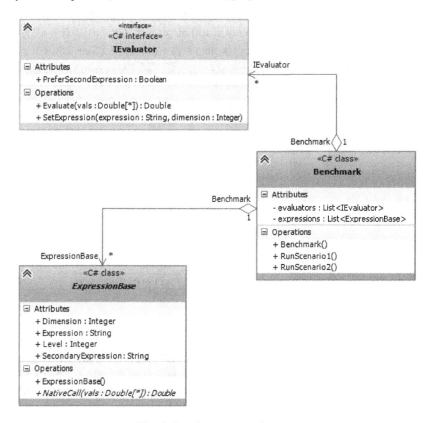

Fig. 3. Benchmark class diagram

Table 1. List of functions used in the benchmarking process [14]

Category	Function name	Expression
Simple	Constant	$f = 10 + 750$
Simple	Second constant	$f = 10 + \pi + 2^9$
Simple	Sum	$f(x, y) = x + y$
Simple	Linear	$f(x) = 55x - 150$
Simple	Sphere, n = 2	$f(x) = \sum_{i=1}^{n} x_i^2$
Medium	Quadratic	$f(x, y) = 55x^2 - 150x + 44 + 12y^2 - 22 - 4$
Medium	Rosenbrock, n = 2	$f(x) = \sum_{i=1}^{n-1} [100(x_{i+1} - x_i^2)^2 + (x_i - 1)^2]$
Medium	Beale's	$f(x, y) = (1.5 - x + xy)^2 + (2.25 - x + xy^2)^2$
Medium	Booth's	$f(x, y) = (x + 2y - 7)^2 + (2x + y - 5)^2$
Medium	Bukin N.6	$f(x, y) = 100\sqrt{|y - 0.01x^2|} + 0.01|x + 10|$
Medium	Matyas	$f(x, y) = 0.26(x^2 + y^2) - 0.48xy$
Medium	Three-hump	$f(x, y) = 2x^2 - 1.05x^4 + \dfrac{x^6}{6} + xy + y^2$
Medium	Easom	$f(x, y) = -\cos(x)\cos(y)\exp(-((x - \pi)^2 + (y - \pi)^2))$
Medium	McCormick	$f(x, y) = \sin(x + y) + (x - y)^2 - 1.5x + 2.5y + 1$
Complex	Ackley's	$f(x, y) = -20\exp(-0.2\sqrt{0.5(x^2 + y^2)}) - \exp(0.5(\cos(2\pi x) + \cos(2\pi y))) + 20 + e$
Complex	Goldstein-Price	$f(x, y) = (1 + (x + y + 1)^2(19 - 14x + 3x^2 - 14y + 6xy + 3y^2))$
Complex	Lévi	$f(x, y) = \sin^2(3\pi x) + (x - 1)^2(1 + \sin^2(3\pi y))$
Complex	Cross-in-tray	$f(x, y) = -0.0001(|\sin(x)\sin(y)\exp(|100 - \dfrac{\sqrt{x^2 + y^2}}{\pi}|)| + 1)^{0.1}$
Complex	Eggholder	$f(x, y) = -(y + 47)\sin(\sqrt{|y + \dfrac{x}{2} + 47|}) - x\sin(\sqrt{|x - (y + 47)|})$
Complex	Hölder table	$f(x, y) = -|\sin(x)\cos(y)\exp(|1 - \dfrac{\sqrt{x^2 + y^2}}{\pi}|)|$
Complex	Schaffer N.4	$f(x, y) = 0.5 + \dfrac{\cos(\sin(|x^2 - y^2|)) - 0.5}{(1 + 0.001(x^2 + y^2))^2}$

4   Results

The test program was compiled within the .NET 4.5 platform, "Any CPU" platform setting and release configuration. We ran it on a laptop with Intel i7 3517 CPU, 10 GB RAM, SSD disk with Windows 8.1 Pro as the main system, and OpenSUSE 13.1 with Mono 3.12.0 and OSX 10.8.5 with Mono 3.12.0.

Tables 2 and 3 show the results for each mathematical parser against the platform used and the function set used. Each row in the table refers to the result for one mathematical parser (if a parser was unable to parse any expression, the result for the whole category was marked as failed). The three main columns indicate each platform in which the benchmark was evaluated. Each of these columns is divided into three sub-columns representing the function set used (i.e. S – Simple, M – Medium, C – Complex).

For the first scenario, 24 different expressions were used, as shown in Table 1, and each function was evaluated 100,000 times. The measured function evaluation time obtained was summarized for each category and then divided by the total number of functions in the category.

Our parser is called "Roslyn" and its function evaluation performance is the best of all libraries for function evaluation. This claim is also proven on other platforms against other libraries.

Table 2. Benchmark result for scenario 1 in ms

Library	Windows			OpenSUSE			OSX		
	S	M	C	S	M	C	S	M	C
Mathos	7262	26317	failed	9036	36277	failed	11416	45093	failed
native	30	148	162	24	114	178	21	90	145
Roslyn	111	181	202	77	165	235	79	148	225
DE	149	202	241	72	162	281	72	135	212
Sprach	484	785	722	433	853	957	507	981	1173
Flee	766	1048	1068	540	1141	1230	677	1367	1412
EE	1847	1190	1317	1016	1546	1681	1130	1573	1664
Jace	1070	2071	2511	866	3200	6697	933	4144	10351
XFunc	1780	6215	5678	2193	8833	7779	2249	9182	8144
NCalc	1411	4654	5977	2199	6977	10071	2667	8325	12107

In the second scenario, 1,000 functions were used (the function set was created by random choice from 24 function sets as shown in Table 1). Each of these functions was evaluated only once.

In the results in Table 3, our library is also called "Roslyn" and we can see that our approach occupied a place in the middle of the table. This result is due to the .NET compilation time overheating.

Table 3. Benchmark result for scenario 2 in ms

Library	Windows			OpenSUSE			OSX		
	S	M	C	S	M	C	S	M	C
Mathos	43	74	failed	49	80	failed	60	97	failed
native	1	1	1	1	1	1	1	1	1
XFunc	94	183	211	158	339	398	195	428	495
NCalc	92	174	269	99	102	161	107	129	143
Sprach	999	1863	2799	4534	10687	18668	5626	13872	26114
Flee	2602	3119	3646	4584	5373	5996	6445	7347	8245
Roslyn	2417	3145	3749	4975	5300	6164	5673	6134	7154
Jace	1533	3096	4068	583	1103	1632	677	1333	1941
DE	5019	5871	6554	2371	2972	3454	2799	3628	4319
EE	11253	21458	30579	2896	5650	8270	3401	6805	9922

5   Conclusion

The main aim of this work is to provide a multiplatform comparison of a dynamic compilation using Roslyn and mathematical parser libraries in .NET for expression evaluation and to investigate the opportunities of the new .NET compiler platform called Roslyn.

The first part of this work introduced our own implementation of a mathematical parser based on the Roslyn compiler service and using on-fly compiling. We also discussed the high level scheme of its current implementation and also mentioned previous implementations.

Further, we designed a benchmark to measure the performance of our solution. The benchmark is based on parsing string expressions and evaluating them against input data. The benchmark is split into two scenarios. The first is focused on evaluating a large amount of input data against a few functions. The second is focused on a large amount of function expressions and parsing while the number of evaluation of these functions is small.

To make a comparison of the performance of our approach, we chose some other existing .NET mathematical parser libraries and included them in our benchmarking. We also ran the benchmark on different platforms to get a more accurate result. On the Windows platform, we ran our benchmark under .NET 4.5 and we ran our benchmark under Mono 3.12.0 on the Linux platform (represented by OpenSUSE) and the OSX platform.

The result of our benchmark is that, in the first scenario, our parser occupies first place on two of the three measured platforms while, in the second scenario, our parser is not as powerful. This is because we had some overheating problems during the compilation phase. Currently, we are using a beta version of the Roslyn service, so maybe there could be some performance improvements in the final release of the Roslyn service.

The results of this work were used in heat load modeling [17]. Users can use this application as an Excel add-in, and the application is built using C# language and the .NET framework.

Acknowledgments. This work was supported by the Grant Agency of the Czech Republic - GACR P103/15/06700S; further, by financial support from the research project: NPU I No. MSMT-7778/2014, by the Ministry of Education of the Czech Republic; and also, by the European Regional Development Fund under the Project: CEBIA-Tech No. CZ.1.05/2.1.00/03.0089. Further, it was supported by the Internal Grant Agency of Tomas Bata University under Project No. IGA/FAI/2015/62

References

1. Faustino da Silva, A., Santos Costa, V.: Design, Implementation, and Evaluation of a Dynamic Compilation Framework for the YAP System. In: Dahl, V., Niemelä, I. (eds.) ICLP 2007. LNCS, vol. 4670, pp. 410–424. Springer, Heidelberg (2007)
2. Ganz Jr., C.: Runtime Code Compilation. In: Pro Dynamic .NET 4.0 Applications, pp. 59–75. Springer (2010)
3. Compilers Principles Techniques and Tools, 2nd edn. Pearson Education, Inc., Boston (2007)
4. Farrell, J.A.: "Compiler Basics" (1995),
   http://www.cs.man.ac.uk/~pjj/farrell/compmain.html
5. Maeda, K.: Quick Parser Development Using Modified Compilers and Generated Syntax Rules. In: Recent Advances in Applied Mathematics and Computational and Information Sciences, Houston, vol. II (2009)
6. NCalc - Mathematical Expressions Evaluator for .NET (2014),
   https://ncalc.codeplex.com/
7. Sprache.Calc (2014), https://github.com/yallie/Sprache.Calc
8. Fast Lightweight Expression Evaluator (2014), http://flee.codeplex.com/
9. Jace.NET (2014), https://github.com/pieterderycke/Jace
10. Mathos Parser (2014), http://mathosparser.codeplex.com/
11. xFunc (2014), http://xfunc.codeplex.com/
12. C# Expression Evaluator (2014), https://csharpeval.codeplex.com
13. Dynamic Expresso (2014),
    https://github.com/davideicardi/DynamicExpresso
14. S. F. University, Test Functions and Datasets (2014),
    http://www.sfu.ca/~ssurjano/optimization.html.

15. Microsoft, MSDN Forum (2007), `https://social.msdn.microsoft.com/`
    `Forums/en-US/8e839652-894d-4891-911b-621c89f1c7f7/`
    `compilerparametersgenerateinmemory-property-doesnt-really-`
    `work?forum=netfxbcl`
16. Microsoft, .NET Compiler Platform ("Roslyn") (2014),
    `https://roslyn.codeplex.com/wikipage?title=`
    `Overview&referringTitle=Documentation#INTRODUCTION`
17. Capek, P., Kral, E.: A Comparison of the Performance of a Composite Pattern and a Mathematic Expression Parser and Interpreter. In: Recent Advances in Applied Mathematics, Modelling and Simulation, Florence (2014)

Value Analysis and Optimization a Machining Process for Hot-Box Core

Florin Chichernea

Transilvania University of Brasov, Department of Materials Science, Blvd. Eroilor,
Nr. 29, Postal Code 500036, Brasov, Romania
chichernea.f@unitbv.ro

Abstract. This paper presents a modern method for optimizing the selection of a machining process for hot-box core, for maximizing its performance and minimizing its cost. The work strategy involves setting the functions of the "product" in this case a machining process, applying the value analysis method in order to obtain an optimal machining process, to optimize the product's value / cost ratio. For the automation of calculations, the author has developed a computer programs that perform all calculations and draw all the necessary diagrams for Value Analysis iterations.

Keywords: value analysis, modelling, optimizing the machining process, hot-box core.

1 Introduction

Based on our experience, on the relevant examples of applying the Value Analysis method [1-3], the author proposes a method for optimizing the machining technologies for various parts.

The optimization is achieved by setting the functions of the constituent elements, of the operations that lead to changing the structure of the machining technological processes accompanied by a reduction in costs without altering the performances of the machining processes.

The hot-box core are used in the foundry sector in order to obtain the cores for cast semi-products in large series.

The technology of these devices is complex due to the complex shapes of the active part and is carried out mostly manually, thus requiring a large volume of work and highly skilled workers.

The advantages of this process are the following: reduce the consumption of the core sand, reduces labor to obtain cores and mechanical knock-out by vibration, increase quality castings, because the permeability and compressibility of the core are best, increase the dimensional accuracy of castings, producing parts without defects, increase surface quality of cast parts, process may be automated and robotized,

The disadvantage of using cores in hot-boxes, relatively high cost of binders and core boxes is balanced by the production of high volume series.

© Springer International Publishing Switzerland 2015
R. Silhavy et al. (eds.), *Software Engineering in Intelligent Systems*,
Advances in Intelligent Systems and Computing 349, DOI: 10.1007/978-3-319-18473-9_35

2   Technological Processes

The hot-box core are usually manufactured from cast steels (OT 400, OT 450, ..., STAS 600), iron steel (Fc200, Fc300, ..., EN-GJL-200) whose technology and manufacturing difficulty resemble those for the hot-box core [4].

In the manufacture of the hot-box core, the first technology that was based on the manual engraving method using chisels.

Preliminarily there was performed a roughing processing for the hot-box core slot using various technological procedures (milling, drilling, turning, etc.).

This procedure was replaced with other technologies when the copying milling and the electro erosionprocessing machines appeared.

Depending on the equipment existing in the workshop, there can be adopted one of the technological processes presented below.

For the hot-box core there are used the variants of technological processes presented in tables 1, 2, 3 and 4.

Finishing the slot (obtained by copying milling), using electroerosion and then manually, allows for smaller shape and dimension deviations, and a continuous surface is obtained.

Complete machining, roughing and finishing the slot, using electroerosion, offers the highest dimensional accuracy but requires appropriate machines and using at least two electrodes, for roughing and finishing.

The method is used to manufacture small and medium high precision hot-box core [4].

Table 1. Technological process, variant 1

No.	Operation
1	Steel ingot casting
2	Cutting heads (feeders)
3	Free forging in two or three directions
4	Primary heat treatment (annealing)
...	...
14	Manual adjustment of the slot and checking using templates
15	Finishing and polishing the hot-box core slot
16	Closing the two semi hot-box core and obtaining the final control part. Checking the control part. Checking the other dimensions of the hot-box core

Table 2. Technological process, variant 2

Same as first variant, including operation 13, then may follow operations	
14	Processing and finishing using electroerosion, until a machining allowance of 0.1 - 0.25 mm remains to polish the surface
15	Hot-box core slot surface polishing
16	Closing the two semi hot-box core and obtaining the control part. Final check

Table 3. Technological process, variant 3

Same as first variant, including operation 13, then may follow operations	
14	Roughing processing using electroerosion for the hot-box core slot until a machining allowance of 0.3 - 0.5 mm remains
15	Finishing machining using electroerosion in an operating mode appropriate for obtaining the quality of the hot-box core surface (R_a = 0.2 - 0.4 mm)
16	Final control

Table 4. Technological process, variant 4

Same as first variant, including operation 7, then may follow operations	
8	Roughing machining milling, drilling, turning, for easing the electroerosion operation
9	Secondary heat treatment for improvement
10	Roughing machining using electroerosion of the hot-box core slot
11	Manual grinding of the hot-box core slot
12	Final control

3   Selecting the Machining Process Using Value Analysis

There shall be selected the optimal technological process for machining a hot-box core, made of different semi-products and there shall be presented the implications of selecting each semi-product from the point of view of mechanical processing and costs.

The functional shape of the part is represented by: dimensions, shape tolerances and the position of surfaces, dimensional tolerances, and quality of the surfaces, the hardness and the operating conditions.

Product functions are the following:

- the main function is: Ensure hot-box core manufacture,
- service functions: FS1 - Ensure semi manufacturing production, FS2 - Ensure heat treatment, FS3 - Ensure machining,
- constraint functions: FC1 - Ensure resilience, FC2 - Allows easy assembly/disassembly, FC3 - Allows control, FC4 - Resistant to the external environment actions, FC5 - Ensure thermal resistance, FC6 - Provide assemblies, FC7 - Allows restoration, FC8 - Ensure user security, FC9 - Ensure resistance to abrasion,
- estimation functions: FE1 - Ensure the user interface.

The object of the Value Analysis is the activity, product or its subassemblies / components. The product is the only one bearing value and its subassemblies or components contribute to its utility.

In this article the Value Analysis product is considered to be the technological process for machining a hot-box core. The components of this Value Analysis product are the stages / operations of the technological process for machining a hot-box core.

There shall be presented the iterations and conclusions drawn from applying the Value Analysis method in the assumptions described above.

Table 5 shows the classification of the functions starting from the functional analysis of the product, respectively the process of machining a hot-box core.

Table 5. The classification of the functions

*Symbol	Functions
FS1	Ensure semi manufacturing production
FS2	Ensure heat treatment
FS3	Ensure machining
FC2	Allows easy assembly/disassembly
FC3	Allows control
FC4	Resistant to the external environment actions
FC5	Ensure thermal resistance
FC6	Provide assemblies
FC7	Allows restoration
FC8	Ensure user security
FC9	Ensure resistance to abrasion
FE1	Ensure the user interface

*FS - Service Function; FC - Constraint Functions; FE - Estimation Function

4 Iteration 1

In this article the author highlights and present:

- the particular role of applying the Value Analysis approach to the selection of the die machining processes,
- the working mode for optimizing the value / cost ratio,
- a valid and useful guide for specialists to optimize the value / cost ratio of the die machining technological processes.

In this article the Value Analysis product / set is considered to be a hot-box core machining technological process. The components of this product / set are the stages / operations of the hot-box core machining technological process.

4.1 Value Weighting of the Functions

Throughout the iterations of the Value Analysis there shall be kept the 12 functions outlined in table 5.

Table 6 shows the value weighting of the functions.

The authors have developed a software that calculates all the values from the shown tables and draws all the diagrams necessary for presenting the findings, in all the iterations of the Value Analysis method.

The calculus is made using the least squares method.

Value weighting of the functions are the values from the last row of table 6.

Table 6. Value weighting of the functions (*coordinate X). *Percentage %

Functions	FS1	FS2	FS3	FC2	...	FC8	FC9	FE1	Total
FS1	1	0	0	0		0	0	0	
FS2	1	1	0	0		0	0	0	
FS3	1	1	1	0		0	0	0	
FC2	1	1	1	1		0	0	0	
FC3	1	1	1	1		0	0	0	
FC4	1	1	1	1		0	0	0	
FC5	1	1	1	1		0	0	0	
FC6	1	1	1	1		0	0	0	
FC7	1	1	1	1		0	0	0	
FC8	1	1	1	1		1	0	0	
FC9	1	1	1	1		1	1	0	
FE1	1	1	1	1		1	1	1	
Points	12	11	10	9		3	2	1	78
Ratio	0.154	0.141	0.128	0.115	...	0.038	0.026	0.013	1
*P	15.4	14.1	12.8	11.5	...	3.85	2.56	1.28	100

4.2 Economic Dimensioning of the Functions

The allocation of costs to functions is made in economic dimensioning of the functions phase. The allocation of costs to functions was performed in the matrix functions - costs from table 7. In table 7 the cost is distributed on the function / functions it is part of.

Table 7. Cost weighting of the functions (*coordinate Y, **cost $)

Parts	Functions								Cost
	FS1	FS2	FS3	FC2	...	FC8	FC9	FE1	parts**
Alloying elements	10	40	20	0		0	20	10	200
Casting	33.3	14.3	9.5	18.1		0	1.9	1.9	95
Forging	26.3	7.5	7.5	13.5		0	0	3.75	75
Face milling	14	0	9.2	8		6	0	2	40
...									
Test pieces	0	0	0	0		0	0	6	30
Reshuffle	0	0	0	0		0	0	8	40
Total cost	184	130.5	124.4	131.4		32.25	52.7	49.4	1105
Ratio	0.167	0.118	0.113	0.119	...	0.029	0.048	0.045	1
*Cost of functions %	16.7	11.8	11.3	11.9	...	2.92	4.77	4.47	100

In the first iteration the machining process is as follows:

- obtaining semi-product (cast + forged),
- primary heat treatment,
- roughing (milling, drilling, milling copying, electro erosion),
- secondary heat treatment (improvement, Thermochemical nitro - ferrox treatment),
- manual semi finishing, manual finishing (grinding, ...),
- manual check (with template, ultrasonic), obtaining test pieces, repairs.

The check of this identity is performed using regression analysis by determining the linear function (the regression line) that represents the average proportionality.

The regression line passes through the origin, as it is considered that a function with "0" value costs "0".

Cost weighting of the functions are the values from the last row of table 7.

4.3   The Weighting of Functions in Value and Cost

The check of this comparison is performed using regression analysis by determining the linear function (the regression line) that represents the average proportionality.

The regression line passes through the origin, as it is considered that a function with "0" value costs "0". The line has the shape: $y_3 = a * x$ (1), where y is the cost weighting of the functions and x is the value weighting of the functions.

In the case of perfect proportionality all points are on the line (1). In order to simplify, the calculation is tabulated. The coordinates x_i and y_i are given in tables 6 and 7 and based on the data calculated in this table the diagrams from figures 1, 2 and 3 are drawn: the value weighting of the functions (fig. 1), the cost weighting of the functions (fig. 2) and the cost and value weighting of the functions (fig. 3).

The diagram in fig. 1 shows the value ranking, prioritization and weighting of the functions. Regression equation describing this distribution is $y_1 = - 1.282 * x + 16.660$ with R squared value on chart $R^2 = 1$.

The assessment of the functions which is shown in fig. 2 highlights the most expensive functions.

These functions are shown in the example in fig. 2, functions FS1, FS2, FC2 and FC3.

Regression equation describing this distribution is $y_2 = - 1.110 * x + 15.550$ with R squared value on chart $R^2 = 0.813$.

If there is such a distribution, the first functions in the order of costs, representing 20 - 30% from the total number of functions, the functions are considered expensive.

The diagram in fig. 3 shows:

-    the equation line $y_3 = x$ (1) (the first bisector) the line that averages the weighting of functions in value and cost, expresses the ideal situation of the disparity between the two weightings, the weighting of functions in value and costs.

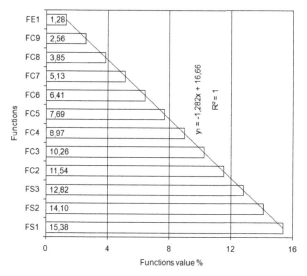

Fig. 1. Value weighting of the functions

- the regression line of equation $y_4 = 0.700 * x + 2.270$, with R squared value on chart $R^2 = 0.3689$, which approximates the arrangement of the points, expresses the real situation of the disparity between the two weightings, the weighting of functions in value and costs,
- functions FS1, FC3, FC9 and FE1 are situated above the lines aforementioned. The weighting of the cost is larger than the weighting of the value of these functions. These functions are deficient and attention should be focused on them. The cost of these functions should be reduced.

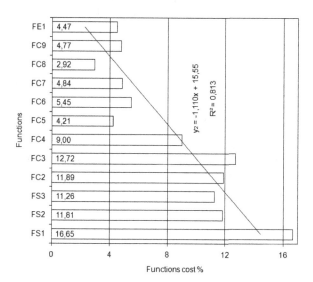

Fig. 2. Cost weighting of the functions

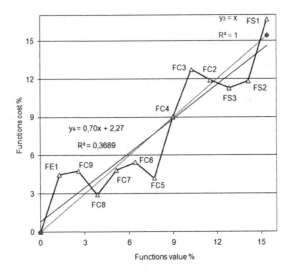

Fig. 3. Weighting of the functions in value and cost

4.4   Iteration 2. Economic Dimensioning of Functions

In the second iteration of the Value Analysis method there shall be considered the functions situated above the ideal regression line (1): FS1 - Ensure semi manufacturing production, FC3 - Allows control, FC9 - Ensure resistance to abrasion, FE1 - Ensure the user interface.

For the second iteration there shall be presented only the results in tabulated form. The weighting of the functions in value is the same as for the first iteration as no functions were added or removed from the system (table 5).

As the functions that cost more are highlighted in fig. 3, solutions shall be suggested for reducing the cost of these functions.

The cost of these functions can be reduced by answering the following questions:

- can there be used less expensive semi-products?
- can the thermal regimes of the heat treatments be reduced (duration and temperature)?
- can there be eliminated / used another a heat treatment operation?
- can the hot-box core be milled at a lower cost?

In the second iteration, actions were taken for the following cost elements, manufacturing process are the follows:

- obtaining semi manufactured (cast),
- primary heat treatment in stages,
- grinding (3D milling),
- secondary heat treatment (improvement),
- hand finishing (grinding, ...),
- 3D control (ultrasonic),
- obtaining test pieces and repairs.

The allocation of costs to functions was performed in table 8. In table 8 the cost is distributed on the function / functions it is part of.

Table 8. Cost weighting of the functions (*coordinate Y, **cost $)

Parts	Functions								Cost
	FS1	FS2	FS3	FC2	...	FC8	FC9	FE1	parts**
Alloying elements	10	40	20	0		0	20	10	200
Casting	33.3	14.3	9.5	18.1		0	1.9	1.9	95
Forging	26.3	7.5	7.5	13.5		0	0	3.75	75
Improvement heat treatment	7.5	15	7.5	13.5		0	7.5	0	75
...									
3D control	0	0	0	0		0	0	10	100
Test pieces	0	0	0	0		0	0	7	35
Reshuffle	0	0	0	0		0	0	4	20
Total cost	138.8	112.3	91.45	93.55		17.25	44.4	42.4	905
Ratio	0.153	0.124	0.101	0.103		0.019	0.049	0.047	1
*Cost of functions %	15.3	12.4	10.1	10.3		1.91	4.91	4.69	100

Cost weighting of the functions are the values from the last row of table 8.

4.5   The Weighting of Functions in Value and Cost

Coordinates x_i and y_i are given in tables 6 and 8 and, based on the data calculated and presented in this tables, the diagrams from fig. 4 and fig. 5 are drawn:

- the diagram of weighting functions in value (identical with figure 1 – iteration 1),
- the diagram of weighting functions in cost (fig. 4),
- the diagram of weighting functions in value and cost (fig. 5).

The critical assessment of the functions presented in fig. 4 highlights the most expensive functions.

Regression equation describing this distribution is $y_5 = -1.045 * x + 15.12$ with R squared value on chart $R^2 = 0.734$.

The diagram in fig. 5 represents the regression line drawn using the least squares method and presents the comparison of functions in terms of value and cost.

In the diagram from fig. 5 there can be seen the following lines:

- the equation line $y_3 = x$ (1) (the first bisector), the line that averages the weighting of functions in value and cost, expresses the ideal situation of the disparity of the two weightings, the weighting of functions in value and cost,
- the regression line, of equation $y_6 = 0.855 * x + 1.115$, with R squared value on chart $R^2 = 0.789$ which approximates the arrangement of the points, expresses the real situation of the disparity of the two weightings, the weighting of functions in value and cost,

- functions FC3, FC4, FC9 and FE1 are situated above the lines aforementioned. The weighting of the cost is larger than the weighting of the value of these functions.

These functions are deficient and attention should be focused on them. The cost of these functions should be reduced, using engineering methods.

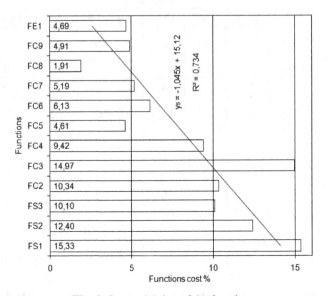

Fig. 4. Cost weighting of the functions

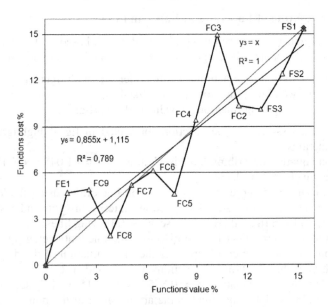

Fig. 5. Weighting of the functions in value and cost

5   Results

Following the second iteration there can be seen in fig. 3 and fig. 5 that the value of some functions increased, of other functions decreased, but the cost of those that increased eventually decreased due to the decrease of the cost of the "product", in this case a machining process.

In the second iteration the functions FC3, FC4, FC9 and FE1 are situated above the regression line.

There can be seen comparatively to fig. 3 (iteration 1) in fig. 5 (iteration 2) that the functions are grouped closer to the ideal regression lines.

Below are presented comparatively the equations of the regression lines (the real situation):

- iteration 1: $y_4 = 0.700 * x + 2.270$ $R^2 = 0.3689$,
- iteration 2: $y_6 = 0.855 * x + 1.115$ $R^2 = 0.7890$.

There can be seen an increase in the value of the correlation coefficient R^2 in the second iteration as compared to the first iteration, thus resulting that the dispersion of the points decreased in relation with the regression line.

The iterations continue until the correlation coefficient R^2 tends to value 1 and the regression line (the real situation) tends to $y = x$ (the ideal situation).

In two iterations of the Value Analysis study the machining technology for the hot-box core was redesigned and optimized from the point of view of:

1 - engineering:
- the hot-box core machining process was changed,
- the primary continuous heat treatment was replaced with a heat treatment in stages,
- the classic milling process (less expensive) of the hot-box core was replaced with high velocity milling, using 3D processing machines (workmanship, more expensive equipment but higher productivity and precision),
- the manual control process, using templates (less expensive) was replaced with a 3D machines control process (workmanship, more expensive equipment but higher productivity and precision),
- as the machining processes are more precise in the 2-nd iteration, the final remedies are fewer than in the first iteration,

2 - economics:
- the cost of the product decreased from 1105 $, in the first iteration to 905 $ in the second iteration, a 18.09 % decrease,
- the cost of functions FS1, FS3, FC2 and FC8 decreases in the second iteration compared to the first iteration (table 9).

Table 9. Evolution of the cost functions in two iterations

Functions	FS1	FS2	FC2	FC8
First iteration (%)	16.65	11.26	11.89	2.92
Second iteration (%)	15.33	10.10	10.34	1.92
Decreases (%)	8.61	11.40	15.00	53.10

In the third iteration of the Value Analysis study there shall be analysed the functions situated above the regression line $y_3 = x$ (1), (FC3 - Allows control, FC4 - Resistant to the external environment actions, FC9 - Ensure resistance to abrasion, and FE1 - Ensure the user interface), there shall be analysed the "components" participating to achieving these functions and solutions shall be proposed for reducing the costs.

6   Conclusions

The Value Analysis is a work whose objective is to find the "compromise" between cost and functionality of a product, all ensuring a level of quality necessary and sufficient.

With this method are identified:

- very expensive functions,
- they highlight the technical components that cost very much,
- show where a technical solution must be changed,
- in this way does not go on attempts, in engineering design,
- it goes directly to solutions that cost very much, all this constituting the Value Analysis benefits.

A major difficulty of Value Analysis methodology, even in consecrated application (designing, redesigning of industrial products, ...) is cost weighting of the components on the function / functions attended by (table 7 and table 8). This point is very sensitive and difficult in Value Analysis method, especially in the field studied in this work where the cost weighting of the functions in cost should be made taking account of the phases of the technological process of manufacturing.

In future applications and researches author will have to focus on this aspect, insisting on closer collaboration with industry specialists and using the work in interdisciplinary groups.

This method can be used for optimizing the value / cost ratio for different types of machining technological processes for various parts.

Designing an economic model for making decisions regarding the optimization of the technological processes for machining parts in terms of the Value Analysis method is an absolute novelty in the field.

References

1. Chichernea, F., Chichernea, A.: Value Management. Editura Universității Transilvania Brașov (2011) ISBN 978-973-598-852-4
2. Chichernea, F., Chichernea, A.: Iterations of the Value Analysis. Editura Universității Transilvania Brașov, ISBN 978-606-19-0040-4
3. Chichernea, F., Chichernea, A.: Value Analysis from theory to practice. Editura Universității Transilvania Brașov (2012) ISBN 978-606-19-0039-8
4. Bejan, V.: The technology for manufacturing and repairing technological equipment, Bucharest, vol. I, II (1991) ISBN 973-95458-7-4

MASCM: Multi Agent System for Cloud Management

Tamer F. Mabrouk

Computer Engineering Department,
Alexandria Higher Institute of Engineering and Technology, AIET
tamer_fm@yahoo.com

Abstract. Cloud computing has been designed and implemented to create a large scalable computing infrastructure that offers service-oriented solutions for individuals and organizations. The current research activities have focused on the implementation of the cloud infrastructure, while a little attention has been directed towards introducing innovative solutions for cloud management. This paper introduces the integration of cloud computing and the Multi-Agent Systems (MAS) as a complementary parts to create an Agent-based Cloud Computing where agents can manage and configure the cloud resources and services in an autonomous and intelligent way to provide a higher degree of flexibility and interoperability for both cloud providers and consumers.

Keywords: Cloud Computing, Multi-Agent Systems, Service Level agreement.

1 Introduction

The fast development technologies have introduced cloud computing as a service over the internet due to its scalability, reliability and cost-effectiveness. Although virtualization and cloud computing have helped organizations in breaking the physical bonds between users and IT infrastructures, the adoption of cloud computing is a challenge that enterprises have to face due to the following reasons: (a) The drastic growth of data volumes and software applications. (b) The organizational changes that will affect peoples' work. (c) The corporate governance issues regarding the use of cloud computing. (d) The security and privacy considerations regarding the outsource of organizational data, applications, and other resources to the cloud environment.

In spite of the ambiguity and uncertainty exists regarding the actual realization of the cloud adoption, cloud computing is likely to have a significant effect on the way software is procured, developed and deployed in the near future.

We believe that the success of cloud adoption depends on the enterprise' degree of awareness of the cloud's benefits, risks and effects. One of the most promising solutions is to integrate MAS within the cloud platforms. The advantage of this approach is to empower any cloud platform with a practical intelligent adaptation mechanism for cloud management. At the same time, cloud platforms itself can offer an ideal platform to host the MAS where agents can operate, coordinate and collaborate due to the cloud large amount of processing and memory resources that can be dynamically configured for executing agent-based systems at a significant scale.

© Springer International Publishing Switzerland 2015
R. Silhavy et al. (eds.), *Software Engineering in Intelligent Systems*,
Advances in Intelligent Systems and Computing 349, DOI: 10.1007/978-3-319-18473-9_36

In this paper, we propose an architectural model for a MAS for Cloud Management (MASCM) as an intelligent and autonomous system that can improve the cloud resources management, services discovery and Service Level Agreement (SLA) negotiation.

2   Background

The main technologies involved in this paper are cloud computing and Multi-Agent System. A brief review of these technologies is provided next.

2.1   Cloud Computing

Cloud Computing combines several existing technologies which have evolved during the last decades such as operating systems, databases, servers, networks, middleware and virtualization. The US National Institute of Standards and Technology (NIST) have published a definition that represents the most commonly agreed aspects of cloud computing [1]. This definition describes cloud computing as "a model for enabling convenient, on-demand network access to a shared pool of configurable computing resources (e.g., networks, servers, storage, applications and services) that can be rapidly provisioned and released with minimal management effort or service interaction".

On-demand, self-service and pay-by-use form the global nature of cloud computing [2]. The on-demand nature of cloud computing helps in supporting the performance and capacity aspects of service-level objectives, the self-service nature of cloud computing helps in creating elastic environments that expand and contract based on the workload and target performance parameters and the pay-by-use nature of cloud computing may take the form of equipment leases that guarantee a minimum level of service from a cloud provider.

The use of virtualization in clouds has created a set of layers [3], which can be classified and organized into: (a) Software as a Service Layer which provides the capability of providing users with an integrated service comprising hardware, development platforms, and applications. (b) Platform as a Service Layer which provides the capability of providing users with an application or development platform in which they can create their own application that will run on the Cloud. (c) Infrastructure as a Service Layer which provides the capability of delivering IT infrastructure based on virtual or physical resources as a commodity to users.

2.2   Multi-agent System

An intelligent agent is an entity that has a certain amount of intelligence and can perform a set of tasks on behalf of the users, due to agent's autonomy, reactivity and mobility [4]. Today's agent-based applications often involve multiple agents in order to maintain complex distributed applications.

Since intelligent agents should be able to play on behalf of the users, it is more likely to depend on roles which help designers and developers in modeling the intelligent agents to perform more tasks than those they are developed for, which is more likely to what happens in the real life, where people learn how to do things and expand their knowledge [5].

Intelligent Agents are designed to interoperate and interact for information gathering and processing tasks. This may include: (a) Locating and accessing information from various on-line sources. (b) Filtering any irrelevant information and resolving inconsistencies in the retrieved information. (c) Integrating information from heterogeneous information sources. (d) Instantaneously adapting to the user's needs. (e) Decision making.

3   Approach

Cloud Computing depends on virtualization where different host computers might run multiple virtual machine (VMs). The Cloud Computing environment becomes much more complicated as the number of hosts, VMs and client applications continue to grow. In recent years there has been a lot of research concerning the management of VMs to ensure that the cloud providers allocate enough resources to VMs according to load and resource usage based on SLA.

In typical cloud architecture, the IaaS is composed of datacenters containing thousands of physical machines organized in clusters. Each physical machine runs several VMs which is responsible of executing the applications and services of different customers with different service level requirements. The customers can mutually agree upon a set of SLAs with different performance and cost structure moreover, they can switch between different SLAs at runtime.

Therefore, the main aim of Cloud Computing is the efficient use of the computing infrastructure, service delivery, data storage and scalable virtualization techniques at reduced costs in order to make Cloud Computing Systems more adaptive, flexible and autonomic in resource management, service provisioning and in running large scale applications.

Based on the previous facts, a Multi-Agent System proposed which emphasis on cloud management where agents can be used as basic components for implementing intelligence in Cloud Computing Systems.

3.1   MASCM Hierarchical Model

MASCM aims to create a consistent environment where each agent has its own knowledge and capabilities to interact and operate through the different cloud layers. The hierarchy of agents is illustrated in Fig.1.

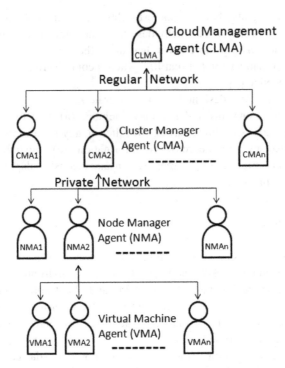

Fig. 1. Agent Hierarchy in MASCM based on the Cloud Architecture

In our approach, each agent has its own knowledge and capabilities to interact and operate through the different cloud layers. It worth noting that the cloud architecture is composed of hierarchical layered elements, each has a certain responsibility, therefore assigning a single centralized manager with all the management capabilities is almost impossible.

In IaaS layer, agents are responsible for the intelligent provisioning of basic resources to user applications. PaaS layer, agents play a role in the efficient deployment and execution of programming environments that developers use for application implementation. Finally, in SaaS layer, agents are programmed to optimize the use of applications provided as services and the management of the underlying hardware/software infrastructure taking care of its efficient utilization and, at the same time, for maintaining the declared QoS.

The proposed MASCM agents are: (a) *Cloud Manager Agent (CLMA)* which is responsible for executing the users' requests as monitoring and managing all clusters. (b) *Cluster Manager Agents (CMA)* is responsible for managing all nodes in the cluster, allocates new instances of VM and decides which Node will run each VM instance. (c) *Node Manager Agents (NMA)* which runs on the physical machine and is responsible for node monitor and management, through interacting with the OS and supervising the node to start, stop, deploy and destroy the VM instances. (d) *Virtual Machine Agents (VMA)* who is responsible for VM management.

3.2   MASCM Role Concept and Assignment

During the operational phase, complex tasks are divided into smaller tasks and delivered to different responsible agents at the lower levels. Each agent operates based on a set of roles to ensure that the SLAs are met. These roles are classified into two types: (a) Static roles which are performed on regular base independently without the need to get a permission to do so. (b) Dynamic roles which are acquired and released based on event driven schema through sending a direct request from the higher level agents based on the proposed hierarchy. These kinds of roles will help higher level agents in defining what is expected to happen in such a dynamic environment based on demand for each instance.

For example, VMA agent should always keep a record containing the necessary information about the managed object, which might change more often and hence should be updated through performing some static roles on regular base to set the VM operating system, CPU, memory utilization and allocation, etc, or through performing some dynamic roles that are acquired and released at runtime to activate/deactivate VM, get the VM IP, etc.

It is important to note that the intelligent component of MASCM is a Rule Engine to assume roles in an intelligent dynamic way. These rules act as policies for self-configuration or as actions to ensure that SLAs are met.

It is recommended that the rule-based knowledge representation scheme for role-assignment is constructed in the form of traditional production rule "If ... Then" Clauses. This grants several advantages, such as:

- Rules for role-assignment can be changed during the working environment without static dependencies defined in the code, which grant high flexibility.
- Permitting the reuse of solutions and experiences to allow the extending of an existing role to produce new roles.
- Allowing separation of concerns between the algorithmic issues and the interaction issues in developing agent-based applications.
- Taking the advantage of local processing in reducing the communication latency in roles that acquire high data transfer rate and making decisions based on the real time situation.

The intelligent component in MASCM is performed in memory as shown in the following algorithm [6]:

1. *The user and/or agent request is passed to the MASCM Rule Engine.*
2. *MASCM Rule Engine fetches the corresponding rules for role assignment.*
3. *MASCM Rule Engine fires the fetched rules to translate the user and/or agent request into the Role which is assigned to the proper agent to fulfill it.*

After acquiring a role, agents may successfully perform the role, or send help message if needed, or send a fail message back as shown in Fig.2.

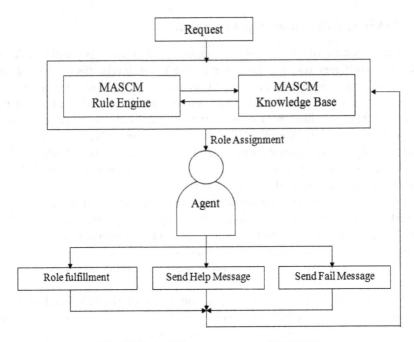

Fig. 2. The intelligent component of MASCM

3.3   MASCM Communication Model

Interaction among agents in a MAS is mainly realized by means of communication. The major challenges concerning agents' communication are: (a) How do agents communicate, which should be performed in a reliable, asynchronous and loosely coupled way. (b) When to send/receive a message between agents in the hierarchy, which might be a request that implies a role acquirement for agents in the lower levels or simply a message that contain information to help higher level agents in decision making process. (c) What information to exchange and this requires a predefined message format to minimize any possibility of extra overload on the communication process [7].

Agents in the proposed hierarchy can communicate through a message based schema. The proposed message format shown in Fig. 3.

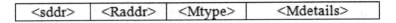

Fig. 3. The proposed message format

- <Sddr> field: Sender Address.
- <Raddr> field: Receiver Address.
- <Mtype> field: Message Type.
- <Mdetails> field: Message Details.

Messages between agents can be configured through the <Mtype> field according to the following schema:

- If it was a Request message for Role Acquire, Role Release or Help Request then the field <Mtype> is set to <R>.
- If it was an Information message, then the field <Mtype> is set to <I>.
- If it was a status *Update* message, then the field <Mtype> is set to <U>.

By this way, agents can take the advantage of working in hierarchical organization to improve the scalability by reducing communication overhead that only happens between parent and child. Therefore, the communication overheads will only affect the local network where agents need to send messages based on how they are networked together.

4   Conclusion

In this paper, we described some details towards the use collaborating agents for the management of cloud environments. The merger between cloud computing and agents can be convenient for both parties.

Cloud Computing can offer a powerful, reliable, predictable and scalable computing infrastructure for the execution of MAS applications and it will create new opportunities to the agents computing area and expanding their knowledge beyond the possibilities offered by traditional computing platforms. On the other hand, Agents will be the basic components for implementing intelligence in Cloud computing systems, making them more adaptive, flexible and autonomic in resource management, service provisioning and in running large-scale applications.

References

1. Mell, P., Grance, T.: The NIST Definition of Cloud Computing (2009)
2. Weiss, A.: Computing in the clouds. NetWorker 11(4), 16–25 (2007)
3. Marks, E.A., Marks, B.L.: Cloud Computing: A Practical Approach. McGraw-Hill (2010)
4. Jennings, N.R., Wooldridge, M.: Agent Technology: Foundations, Applications, and Markets, pp. 29–40. Springer (2002)
5. Cabri, G., Ferrari, L., Leopardi, L.: Role-based Approaches for Agent Development. In: Proceedings of the 3rd Conference on Autonomous Agents and Multi Agent Systems (AAMAS), pp. 1504–1505 (2004)
6. Mabrouk, T.F., Sherbiny, M.M., Guirguis, S.K., Shawky, A.Y.: A Multi-Agent Role-Based System for Business Intelligence. In: Innovations and Advances in Computer Sciences and Engineering, pp. 203–208. Springer, Netherlands (2010)
7. Mola, O., Bauer, M.A.: Collaborative policy-based autonomic management: In a hierarchical model. In: 7th International Conference on Network and Service Management (CNSM), pp. 1–5 (2011)

Author Index

Printed in the United States
By Bookmasters